Environment Climate Change and Sustainable Development

Environment Climate Change and Sustainable Development

Edited by:
Dr. Vinita Katiyar
Dr. S.K. Verma

Co-Editor:
Vikram Singh

RADHA PUBLICATIONS
New Delhi-110002

First Edition 2012
ISBN: 978-81-7487-760-4

ISBN 81-7487-760-6

RADHA PUBLICATIONS
4231/1, Ansari Road, Daryaganj
New Delhi-110002
Phone : 23254306, 23247003

Opinions expressed in this book do not reflect the policies or views of this organization, but of the individual contributors. The authors are solely responsible for the details and statements in their research paper.

Published by Nitin Garg for Radha Publications and printed by Asain Offset Printers, Moujpur, Delhi-110053

Preface

Development depends upon the environment of that economy because economic development depends upon available natural resources. Production of goods and services depend upon natural recourses and the technology of that time and place. Trade takes place according the difference of environment and availability of natural resources between two economies, so the environment plays an important role in economic development, production, pricing of products and trade. Thus environment is the base of economic development. Nature provides environment and life on earth. Environment consists of the living organism and physical component. Living organism includes plant, animal, microbes, and human being. Nature also provides us many natural resources as environment. Human being gets many substances from the nature to meet their basic needs are called natural resources. This environment decides the existence, growth and development of the economy and all its related activities.

Economic development affects the environment and environment affects the economic development, because environment provides raw material for production; and economic development links with environmental degradation. Economics has engaged in exploitation of nature and its resources for the development. Economic development changes human life in positive manner and make comfort but on other hand it also involve environmental degradation and depletion of natural resources, and exploited the nature and caused serious damage to the environment, so environment has lost its regenerative capacities and bringing deterioration day by day.

Continuous degradation of natural resources leads adverse affects on environment, climate change and also hindrance the future possibilities of development. For meets the needs of present and future we must save our environment and shift the paradigm of mal-development to sustainable development by changing the way of development. Achieving progress towards sustainability implies maintaining improvement in both, human and the ecosystem. For sustainable development we need some changes in our present development models and make a balance between environment and economic development. For sustainable economic development policies are developed for minimized the resources consumption, damage and destroy.

Climate change, global warming, environment degradation and sustainable development have been main issue of debate in recent time on global concern. For awareness these concern in society, we took up the work of editing the paper for theme "Environment, Climate Change and Sustainable Development". This theme covers all aspects related environment and development. Several contributors sent their views related to the theme in different manner. To provide a systematic view of idea and analysis contained in 38 articles, the present edit book was distributed over six sub-themes:

I. Sustainable Development and Climate Change
II. Sustainable Development and Water Management
III. Sustainable Development and Land Management
IV. Sustainable Development and Energy
V. Sustainable Development and Environment Pollution
VI. Sustainable Development and Miscellaneous Issues

In view of vast nature of subject, the book "Environment, Climate Change and Sustainable Development" has become slightly bigger in size than we originally planned it to be. we could not resist the temptation of including very useful contributions. However, it can be reasonably hoped that this book will serve as a useful addition to the literature on Environment, Climate Change and Sustainable Development.

Shikohabad

Vinita Katiyar
S.K.Verma
Editors

Contents

List of Contributors

1. **Dr. Addala Narayana Swamy**, Department of Geophysics, Andhra University, Visakhapatnam, India.
2. **Alia Yusuf**, Research Scholar, Department of Geology, Aligarh Muslim University, Aligarh, UP, India.
3. **Dr. Alok Chantia**, Assistant Professor, Department of Anthropology, SJNPG (KKC) College, Lucknow, UP.
4. **Dr. Amar Prakash,** Assistant Professor, Department of Psychology, S.R.K. (P.G.) College, Firozabad, UP.
5. **Ambica Prasad**, Research Scholar, Department of Economics, S.R.K. (P.G.) College, Firozabad, Uttar Pradesh.
6. **Amita Agrawal,** Assistant Professor, Department of Chemistry, K.P. Engineering College, Etmadpur, Agra, Uttar Pradesh.
7. **Dr. Anupam Pathak**, Assistant Professor, K.A. (P.G.) College, Faculty of Geography, Kasganj (Kanshiram Nagar), Uttar Pradesh.
8. **Dr. Arun Kumar Singh**, Associate Professor, Department of Geography, MMV, BHU, Varanasi, Uttar Pradesh.
9. **Dr. A. K. Rao**, Associate Professor, Department of Geology, Aligarh Muslim University, Aligarh, India.
10. **Dr. C.S. Verma,** Associate Professor, Department of Economics, M.B.P. Government Post Graduate College, Aashiana, Lucknow, Uttar Pradesh.
11. **Dr. Deepali Singh,** Dy Director & Co-ordinator IGNOU Dau Dayal Mahila (P.G.) College, Firozabad, Uttar Pradesh.
12. **Dr. Dattatraya Harpale**, Department of PGSR, S. N. D. T. Women's University Karve Road, Pune, Maharastra.
13. **Dushmanta Dutta,** School of Applied Sciences and Engineering, Monash University, Churchill, Australia.

14. **Dr.Helen**, Associate Professor, Communication Centre, Kerala Agricultural University, Mannuthy ,Thrissur Dt, Kerala.
15. **Dr.H.Kamala** , Assistant Professor, Department of Geography, Maharaja's College, University of Mysore, Mysore, Karnataka.
16. **Dr. H.P. Malonia**, Assistant Professor, Department of Commerce , S.R.K. (P.G.) College, Firozabad, Uttar Pradesh.
17. **Inder Jeet Singh**, Research Fellow, Department of Geography, Punjab University, Chandigarh, Punjab.
18. **Dr. Jagadish Singh**, Assistant professor & Head, Deptartment of Economics, Harishchandra P.G.College, Varanasi, Uttar Pradesh.
19. **Joy Jacqueline Pereira**, Associate Professor, School for Social Development & Environmental Studies, FSSK. University, Kebangsaan, Malaysia.
20. **Dr.Komala Bhat**, Head, Department of Geography, M.M.Arts and Science College Sirsi, Uttarakannada, Karnataka.
21. **K.Shimola**, Research Scholar, Centre For Water Resources, Anna University, Chennai, Tamil Nadu.
22. **Manoj Sharma**, Senior lecturer in Hotel Management, College of Hotel Management, Agra campus, Dr. MPS Group of Institutions, Agra.
23. **Mohd Ekhwan Toriman,** Associate Professor, School for Social Development & Environmental Studies, FSSK. University, Kebangsaan, Malaysia.
24. **Mohammad Sadique,** Research Scholar, Department of Geology, Aligarh Muslim University, Aligarh, India.
25. **Dr. Mridul Kumar Verma**, Associate Professor, Faculty of Education, A. K. (P.G.) College, Shikohabad, Distt. – Firozabad, Uttar Pradesh.
26. **Muhammad Barzani Gasim**, School for Social Development & Environmental Studies, FSSK. University, Kebangsaan Malaysia.
27. **Mushrifah Idris**, School for Social Development & Environmental Studies, FSSK. University Kebangsaan, Malaysia.
28. **Dr.M.Krishnaveni**, Associate Professor, Centre for Water Resources, Anna University, Chennai, Tamil Nadu.
29. **Dr.M.C. Dwivedi**, Sher-e-Kashmir University of Agricultural Sciences &Technology, Jammu, Jammu & Kashmir
30. **Dr. M.K. Agrawal**, Associate Professor, Department of Commerce, S.R.K. (P.G.) College, Firozabad, Uttar Pradesh.

31. **Neha Sharma**, Research Scholar, Department of Economics, S.R.K. (P.G.) College, Firozabad, Uttar Pradesh.
32. **Dr.Niharika Chaturvedi**, Assistant Professor, Department of Sanskrit, S.R.K. (P.G.) College, Firozabad, Uttar Pradesh.
33. **Nor Azlina Abdul Aziz**, Associate Professor, School for Social Development & Environmental Studies, FSSK. University, Kebangsaan, Malaysia.
34. **Nor Rohaizah Jamil**, Associate Professor, School for Social Development & Environmental Studies, FSSK. University, Kebangsaan , Malaysia.
35. **Dr. Pawan Kumar Dubey**, Department of Economics, K.A. (P.G.) College, Kasganj , Uttar Pradesh.
36. **Mr. Pradeep Kumar Singhal**, Research Scholar, Department of Commerce, S.R.K. (P.G.) College, Firozabad, Uttar Pradesh.
37. **Dr. Prakash Pandey**, Lecturer, Advance Institute of Management, Ghaziabad, Uttar Pradesh.
38. **Dr. Praveen Sood,** KVMM, Kanpur, Uttar Pradesh.
39. **Preeti Misra**, Associate Professor, Department of Human Rights, Babasaheb Bhimrao Ambedkar University, Lucknow, UP, India.
40. **Prerana Bansal**, Research Scholar, Singhania University, Pilani, Rajasthan, India.
41. **Dr.P. Suseela**, Associate Professor, Communication Centre, Kerala Agricultural University, Mannuthy, Thrissur Dt, Kerala.
42. **Rajesh Kumari**, Assistant Professor, Department of Geography, Govt. College, Bahadurgarh (Jhajjar), Haryana, India.
43. **Dr. Ranju Narang**, Sr. Lecturer and Head, Department of Economics, SSSS (P.G) College, Rasna, Meerut, Uttar Pradesh.
44. **Rakhi Sharma**, Research Scholar, Department of Economics, S.R.K. (P.G.) College, Firozabad, Uttar Pradesh.
45. **Rashmi Rani Anand**, Research Scholar, School of International Studies, Jawaharlal Nehru University, New Delhi, India.
46. **Dr. Reeta Rautela**, Sr.Lecturer, IMS Dehradun, Uttrakhand, India.
47. **Dr. Ruchi Sharma**, Sr. Lecturer, Advance Institute of Management, Ghaziabad, Uttar Pradesh.
48. **Dr. R.C. Bharti**, Assistant Professor, IT, IMS, Dehradun, Uttrakhand, India.
49. **Saikiran Tharimena**, Department of Civil Engineering, Visvesvaraya National Institute of Technology, Nagpur India.

50. **Sangyu Yaden**, School of Sciences, Department of Geography, Nagaland University, Lumami, Nagaland.

51. **Sharad Chandra Dwivedi**, Center for the study of Regional Development, Jawaharlal Nehru University, New Delhi.

52. **Dr.Shikha**, Assistant Professor, Department of Economics, S.R.K. (P.G.) College, Firozabad, Uttar Pradesh.

53. **Simran Kaur**, Department of Geography, Punjab University, Chandigarh, Punjab.

54. **Dr. Shivani Kapoor,** Dayanand Academy of Management Studies Kanpur, Uttar Pradesh.

55. **Soyhunlo Sebu**, School of Sciences, Department of Geography, Nagaland University, Lumami, Nagaland

56. **Dr. Satendra Bhardwaj**, Sr. Lecturer, Advance Institute of Management, Ghaziabad, Uttar Pradesh.

57. **Dr.Subodh Kumar**, Assistant Professor, IT, IMS, Dehradun, Uttrakhand, India.

58. **Dr. Sudhakar Dwivedi,** Sher-e-Kashmir University of Agricultural Sciences &Technology, Jammu, Jammu & Kashmir

59. **Dr. S. K. Mishra,** Secretary, UPUEA and Reader, P.G. College, Sakaldiha, Chandauli , Varanasi, Uttar Pradesh.

60. **Dr. S. K. Verma,** Associate Professor and Head, Department of Economics, S.R.K. (P.G.) College, Firozabad, Uttar Pradesh.

61. **Dr. U. S. Pandey**, Associate Professor, Department of Sociology, S.R.K. (P.G.) College, Firozabad, Uttar Pradesh.

62. **Vikram Singh**, Research Scholar, Department of Geography, S. C. (P. G.) College, Mainpuri, Uttar Pradesh.

63. **Dr. Vinita Katiyar**, Assistant Professor, Department of Economics, Narain (P. G.) College, Shikohabad, Uttar Pradesh.

64. **Dr. Vipin Agrawal**, Principal S.R.K. (P.G.) College, Firozabad, Uttar Pradesh.

65. **Dr. Virendra Nagarale**, Department of PGSR, S. N. D. T. Women's University, Pune, Maharastra, India.

66. **Vishal Agrawal,** Research Scholar, Department of Commerce, S.R.K. (P.G.) College, Firozabad, Uttar Pradesh.

67. **Y B Katpatal,** Department of Civil Engineering, Visvesvaraya National Institute of Technology, Nagpur, India.

68. **Dr. Y. P Chandrashekar**, Lecturer, Department of Geography, Karnataka State Open University, Mysore, Karnataka.

69. **Ziaur Rehman,** Research Scholar, Department of Geology, Aligarh Muslim University, Aligarh, U.P., India.

Section-I
Sustainable Development and Climate Change

1. Sustainable Development, Climate Change and India

S.K. Verma
Vinita Katiyar

Sustainable development, environmental degradation and climate change are interrelated concern. Un-sustainable economic development links with environmental degradation and emission of Green House Gases, which leads to variations in climate. Climate change is one of the most important global environmental challenge, climate variations lead to increase in temperature, alteration in precipitation, increase in river flow, rise in sea level, melting in glaciers and increase in frequency and intensity of natural disaster such as drought, flood, cyclone etc., climate change with implications for food production, water supply, health, energy, and loss of bio-diversity. This paper addresses the impact of economic development on climate change. The efforts made by the UNFCCC and the Kyoto Protocol provisions are clearly inadequate to address the climate change challenge. This paper also highlight the impact of climate change on economic development specially agriculture, human resource and bio- diversity. The most effective way to address climate change is to adopt a sustainable development pathway by shifting to environmentally sustainable technologies and promotion of energy efficiency, renewable energy, forest conservation, reforestation, water conservation, etc.

Key Words: Eco-domestic product, Genuine savings, Global warming, Green House Gases.

Introduction

Climate change is one of the complex problems facing mankind today. The complexity of this problem is attributed to its deeper global ramifications on a vast range of issues impacting the very survival of life on Earth. There are varieties of perceptions and global warming regarding the exact size and consequences of climate change. Climate change is one of the most important global environmental challenges facing humanity with implications for food production, natural ecosystems, freshwater supply, health, etc. According to the latest scientific assessment, the earth's climate system has demonstrably changed on both global and regional scales since the preindustrial era. Further evidence shows that most of the warming (of 0.1°C per decade) observed over the last 50 years, is attributable to human activities[1] Today, it has been scientifically established that significant global warming is occurring. Warming of the climate system is unequivocal, as is now evident from observations of increases in global average air and ocean temperatures, widespread melting of snow and ice and rising global average sea level[2].

The climate change is large challenge for sustainable development. Thus, climate policies can be more effective when consistently embedded within broader strategies designed to make national and regional development paths more sustainable. The impact of climate variability and change, climate policy responses, and associated socio-economic development will affect the ability of countries to achieve sustainable development goals. The pursuit of these goals will in turn affect the opportunities for, and success of, climate policies. In particular, the socio-economic and technological characteristics of different development paths will strongly affect emissions of Green House Gases, the rate and magnitude of climate change, climate change impacts on economy, the capability to adapt, and the capacity to mitigate.

Methodology and Objectives

This paper widely covers concept of climate change, sustainable development and interrelation between sustainable development, environmental degradation and climate change. This paper also enlightened share of emission of green house gases the country-wise and sector –wise. Paper also focuses adaptation towards climate change and sustainable development. Mainly secondary data are used in paper for the study. Data are collected from different publications of Government of India, NSSO, World Development Report, Various IPCC Assessment Reports etc. The main objectives of this paper is -

- To examining the impact of economic development on climate change.
- To examining the impact of climate change on economic development.

Climate Change

Climate change refers to the variation in the global climate or in regional climates over time. Climate change is defined as "a change which is attributed directly or indirectly to human activity that alters the composition of the global atmosphere and which is in addition to natural climate variability observed over comparable time periods."[3] Human activities for economic development is responsible for alteration in environment through environment degradation and pollution by increase amount of in Green House Gases which leads to climate change. The Earth's climate is dynamic and always changing through a natural cycle. The solar energy, passing through the atmosphere, is absorbed by the Earth's surface and a significant part of it is reflected back into the atmosphere. However, the atmosphere of the Earth contains small quantities of carbon dioxide, methane and nitrous oxide (collectively called greenhouse gases (GHGs)) that traps some of the outgoing infra red radiation and reflects it back to Earth thus keeping the surface warmer. In the absence of this greenhouse effect (trapping by GHGs) the Earth's mean temperature would be 30°C lower than normal temperature. Thus, most of the present life forms on the Earth depend on the natural greenhouse effect for their existence. However, increase in the emission of these GHGs due to human activities causes the enhanced greenhouse effect. Global GHG emissions due to human activities have grown since pre-industrial times, with an increase of 70% between 1970 and 2004[4]. Apart from the three natural GHGs (carbon dioxide, methane and nitrous oxide), the increased emission also includes several "man-made" gases including chlorofluorocarbons (CFCs), hydrofluorocarbons (HFCs), perfluorocarbons (PFCs) and sulphur hexafluoride (SF6). Increase in the concentration of these GHGs tends to increase the surface temperatures. This rise in the average temperature of the Earth is called global warming, which is likely to lead to climate changes and threatening for the ecosystems of the entire world.

Sustainable Development

The meaning of term 'sustainable' is relatively easy to interpret: it means 'enduring' and 'to keep in being'. Development could be narrowly defined in traditional terms as real GNP per capita, or real consumption per capita, but in broad term it could be to include other indicators such as education, health and some measure of the 'quality of life' including human freedom. On similar ways, Sustainable Development has been defined over a wide spectrum, starting with the narrow definition of 'increasing GNP and per capita income over time' to a wider conception of 'environmental sustainability'. World Commission on Environment and Development's (popularly known as the 'Brundtland Commission') defines sustainable development as "development that meets the needs of the present generation without compromising the ability of future generations to meet their own needs" (WCED, 1987). Sustainable development has come to settle as a desirable goal for the society due to the emphasis and endorsements of the term on the nexus between environment and economic development. According to Perman *et al.* (1999), the rationale for sustainable development arises from the 'desirability of cautious behaviour' with an objective to 'minimize ecological disruptions'. Such desirable behaviour, has moral, social, ecological, and economic grounds. From a moral stand point, they argue that actions of present generation must not jeopardize equal opportunities for future generation as enjoyed by the present. From angle of social sustainability, social equality is important factor for sustainability and on ecology ground bio-diversity is an important objective in its own right, any activity that threatens such diversity is undesirable. Lastly, from the point view of economics sustainable economic behaviour is more efficient than non-sustainable behaviour. Measuring sustainability is one of the most difficult tasks. It can be measure by two following approaches-economic approaches and accounting approaches. Although both the approaches provide some guidelines towards valuing the environment and thus measuring sustainability but none of the approach is complete in any sense.

Sustainability can be measure in economic terms through intergenerational equity in welfare. Sustainable Development ensures that the future generations of individuals are at least at the same level of welfare as enjoyed by today's generation. Hence, intergenerational equity in welfare of the future generations lies at the heart of sustainable development

Sustainability can be measure in accounting terms through impact of our economic activities on the environment. It is important to consider the role of the environment in our economy and the impact of our economic activities on the environment. Hence, a true national income accounting system must consider the role of environment and the impact of economic activities on the environment. Hence, the value of such impact needs to be subtracted from the conventional concept of income to arrive at a true figure of net income. Such accounting measure which is a true indicator of sustainable development is called Green Accounting. Green Accounting estimates the prices for all national assets, including natural and human capital assets, and includes those estimates in the 'financial statement' of the nation. Green Accounting involves an array of quantitative estimations, which involves modeling and valuing the non marketed services of environmental assets such as forests, calculating the value of education as a generator of future incomes, valuing future liabilities in the form of pollution abatement costs and healthcare costs, etc. One of the important indicators of Green National Accounting system is the Environment Adjusted Domestic Product or Eco-Domestic Product, also popularly known as the 'Green GDP'. The handbook on a System of Integrated **'Environmental and Economic Accounts'** published by the United Nations (1993) defines eco-domestic product as follows:

Eco-domestic product (EDP) = Gross domestic product (GDP) – [depreciation of produced assets + depletion of natural resources + degradation and pollution damage]

However, in term of sustainability, eco-domestic product does not directly measure whether this true income can be sustained. This leads to the concept of genuine savings, which can be conceived as the net change in wealth over an accounting period as new assets are invested, old assets depreciate, and natural resources are augmented or depleted. Since wealth represents the potential to generate income and induce well-being, changes in wealth or genuine savings are intimately linked with sustainability. The accounting definition of genuine savings is as follows:

Genuine savings = Net savings – [depletion of natural resources + pollution damages]

Where,

Net savings = gross savings - depreciation of produced assets.

Therefore, using the accounting measure, one can infer a country as sustainable, if the genuine savings are non-declining over time. In other words, a country is sustainable if the growth in its genuine savings is persistently positive over time or environment degradation is minimizing during development process.

Linkage between Environmental Degradation and Sustainable Development and Climate Change

Sustainable development deals with the linkage between economic development and environment. It has been observed that the economic development and environment are inseparably connected. This inter-relationship is essential and exists because-

a) Economic development affects the environment.

b) Environmental quality affects the economic development

There is close linkage between Environment and development, Economic Development of any country is primarily depends upon available natural resources. Environment supplies raw materials for economic development of the economy, raw materials transformed into consumer product through the production process. It also clean up waste material and recycle them for renewable and reuse. Environment also affects the productivity and efficiency of human resource.

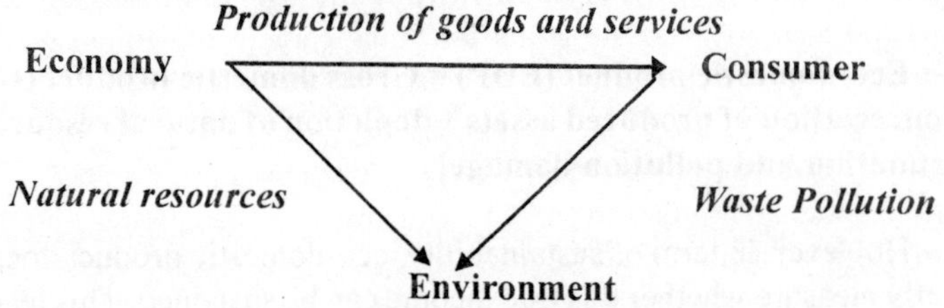

Figure -1

Not only environment affects economic development but economic development also affects environmental degradation. Environmental degradation occurs as the result of our production and consumption related activities and environmental degradation has an impact on green house gases emission that leads to global warming. Climate change is subsequence result of global warming. Thus Economic development links with environmental degradation, environmental degradation not only itself effects economic development leads climate change and climate change also has an impact on economic development of a country. There is a cyclic trend from environmental degradation to economic development.

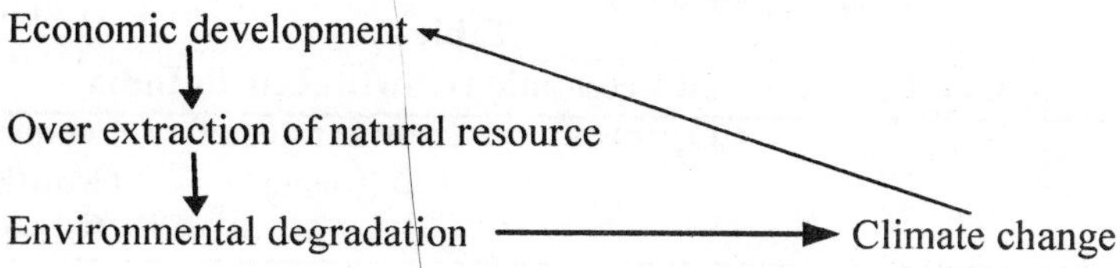

Figure -2

(I) Impact of Economic Development on Climate Change

Developing countries are faced with immediate concerns that relate to forest and land degradation, freshwater shortage, food security and air and water pollution. Climate change will exacerbate the impacts of deforestation and other economic pressures, leading to further water shortages, land degradation and desertification.

India is also a developing country, the process of industrialization; urbanization; transportation and utilization of latest and modern technique for development in different sector including agriculture increase the concentration of Green House Gases in atmosphere. Affects of green house gases emits during the development process is not regional but it affects whole glob. Development planning in India has increasingly incorporated measurable goals for enhancement of human well being, beyond mere expansion of production of goods, services and the consequent growth of per capita income recent years. India has many future development goals, which are directly or indirectly linked to environment degradation and energy consumption and therefore to GHG emission. Indian economy is growing economy. There is mass development of industrialization. The development of a diversified industrial structure, base on a combination of large scale industries, along with growing population has contributed to the growing incidence of air pollution. Due to large size of population and high growth rate of population in India demand for food grain increased, and for meets the need of demand, new technique for food production called 'Green Revolution' has introduced, which also main cause of environmental damage. Industrialization and urbanization damage the land, desertification leads to climate change. Table -1 shows the growth rate of CO_2 in India (in tones) and economic growth rate of India, as the economic growth increase emission of CO_2 also increased. There is positive correlation between economic growth rate and growth rate of CO_2 emission in India (in tones).

Table -1

CO_2 Emission and Economic Growth Rate in India

Year	CO_2 (tones)	Growth rates of CO_2(tones)	Economic Growth rate (Percent)
2000-01	78055.80	-	5.15
2001-02	80147.64	2.68	4.10
2002-03	83097.36	3.68	8.60
2003-04	86739.47	4.38	6.90
2004-05	915535.21	5.53	7.30
2005-06	96662.77	5.60	7.70
2006-07	104290.67	7.89	8.10
2007-08	111223.17	6.65	7.50
2008-09	118355.70	6.41	7.41

Source: Growth rate Compile from different volumes of Economic Survey, Government of India and Bureau of Energy Efficiency.

Country -wise Green House Gases Emission

Developed countries have largest historical and current share in emissions of GHGs than the developing countries. Present trend shows, the developed countries emit more GHGs than developing countries as shown in table-2.

Table – 2

Per Capita CO_2 Emission and Percentage Share of Global CO_2 Emission of Different Countries

Country	Per Capita CO_2 Emission (tones per Person)	Percentage share of global CO_2 Emission(2007)	Percentage change in GHG Emissions (1992-2007)
China	4.75	21	154.4
USA	19.94	20.1	18.1
India	1.25	4.68	110.99
Brazil	2.05	1.33	67.22
South Africa	9.35	1.51	40.11
Japan	9.91	4.22	17.13
Australia	21.99	1.53	67.8
Spain	9.47	1.28	50.8
Germany	10.13	2.77	-7.1
UK	9.28	1.89	-2.62

France	6.36	1.35	5.84
Global	4.52	100	39.22

Source: *Energy Information Administration (EIA) U.S. Department of Energy.*

Table -2 shows per persons and percentage CO_2 emission of different countries. Share of China and USA in total global pollution is highest. Germany Sweden and UK have drastically reduced CO_2 emission level during fifteen years, and fastest developing countries like India, Brazil, South Africa, China, are likely increase their GHGs emission level, but because of their higher population and economic growth rates, the fossil fuel CO_2 emissions from developing countries are likely to soon match or exceed those from the developed countries. When fossil fuel CO_2 emissions alone are considered, due to population and economic growth in the coming decades, the contribution of developing countries as a group will soon overtake the developed countries. Historically, the responsibility for emissions increase lies largely with the industrialized world, though the developing countries are likely to be the source of an increasing proportion of future increases.

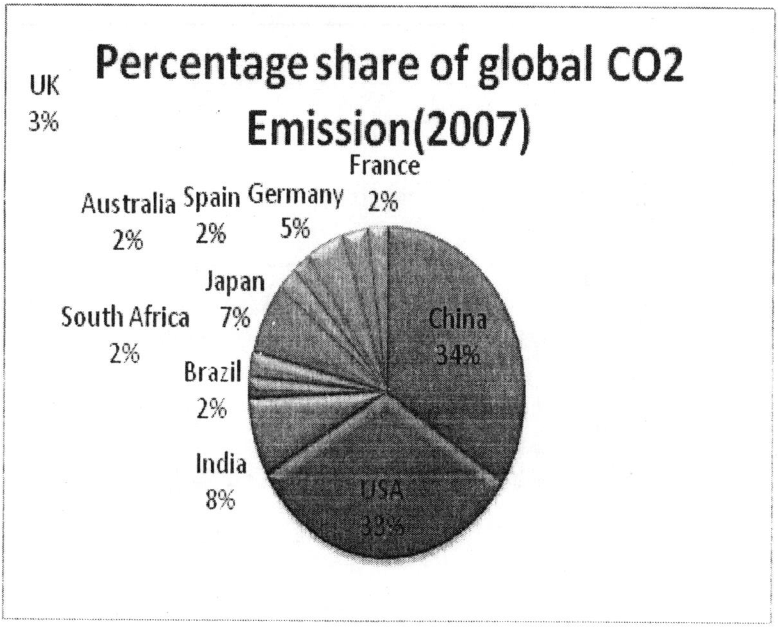

Figure – 1 Percentage share of global CO_2 Emission of different countries ***(Compile by Table-2)***

Table -3 shows the trend of per capita CO_2 emission of some selected countries. Developing country like India, China shows increasing trend 1990 to 2008, but per person CO_2 Emission is low then developed countries. Developed country like USA, UK and Russia shows decreasing trend 1990 to 2008, but per person CO_2 Emission is very high than developed countries. US emits twenty times more, Most country in Western Europe and Japan emit ten times more per capita carbon. Composition of different gases in GHG emission is as shown- highest emission of CO_2 (65 percent), followed by CH_4 (31percent), followed by N_2O only (4 percent).

Table -3

Trend of per capita CO_2 Emission of Some Selected Countries (tones /person)

Country	1990	2003	2004	2008
USA	19.3	19.5	20.6	18.5
UK	10.2	10.6	9.8	8.7
Russia	15.7	10.8	10.9	11.9
China	2.2	3.2	3.8	5.7
India	0.8	1.0	1.2	1.4

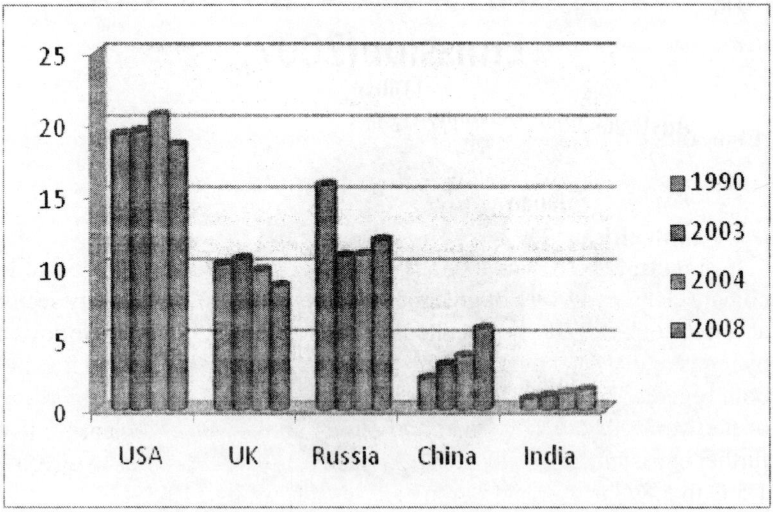

Figure – 2 Trend of per capita CO_2 Emission of some selected Countries (tones /person) (*Compile by table -3*)

CO_2 Emission Sector-wise

The unimpeded growth of GHG emissions is raising the earth's temperature. The consequences include melting glaciers, more precipitation, more and more extreme weather events, and shifting seasons, as mentioned earlier. IPCC has shown that the earth temperature has increased by 0.74° between 1906 and 2005 due to increase in anthropogenic emissions of greenhouse gases.

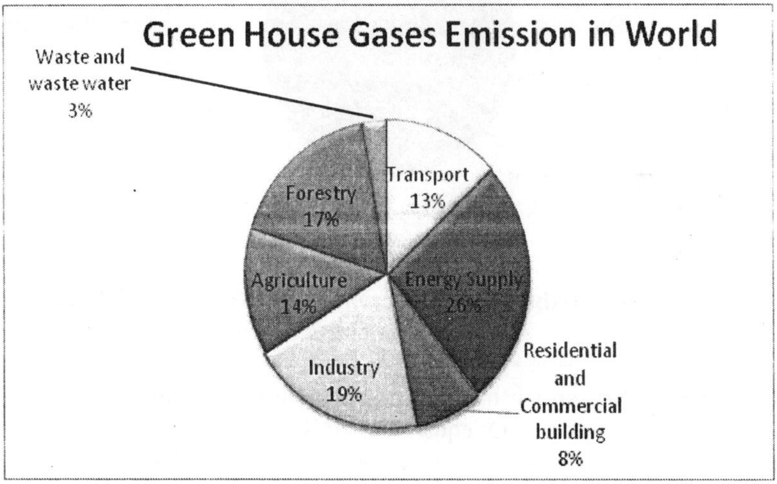

Figure- 3 Contribution of Different Sectors of the World in Green House Gases Emissions.

Source: World Development Report 2010.

Figure-3 presents the contribution of different sectors in world to climate change (sources of greenhouse gas emissions). The energy sector accounts highest share in greenhouse emissions of 25.9 per cent, followed by industry (19.4 per cent) and forestry (17.4%). Agriculture and forestry both together accounts for 30.9 per cent share in total GHGs emissions of the world. There is a broad consensus among climate scientists that further emissions of greenhouse gases will cause temperatures to increase 1.5°C to 5.8°C and precipitation patterns to shift by 2100[6].

India's share in global CO_2 emission is only approximate 5 percent. Historically and at present share of India in carbon stock in the atmosphere is relatively very small in terms of per capita emission.

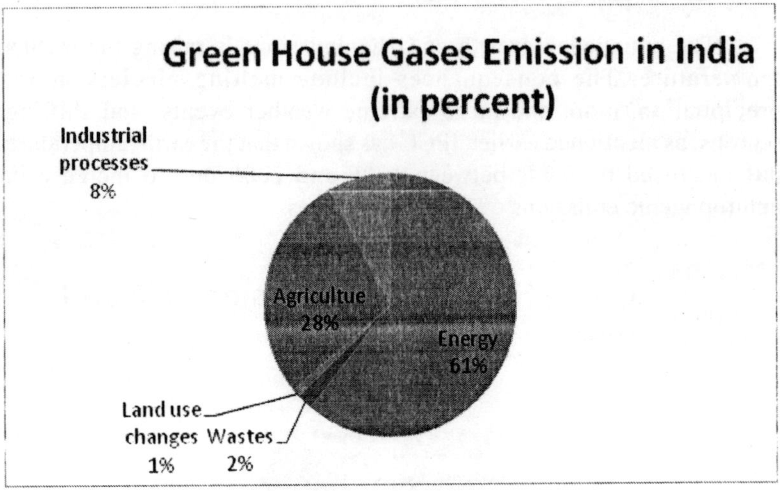

Figure- 4 Contribution of Different Sectors of India in Green House Gases emissions.

The sector- wise distribution of greenhouse gases emission in India shows that the highest CO_2 equivalent emission contribution is in from the energy sector (61%) and lowest from land use, land-use change and forestry 1%. Agriculture contributes 28% of the Indian greenhouse gases emissions, primarily due to methane emission from rice, enteric fermentation in ruminant animals, and nitrous oxides from application of manures and fertilizers to the soil.

Kyoto Protocol

In 1997, parties to the UNFCCC adopted the Kyoto Protocol to strengthen the commitment of the developed countries towards carbon emissions reduction. Kyoto Protocol is viewed as one of the significant measures in the reduction of carbon emissions of the developed countries. Under Kyoto protocol, even though there is an obligation on the part of the developed countries to commit them to a second round of quantified emission reductions for the next decade (the present commitments expire in 2012), they have maintained a deafening silence over this issue. Rather, their intention seems to escape from the quantitative emission reduction commitments that are obligatory for the developed countries under the

Kyoto protocol and to introduce, at the same time, new commitments for the developing countries.

Kyoto Protocol broke new ground by defining three innovative 'flexibility mechanisms' for reducing emissions, and achieving the goals of the Climate Convention. These mechanisms enable Parties to access cost-effective opportunities to reduce emissions or to remove carbon from the atmosphere in other countries. While the cost of limiting emissions varies considerably from region to region, the benefit for the atmosphere is the same, wherever the action is taken. Much of the negotiations on the mechanisms have been concerned with ensuring their integrity. The three Kyoto mechanisms are as follows:

(i) Joint Implementation (JI) under Article 6 provides for Annex I Parties to implement projects that reduce emissions, or remove carbon from the atmosphere, in other Annex I Parties, in return for emission reduction units (ERUs).

(ii) Clean Development Mechanism (CDM) defined in Article 12 provides for Annex I Parties to implement projects that reduce emissions in non-Annex I Parties, or absorb carbon through afforestation or reforestation activities, in return for certified emission reductions (CERs) and assist the host Parties in achieving sustainable development and contributing to the ultimate objective of the Convention.

(iii) Emissions Trading (ET), as set out in Article 17, provides for Annex I Parties to acquire certified emission reduction units from other Annex I Parties.

Among the above three mechanisms, only Clean Development Mechanism is relevant to developing countries such as India. Developing countries could view Clean Development Mechanism as an opportunity not only to attract investment capital and Environmentally Sustainable Technologies (ESTs) but also to implement innovative technical, institutional and financial interventions to promote energy efficiency, renewable energy and forestry activities that contribute to sustainable development. Projects specially designed and implemented in developing countries under Clean Development Mechanism, leading to carbon emission reduction or sequestration will receive payments called carbon price from institutions and agencies in Annex B (Annex I countries with commitment to reduce GHG emissions) countries for every tonne of carbon emission avoided or sequestered. Clean Development Mechanism provides an opportunity for developing countries to access modern

Environmentally Sustainable Technologies and receive financial incentives to overcome the barriers.

Projected Climate Change in India

Projection of future climate is a complex and uncertain exercise. Future projections are needed to assess possible future risk, hazard, and opportunities. Future climate projections are necessary for adaptation towards future climate change and develop a coherent risk management program. A large number of future climate projections have been done, most of them based on Hadley centre regional climate model. Climatologists have projected an increase in the intensity and frequency of extreme events such as drought, floods and cyclones in India.

The IPCC, in its 2007 report, predicts that global temperatures will rise by 2-4.5°C by the end of this century, with a 2.7-4.3°C increase over India by 2080s. The panel also predicted an increase in rainfall over the Indian sub continent by 6-8% and that the sea level would rise by 88cm by 2100. Another projection predicted region-wise temperature and precipitation change for the period 2070-2099 with reference to the base period 1960-1990, is shown in table- 4.

Table-4

Projected Climate Changes in India: 2070- 2099

Region	Jan.- March	April- June	July- Sept.	Oct.- Dec.
Temperature changes(°C)				
Northeast	4.95	4.11	2.88	4.05
Northwest	4.53	4.25	2.96	4.16
Southeast	4.16	3.21	2.53	3.29
Southwest	3.74	3.07	2.52	3.04
Precipitation change (%)				
Northeast	9.3	20.3	21.0	7.5
Northwest	7.2	7.1	27.2	57.0
Southeast	32.9	29.7	10.9	0.7
Southwest	22.3	32.3	8.8	8.5

Source: Cline(2007)

(II) Impact of Climate Change on Economic Development:

Climate change adversely affected economic development, it affects mainly agriculture which is the base of human livelihood and provide raw material for most of the industrial processes. It impacts on agriculture and livestock production many ways like degree of rise temperature, amount and time of precipitation, availability of water etc..Climate change affects on human resource development by affecting the human health. Climate changes also affect environmental balance and show adverse affect on ecosystem and bio-diversity. Climate change will exacerbate the impacts of deforestation and other economic pressures, leading to further water shortages, land degradation and desertification. Increasing global temperatures will result in rising sea levels. Populations that inhabit small islands and/or low-lying coastal areas are at particular risk of severe social and economic disruptions from sea-level rise and storm surges that could destroy cities and disrupt large coastal livelihoods

Impact of Climate Change on Agriculture Sector

Climate Change is projected to have significant impacts on conditions affecting agriculture, including temperature, precipitation and glacial run off. It affects agriculture many ways. It can affect crop yield as well as cropping pattern that can be grown in certain areas. It is also impacted agricultural inputs such as water for irrigation, amounts of solar radiation that affect plant growth, as well as the prevalence of pests. Rise in temperatures caused by increasing green house gases is affected crops differently from region to region. For example, increase of 1 to 3°C in mean temperature(moderate warming) is expected to benefit crop yields in temperate regions, while in lower latitudes especially seasonally dry tropics, even moderate temperature increases 1 to 2°C are likely to have negative impacts for major cereal crops. Warming of more than 3°C is expected to have negative effect on production in all regions[8]. The Third Assessment Report of the IPCC, 2001 concluded that climate change would hit the poorest countries severely in terms of reducing the agricultural products. The Report claimed that crop yield would be reduced in most tropical and sub-tropical regions due to decreased water availability, and new or changed insect/pest incidence. In South Asia losses of many regional staples, such as rice, millet and maize could top 10 per cent by 2030[9]. Agriculture is important for food security in two ways: it provides the food and also the primary source of livelihood for 38.7 percent of the world's total workforce. In Asia and the Pacific, this share accounts for approximately 50 per cent and in sub-Saharan Africa, nearly two-thirds (63 per cent) of the working population still make their living from agriculture.[10]

India is agro based country; Most of the people of India lived in rural areas and depend on agriculture. If we see the impact of climate change on Indian agriculture, we find productivity of most crops may decrease due to increase in temperature and decrease in water availability, especially in Indo-Gangetic plains. This apart, there would be a decline in the productivity of rabi as compared to kharif season crops. Rising temperature would increase fertilizer requirement for the same production targets and result in higher GHG emissions, ammonia volatilization and cost of crop production[11]. Increased frequencies of droughts, floods, storms and cyclones are likely to increase agricultural production variability. Therefore, we have to place equal emphasis on saving lives and sustaining Livelihoods[12].

Table-5
Impact of climate change on Indian Agriculture (agronomic assessment)

Source	Area	Crop	Yield Change (%)	Scenario	Model
Kalra,et al.,2007	DEFRA Study	Rice	-5 to -8	$+1^{0}C$; No change in CO_2	CERES-Rice
			-10 to -16	$+2^{0}C$; No change in CO_2	
			-21 to -30	$+4^{0}C$; No change in CO_2	
		Maize	-10 to -30	$+1^{0}C$ to $+4^{0}C$; 350ppm CO_2	CERES-Maize
		Jowar	-7	$+1^{0}C$	CERES-Sorghum
			-12	$+2^{0}C$	

Table-5 shows that 1^{oC} rise in temperature decrease the yield of Rice 5 to 8 percent, and 2^{oC} rise in temperature decrease the yield of Rice 10 to 16 percent., temperature is increase 4^{oC} decrease the yield of Rice 21to 30 percent. Incensement in temperature of atmosphere also decreases the yield of Maize and Jawar.

Rain Precipitation

Agriculture will be adversely affected not only by an increase or decrease in the overall amounts of rainfall but also by shifts in the timing of the rainfall. Any change in rainfall patterns poses a serious threat to agriculture, and therefore to the economy and food security. Summer rainfall accounts for almost 70 per cent of the total annual rainfall over India and is crucial to Indian agriculture. However, studies predict decline in summer rainfall by the 2050s[13]. Semi arid regions of western India are expected to receive higher than normal rainfall as temperatures soar, while central India will experience a decrease of between 10 and 20 per cent in winter rainfall by the 2050s.[14] Relatively small climate changes can cause large water resources problems particularly in arid and semi arid regions such as northwest India. Climate change adversely affects the precipitation and total precipitation in India decrease at present time.

Table-6

All India Rainfall Distribution and percent Departure from 1992-93 to 2009-10

(in Millimeters)

D- Stands for Departure here.

Year	Monsoon Season (June- Sept.)			Post Monsoon Season (Oct.-Dec.)	Winter Season (Jan.-Feb.)	Pre-Monsoon Season (March-May)	Over all Rain fall (June-May)		
	Actual	Normal	%D	%D	%D	%D	Actual	Normal	%D
1992-93	830.7	899.2	-7.6	-6.7	-7.6	-4.0	1091.6	1175.6	-7.1
1995-96	904.5	904.7	0.0	-1.8	-8.3	-23.4	1154.6	1189.3	-2.9
1999-00	866.9	904.7	-4.0	18.8	1.4	-0.5	1183.5	1197.0	-1.1
2005-06	879.3	892.2	-1.0	10.0	-37.0	3.9	1185.4	1196.8	-1.0
2006-07	886.6	892.2	-0.6	-21.1	-21.7	-15.6	1133.0	1195.5	-5.2
2007-08	936.9	892.2	5.0	-32.2	-1.4	-13.6	1180.2	1194.8	-1.2
2008-09	873.2	892.2	-2.1	-30.7	-46.1	-32.3	1075.0	1196.4	-10.1
2009-10	689.8	892.2	-22.7	7.6	-43.8	-8.1	972.8	1195.6	-18.6

Sources: *Directorate of Economics and Statistics, Department of Agriculture and Cooperation*

Table-6 shows trend of rain precipitation last 18 years. Monsoon satiation has worsened across the country with deficiency in the cumulative rainfall. Delhi, Utter Pradesh, Punjab, Haryana, Bihar, Orissa, Andhra Pradesh, and Maharashtra continue to be worst hit by climate change.

Impact of Climate Change on Ecosystems and Bio-diversity

Climate Change is the main cause of immense biodiversity loss, affecting both individual species and their ecosystems that support economic growth and human well being. It is difficult to predict the overall result of climate changes on animal and plant kingdom and ecosystem. Global warming has devastating effects on the native habitats of many animals and plants, and drives a considerable number of animal and plant species to extinction. Mass extinctions of the earth's flora and fauna have occurred before also but those were driven by natural factors. World Wildlife Fund (WWF) asserted that one-fifth of the world's most vulnerable natural areas may be facing a "catastrophic" loss of species[15]. Studies predict that global warming will also lead to extinction of insects in the tropical zone by the end of the century while insects in the temperate zones and the poles could experience a dramatic increase in numbers. It will also have catastrophic impact on the marine ecosystems.

India has mega-biodiversity, encompassing features from glaciers to deserts. However, climate change is posing grave threat to its ecosystems. Temperature increase due to climate change and human activities are causing fragmentation and degradation of mountain biodiversity. **The Himalayan Ecosystem** is considered as the lifeline to India and neighbouring countries such as China, Pakistan, Nepal, owing to the perennial rivers that arise out of the melting glaciers. However, climate change is threatening this **Ecosystem.** It is also predicted that there will be an increase in the phenomenon of the melting glaciers of the Himalayas have a serious impact given the fact that they give rise to the perennial rivers that further flourishes the agriculture. The Himalayan rivers are closely interlinked with the **Indo-Gangetic Ecosystem,** which is primarily an agricultural ecosystem, nearly 65-70% of Indians having agriculture as their primary occupation[16]. The National Environment Policy, 2006 states that the Indian **Desert Ecosystems** (arid and semi-arid region) occupies 127.3 Mha (38.8%) of the country's geographical area and spreads over 10 states. The Indian desert fauna is extremely rich in species diversity of mammals and winter migratory birds. Recent studies have shown that deserts have shown signs of expansion, this process called desertification. The climate patterns have altered the natural attributes of a desert region; for example the floods in the desert district of Barmer in Rajasthan in 2006. **Coastal and Marine Ecosystem** is one of the assets of India. Themangrove forests (wetlands) of the rivers and the coasts acts as carbon sink as well as a habitat for a unique and diverse species of plants and animals. The wetlands act as a natural barrier to cyclones and flooding that may be caused by the rising sea levels. The climate change also affects the marine ecosystem for example of coral bleaching. In the Peninsular India, even the rivers of the Peninsula are dependent on the monsoons, thus the **Peninsular Ecosystem** is basically a monsoon dependent ecosystem. India is heavily dependent on the monsoon to meet its agricultural and water needs, and also for protecting and propagating its rich biodiversity. Climate change is linked with the changing patterns observed in the monsoons of India.

Impact of Climate Change and Health

Climate change poses a threat to the survival of mankind; it has catastrophic effects on human health. Each year, about 800,000 people die from causes attributable to air pollution, 1.8 million from diarrhea resulting from lack of access to clean water supply, sanitation, and poor hygiene, 3.5 million from malnutrition and approximately 60,000 in natural disasters[17]. A warmer and more variable climate would result in higher levels of some air pollutants, increased transmission of diseases through unclean water and through contaminated food. Climate change has a direct impact on human health. The warmness of the climate impact on likelihood and human health becomes worse. It is anticipated that there will be an increase in the number of deaths due to greater frequency and severity of heat waves and other extreme weather events. Climate change and the resulting higher global temperatures are causing increasing frequency of floods and droughts leading to the risk of disease infections. Areas affected by frequent floods and drought conditions also witness large scale migration of populations to relatively stable regions leading to overcrowding and unhygienic conditions resulting in transmission of diseases. Climate change is a main cause of the spread of infectious diseases. The World Health Organization (WHO) in their studies have indicated that due to rising temperatures, malaria cases are now being reported for the first time from countries like Nepal and Bhutan. Dengue fever is already in evidence at higher levels of elevation in Latin America and parts of East Asia. Climate change could further expand the reach of the disease[18] Rising temperatures and changing patterns of rainfall are projected to decrease crop yields in many developing countries, stressing food supplies. This will ultimately translate into wider prevalence of malnutrition/ under nutrition. In some African countries, yields from rain-fed agriculture could be reduced by up to 50 per cent by 2020.[19] Emission of the Green House Gases have been responsible for the depletion of ozone layer, which protects the Earth from the harmful direct rays of the sun. Depletion of stratospheric ozone results in higher exposure to the ultra violet rays of the sun, leading to an increase in the incidents of skin cancer. The projections by WHO and IPCC suggest that the negative effects of climate change on health are greater on poor populations. The negative health impact will vary from one location to another.

Measures for Sustainable Economic Development

As we Know Economic growth directly links with environment degradation. Goal of high economic growth and development are related with some important environmental issues such as irreversible damage with implementation of food production, water supply, health, energy and bio-diversity. To minimize the environment degradation, most effective way is to adopt a sustainable development pathway by shifting to environmentally sustainable technologies and promoting energy efficiency, renewable energy, soil conservation, water conservation, forest conservation, afforestation, etc. Some important strategies are -

- **Sustainable Management of Natural Resource:** To avoid the over exploitation of natural resource, improve the efficiency of natural resources, manage the value of ecosystem and minimize the environmental degradation we can achieved goal of sustainable development and also save bio diversity.
- **Kyoto Protocol:** To limit the effect of climate change by meeting the commitment under Kyoto Protocol.
- **Combat Poverty:** Poverty is the major cause of un-sustainability, by eradicating poverty we can achieved sustainable economic growth.
- **Adopt Clean Development Mechanism:** The Clean Development Mechanism has accelerated the diffusion of renewable energy and energy efficient technologies by mitigating some of the risk associated with the adoption of new technologies.
- **Conservation of Natural Resources:** Adopt conservation of natural resources like soil, water and forest. Reforestation, aforestation and sustainable forest management can contribute to conserve bio diversity.

- **Minimize Pollution and Green House Gases Effect:** By adopting sustainable technique and clean energy sources as fuel we can be minimize the pollution and emission of GHGs and reduce the effect of climate change.
- **Social Equality and Human Resource Development:** Limited major threats to public health, ensuring food security, equality and sanitation.
- **Awareness:** Education and training can be given to common man to safe guard and adapt to climate change.

India's Eco-friendly Measures for Sustainable Development

Despite the fact that India's contributions to greenhouse gas emissions are very small; the Government of India has taken many measures to improve the situation in this regard. The Ministry of Environment and Forests is the nodal agency for climate change issues in India. India has initiated several climate-friendly measures, particularly in the area of renewable energy. It has only dedicated Ministry for non-conventional energy sources in the world named Ministry of New and Renewable Energy.

India had adopted the National Environment Policy 2006 which provides for several measures and policy initiatives, to create awareness about climate change and help capacity building for taking adaptation measures. The National Forest Policy also provide active measures for expanding carbon sinks through increase in cover area under forest up to 33 per cent by 2012. A major afforestation programme covering 6 mn hectares has been launched under the XIth Plan for this purpose. National Action Plan on Climate Change (NAPCC) has adopted on 30th June 2008, with a view to lay down the priorities and future actions of the Government for addressing climate change and updating India's national programme relevant to addressing climate change. The National Action Plan identifies measures that promote our development objectives and yielding co-benefits for addressing climate change effectively. Eight national missions like –solar energy mission, energy efficiency, sustainable habitat, water, Himalayan ecosystem, green India, Eco-green agriculture and knowledge about climate change, have been adopted specifically for advance India's development and climate change related objectives of adaptation and GHG mitigation. However, India has not set any quantitative goals towards emission reduction. The Government has set up an "Expert Committee on Impacts of Climate Change" on 7 May 2007 under the chairmanship of Dr. R. Chidambaram, Principal Scientific Advisor to the Government of India, to study the impacts of climate change on human being in India and also to identify the measures that India may have to take in the future in relation to addressing vulnerability to climate change impacts on human being. Moreover, a Council has also been set up under the Chairmanship of the Prime Minister of India on 6 June 2007 constituting eminent persons to evolve a coordinated response to issues relating to climate change at the National level and provide oversight for formulation of action plans in the area of assessment. Besides, the Indian Government has initiated "Green India" programme which envisages undertaking massive afforestation of degraded forests land in the country. The "Green India" programme will cover about six million hectare in the country in about 10 years.

Conclusion

Climate change and environment degradation have been main issue of debate in recent time on global concern. It is the greatest challenge to sustainable development. It should be addressed by all countries with a shared perspective and agenda. The developed countries need to look beyond their narrow self interests and work jointly with the developing countries to evolve cooperative strategies on the issue of climate change, which is deeply related to the future of mankind. However, the efforts so far in the direction of meeting the challenges of climate change are not sufficient; we urgently need a new economic paradigm, which is scientific, global, inclusive, cooperative, and environmentally sensitive. The world's current ecological, demographic and economic trajectory is unsustainable, need to turn it in sustainable manner.

The most effective way to facing climate change is to adopt a sustainable development by shifting the way of environmentally sustainable technology, forest conservation, water conservation, soil conservation reforestation etc. Sustainable development based on addressing the needs of the poor and optimal harnessing of scarce resources of water, air, energy, land, and biodiversity will have to be sustained through more cooperative endeavours. Then alone, we could make some headway in saving our lone planet from the brink of climate disasters.

Foot Notes

1. Climate Change 2001: Synthesis Report, Intergovernmental Panel on Climate Change, Geneva, Switzerland, 2001.
2. Climate Change 2007, Synthesis Report (A Report of the IPCC), p.2
3. Article 1, United Nations Framework Convention on Climate Change
4. Climate Change 2007, Synthesis Report (A Report of the IPCC), p.5
6. Houghton et al., 2001.
7. Franklin Cudjoe and Bright Simons, 'UN climate change plans and world's poor', The Pioneer, 9 September, 2008
8. IPCC Fourth Assessment Report, 2007, p.38
9. Climate Change and Agriculture (http:en.wikipedia.org/wiki/Climate-change-and-agriculture)
10. Global Employment Trends: Brief (ILO, January 2007, p.12)
11. Effect of Global Warming on Crop Productivity, Y.S. Shivay and Anshu Rahal, Kurukshetra, July 2008, p.19
12. M.S. Swaminathan, For an Action Plan for Bihar, The Hindu, 5 September, 2008
13. Climate Change and its Impact on India (http://www.greenpeace.org/india/campaigns/ choose-positive-energy/what-is-climate-change...)
14. Impacts of Climate Change: Western and Central India (www.cseindia.org/programme/geg/ pdf/western.pdf)
15. Bruce E. Johansen, 'Global Warming in the 21st century: Plants and Animals in Peril', p.536
16. India 2008
17. World Health Organization: Protecting Health from Climate Change
18. UNDP Human Development Report 2007-2008
19. Climate Change 2007: Impacts, Adaptation and Vulnerability (Working Group II Contribution to the Fourth Assessment Report of the Intergovernmental Panel on Climate Change Summary for Policymakers and Technical Summary)

References

- Herbert Girardet (ed.), Surviving the Century, Earthscan, London, 2007.
- The Day After Tomorrow: Impact of Climate Change on the World's Water, Dr. Pradipto Ghosh (Terragreen, 2008)
- Coping with Climate Change: Gautam Dutt and Fabiina Gaioli (Economic & Political Weekly, 20 Oct., 2007).
- Effect of Global Warming on Crop Productivity, Y.S. Shivay and Anshu Rahal (Kurukshetra, July 2008)
- M.S. Swaminathan, For an Action Plan for Bihar, The Hindu, 5 September, 2008 Eric Neumayer, Greening Trade and Investment: Environmental Protection Without Protectionism, Earthscan Publications Ltd., 2001.

- Bruce E. Johansen, Global Warming in the 21st century: Plants and Animals in Peril (www.greenwood.com).
- P.P.Sangal, India's Climate Change Action Plan, The Economic Times, New Delhi, 29 July, 2008.
- An Assessment of the Intergovernmental Panel on Climate Change: Climate Change 2007: Synthesis Report.
- Climate Change 2007: Impacts, Adaptation and Vulnerability (Working Group II Contribution to the Fourth Assessment Report of the Intergovernmental Panel on Climate Change).
- Third Assessment Report of the IPCC, 2001
- IPCC Fourth Assessment Report,2007.
- United Nations Framework Convention on Climate Change (UNFCCC).
- World Health Organization: Protecting Health from Climate Change.
- Keynote speech by Angel Gurría, OECD Secretary-General, during the World Energy Council Energy Leaders Summit, London, 16 September 2008

2. Environment, Sustainable Development and Globalization:
Challenges and Options for India

C.S.Verma

The impact of globalization on environment and sustainable development needs to be continuously addressed in Indian context which profoundly remains in the transition. International trade arrangements and environmental agreements contain very few provision for harmonizing trade and environment trade and development. The findings of majority of recent reports point out that India's rapid growth is driving equally rapid environmental destruction. An argument often put forth in developing countries is that it is unfair to ask people to make environmental sacrifices during a period of growth and industrialization when Western countries did not have to make the same choices.

This paper attempts to analyse the policy options available for India and suggests the targets it may set for itself, along with some developing countries including China for post-2012 which is not binding under the Treaty but on which they will report to the UNFCCC for the next commitment period till, say 2030, India would simultaneously adopt measures to reduce inequality in per capita energy consumption within the country ,monitored through a suitable measure such as a Gini index for energy inequality.

Key Words:-Globalization, Economic development, Environment, Exploitable resource, Ecosystem, Gini Index.

Introduction

Human being acts as a rational and social partner in the environmental actions. Though the ever increasing human activities lead to degradation of the environment but the role of human beings in the working of the environment can never be under emphasized. The activities of human beings towards the whole environment are generally aimed to continue for a long time, longer than a lifetime. It is therefore very important for us to understand that for the humanity to survive, a synergy has to be developed between development and environment. The same is understood as the integration of environmental, social, political and economic development, underpinned by equity. It meets the needs of the present without compromising the ability of future generations to meet their own needs.

Natural resources and ecosystems, on which human life, and quality of life, depend, have been strained in recent decades to, and sometimes beyond, their limits, as demonstrated by climate change and the global loss of biodiversity. The economy is dependent on the ecosystem. Climate change is likely to have a range of impacts on health and society, such as changes in disease patterns, drought and extreme weather events, conflict over access to water, and forced migration.

The impact of globalization on environment and sustainable development needs to be continuously addressed in Indian context which profoundly remains in the transition. In spite of the potential of globalization to economic convergence it paved for an increase in inequality resulting in increased environmental impacts such as climate change, protection of the ozone layer, biodiversity and desertification. These international trade arrangements and environmental agreements contain very few provision for harmonizing trade and environment trade and development. The increasing tendency of tram national corporations to establish global standards for environmental performance enhances the contribution of FDI to sustainable development. The Commission on Sustainable Development urged creditor countries and international financial institutions to implement speedily the enhanced heavily indebted poor countries (HIPC) initiative. It also urged HIPC countries to develop national poverty strategies in a participatory way so that debt relief would be linked with poverty eradication[1]. The Agenda 21 refers to national sustainable development strategies as important mechanisms for enhancing and linking national capacity so as to bring together priorities in social, economic and environmental policies. National sustainable development strategies provide an opportunity to put into practice common principles of strategic planning for sustainable development. The substantial human and institutional resources required for the formulation of national sustainable development strategies may impose a burden on India[2].

Present Status and Challenges

The United Nation's Global Compact initiative, as a partnership between the United Nations, the business community, international, labour and civil society organizations commit to open markets while meeting the socio-economic needs of the world's people and contributing to a more human world[3]. The main principles promoted by the Global Compact are taken from the Universal Declaration of Human Rights, the International Labour Organization's fundamental principles on rights at work, and the Rio Declaration on Environment and Development all of which enjoy universal political support and a strong international legal foundation. Understanding the links between globalization and sustainable development is essential to more integrated and strategically focused Indian legal policy making. It allows Governments, International organizations and other stakeholders in the development process to make proactive adjustments to changing conditions in the enabling environment for sustainable development at the national level and through international cooperation[4].

The new economic policies and the structural adjustment programme often predict environmental impact without accuracy because of the complex interplay of various economic, social, political and ecological factors. These are put into effect in an effort to meet India's severe balance of payments crisis, and to propel its economy into quicker growth and global integration[5]. A part from direct fiscal policies, the major components of the new package include boosting exports to earn foreign exchange, liberalizing industrial production, dropping barriers to the entry of foreign companies and goods expanding privatization and cutting government spending. The drastic nature of the NEP package has understandably underscored natural environment conventional economists advocates an exploitable resource, and sink into which the effluents of affluence can be thrown[6].

The crux of all of these reports is that India's rapid growth is driving equally rapid environmental destruction. An argument often put forth in developing countries is that it is unfair to ask people to make environmental sacrifices during a period of growth and industrialization when Western countries did not have to make the same choices.

A report in early 2009[7] on the current environmental challenges facing the India outlined five major environmental issues facing the country:

1. Water availability in India is "rapidly" running dry and is an issue that needs to be confronted soon before it faces a severe water crisis. Only 67% of rural Indians have access to water in their homes (as opposed to 95% in 2005). Solutions can start with rainwater harvesting for large buildings and fixing distribution losses.
2. Invasive species "are the second biggest threat to biodiversity after deforestation." India loses a great deal of valuable plants and animals because of invasive species, but at the same time, many of the introduced crops, such as soya and wheat, are financially viable and important. Solutions could include micro reserves for native plants
3. The loss of natural habitats creates situations in which lions, leopards, and monkeys, amongst other animals, create major problems for humans in their daily interactions. As animals ruin property and take lives, humans are tempted to start killing important parts of the environment. The main solution here is not ruining the animals' native environments, or creating reserves.
4. India's energy grid is direly overtaxed, resulting in major power shortages for much of the country. Building efficiency measures, such as those suggested by the Obama Administration in America (reconfiguring buildings to make them more sustainable and making sure future construction is more environmentally friendly, including natural cooling techniques and solar panels.
5. Mining causes significant soil erosion and deforestation, in addition to forced relocation of tribal peoples. Mining needs to be regulated more strictly by states to prevent widespread illegal mining and environmental ruin.

Climate is changing rapidly. It has put the fate of the world and living species at stake. To cater to the insatiable greed of human beings, Mother Nature opened up all its veins to the so called civilization and consequently it has been exhausted as well. Global temperature is rising sharply; it has created a whole host of problems. Consequently, people living in Bangladesh and in other small island states are faced with the prospect of losing their homes, which would make them climate refugees. In the backdrop of the frequent floods, drought and natural calamities; a grave danger seems to be impending.

Dangers

There are many sound reasons which cannot help draw our concern about the current rapid degradation of the earth's environment. Global warming, brought about by human-induced increases in greenhouse gases (CO_2, methane, N_2O, etc.) is the major concern for us. It is destabilizing the world's climate with dreadful effects for most species on the planet which may cause extinction of human beings as well. Each decade is warmer than the one before, with 2009 as the second warmest year (2005 was the warmest) in the 130 years of global instrumental temperature records[8]. Climate change does not occur in a gradual, linear way, but is non-linear, with all sorts of amplifying feedbacks and tipping points. There are already clear signs of accelerating problems that lie ahead. These include:

More than halfway to the 2015 deadline to achieve the Millennium Development Goals (MDGs), major advances in the fight against poverty and hunger have begun to slow or even reverse as a result of the global economic and food crises, a progress report by the United Nations has found[9]. The assessment, launched by UN Secretary-General Ban Ki-moon in Geneva, warns that, despite many successes, overall progress has been too slow for most of the targets to be met by 2015[10].

The report shows that some hard-won gains are being eroded by the climate, food and economic crises. But even though the economic crisis took a heavy toll on jobs and incomes around the world, the world is still on track to achieve the MDG target of cutting the rate of extreme poverty in half by 2015, the report notes. Progress on Fifth MDG, improving maternal health, lags behind the others, although preliminary reports suggest progress is being made[11].

Funding for family planning has declined during the first few years of the present decade. And inadequate funding for family planning is a major failure in fulfilling commitments to improve women's reproductive health.

Public Health problems caused by Environmental Contamination and Emerging Infectious Diseases are a growing concern, world –wide[12]. These Public health threats are affected by the relationship between people and the Physical, Chemical and Biological nature of our Natural Environments. Population growth and the associated pressures of development are increasing the difficulties associated with sustaining effective public health practices and policies. Vector- borne and zoonotic diseases, water contamination, air borne contaminants, bio accumulative contaminants in the food chain and environmental threats to public health the world over require marshalling of all our scientific knowledge and know-how to develop new solutions. Understanding environmental and ecological health is a pre-requisite to protecting public health.

Options

So what should India's position be, both with regard to international negotiations and for itself? It could certainly be argued, that Kyoto targets have actually been meaningless since almost nobody has adhered to them, and nobody has incurred any penalty.[13]

- According to James Hansen, director of NASA's Goddard Institute for Space Studies Carbon trading, the approach promoted in the Kyoto process is "worse than nothing," since it allows emissions to rise appreciably, while giving the false impression that the problem is being addressed. A carbon fee, he argues, "is the only internationally viable approach." He points to the fact that China —already a leader in alternative energy technology, such as solar panels, wind technology, etc.—is planning to impose such a fee, as among the most hopeful of current developments. It is therefore suggested that a possible Indian position could be the following:
- India (along with China and other large Developing Countries) would set a target for itself for post-2012 which is not binding under the Treaty but on which they will report to the UNFCCC for the next commitment period till, say 2030;
- India would simultaneously adopt measures to reduce inequality in per capita energy consumption within the country (monitored through a suitable suitable measure such as a Gini index for energy inequality·
- Efforts should be made to promote the sustainable use of finite natural resources [including, but not only, fossil fuels and water] as a major determinant of health; develop healthy environments, noting that environmental inequalities can generate health inequalities.
- Action should be taken within an awareness of the global context, including the real poverty of billions of people, who live with the health consequences of environmental degradation on a daily basis.
- To implement energetically policies on climate change and energy, and on sustainable consumption and production [including more appropriate measures of economic success].

Conclusion

There can be little doubt that major developing countries, particularly China and India, are feeling the pressure to take concrete steps on climate change. India, as a dense country of over 1 billion people, faces unique challenges that need unique responses. Action to solve these problems now, even at the expense of slightly slower growth in the future, will allow development to be sustainable and last longer.

The India today needs to usher in a season of transformation, a season of stewardship to make long overdue constitutional commitment to protection and improvement of environment and security of future generation. Government's organizations and financing institutions need to use their resources effectively to improve their performance and to set up their cooperation, working as one to overcome hunger and to consolidate the primary role of sustainable agriculture and rural development in food and environmental security. Indian legislatures should rise to occasion and act in a manner that fundamentally changes the lives of the poor and instill confidence by implementation of laws and policies to usher a humane, equitable and environment friendly society.

The scenario is a grave one, our fate depends on some rich countries. Well! The planetary ecological emergency is not just an issue of natural science, but depends even more today on the development of a critical social-science response. What are the social and economic and above all political causes of the crisis, and how can they be alleviated? It remains a political question. But one thing is clear: global warming must be tackled immediately in order to save the plane.

Foot Notes

1. Reddy KS, Patel V, JhaP, Paul VK, Shiva Kumar AK, Dandona L for *The Lancet* India Group for Universal Healthcare. Towards achievement of universal health care in India by 2020: a call to action. Lancet 2011 10.1016/S0140-6736(10)61960-5. published online Jan 12. *PubMed*
2. National Coordination Committee, Jan Swasthya Abhiyan. Globalisation and health, towards the national health assembly II, booklet 1. New Delhi: Jan Swasthya Abhiyan, 2006 . *http:// www.esocialsciences.com/data/articles/Document 12512007440. 470791. pdf.* (accessed Sept 9, 2010).
3. National Coordination Committee, Jan Swasthya Abhiyan. Health system in India: crisis and alternatives, towards the national health assembly II, booklet 2. New Delhi: Jan Swasthya Abhiyan, 2006. *http:/*

/sanhati.com/wp-content/uploads/2009/09/health_system_in_india.pdf. (accessed Sept 9, 2010).

4. Climate Crisis: Challenges and Options, D. Raghunandan et el, AIPSC, 2008.
5. National Commission for Enterprises in the Unorganised Sector. Report on conditions of work and promotion of livelihoods in the unorganised sector . New Delhi: Government of India, 2007. *http://nceus.gov.in/ Condition_of_workers_sep_2007.pdf.* (accessed Dec 1, 2010).
6. The Convergence of Health and Sustainable Development. A Manifesto and a Network. January 2008. *http://www.healthandsustainability.net*
7. http://***sepia/archives/www.livemint.com(accessed* 1 *May 2011).***
8. National Geographic Society. 1995. Water: A Story of Hope. Washington (DC): National Geographic Society.
9. Millennium Development Goals Report 2009. UN Department of Economic and Social Affairs.
10. Millennium Development Goals Report 2010. UN Department of Economic and Social Affairs.
11. Millennium Development Goals Report 2010. UN Department of Economic and Social Affairs.
12. *The Hindu* (New Delhi), Sep 23, 2009. Court directs apollo to provide free medicines to poor and needy. *http://www.thehindu.com/2009/09/ 23/stories/2009092354760400.htm.* (accessed Sept 9, 2010).
13. See the Global Commons Institute 's policy of Contraction and Convergence, which is a carbon cap and trade policy designed to reduce CO2 emissions: **www.gci.org.uk/contconv/cc.html**

References

- Citizen's Report, 1982
- Luis Pinguelli Rosa "Sustainability & Equity in Climate Convention"in http//:www.teri.res.in/xcut/Havana/paper6_pinguelli.doc.
- "Indias Initial National Communication to the UNFCCC", New Delhi, 2004 at http//unfccc.int/resource/docs/natc/indnc1.pdf.
- "China 's National Climate Change Programme" at *http://en.ndrc.gov.cn/ newsrelease/P020070645611910006823.pdf*
- Environmental Issues, Law and Technology - An Indian Perspective. Ramesha Chandrappa and Ravi.D.R, Research India Publication, Delhi, 2009, *ISBN 978-81-904362-5-0*
- *Population Reference Bureau_*, 2001
- Russell Hopfenberg and David Pimentel *HUMAN POPULATION NUMBERS AS A FUNCTION OF FOOD SUPPLY_* oilcrash.com Retrieved on- February 2008
- *Country Profile: India* Library of Congress Country Studies. December 2004. Accessed: May 18, 2011. *http://lcweb2.loc.gov/frd/cs/profiles/ India.pdf*.

- *"A special report on India: Creaking, groaning: Infrastructure is India 's biggest handicap"*. The Economist. 11 December 2008.

Web Sources

- *http://www.monthlyreview.org/100301magdoff-foster.php*
- *http://www.monthlyreview.org/nfte101201.php*
- *http://en.wikipedia.org/wiki/Copenhagen*
- *http://en.wikipedia.org/wiki/Greenhouse_gas)*

3. Socio-Economic Analyses for Various Climate Change Scenarios over Mahanadi Delta, India

Dushmanta Dutta
Addala Narayana Swamy

About 40% of Indian population lives along the coast, that too in several large coastal cities. Indian coast line, in general, and east coast of India, in particular, subjected to severe cyclonic storms and surges as well as resulting floods. Such natural hazards are responsible for the loss of millions of lives and damage to the properties. The possible global warming may increase the sea surface temperature and thus increase the frequency and intensity of the cyclonic storms and resulting storm surges and floods on the east coast of India. There is every possibility that the above will have its impact on the socio-economic aspects of the people living in the rural areas as well as cities along the coast. In the present study, flood inundation modeling studies have been carried out over Mahanadi delta in Orissa state of India. Several important cities like Bhubaneswar, Cuttack, Puri etc. are located in the above area and subjected to frequent floods. A Distributed hydrologic Model developed by Dutta et al. (2000) has been utilized in the present study for flood inundation mapping. Model simulation has been carried out for four future climate change scenarios based on the report of Inter Governmental Panel on Climate Change (IPCC). Preliminary socio economic analysis has been carried out and the results are discussed in the paper.

Keywords: Climate change, Sea level rise, Mahanadi Delta, Floods

Introduction

India is one of the most densely populated countries in the world with over one billion people. About half of the Indian boundary (approximately 7,500 km including two groups of islands) is surrounded by ocean and approximately 40% of total population lives within 100 km ocean coast. Presently, Indian coastline is facing increasing stresses like population growth, urbanisation, industrial development, trade and human pressures like overexploitation of marine resources, dumping of industrial and toxic wastes, oil spills and leaks. On the other side, the people living in the coastal regions of India are highly vulnerable to natural hazards such as cyclones and resulting floods. The natural hazards take millions of lives, damage properties and natural resources in coastal areas. In India about 75% of the rainfall occurs over the four monsoon months (June – September) and the Indian rivers carry 90% of the water between June and November. Thus, only 10% of the river flow is available during the other six months. Almost all major rivers in India, including Ganges, Yamuna and Brahmaputra, flow to the Bay of Bengal. These rivers carry vast amounts of silt created by deforestation and erosion. When the silt is deposited in the Bay, the areas around deltas become shallower. This creates ideal conditions for high tidal waves and storm surges that were an important factor in the devastation during cyclonic storms. Sometimes the infrastructure development in the deltaic environments may prevent seawater from receding thus aggravating the situation during severe floods associated with cyclonic storms and surges.

Global warming may result in an increase in sea surface temperatures as a result of which changes in the frequency, intensity or tracks in cyclones hitting the coastal zones may take place. India is frequently affected by tropical cyclones and storms particularly during or around monsoon season. About 6.5% of worlds 80 tropical storms are formed annually in the Indian ocean (Neumann, 1993; Mc Bride, 1995). Frequency of formation of cyclones is 5 to 6 times more in Bay of Bengal as compared to Arabian Sea (India Meteorological Department, 1979). Therefore, the east coast of India in general and the states of Orissa, Andhra Pradesh and West Bengal in particular, are more vulnerable to the fury of cyclones and resulting floods causing damage to life and property almost every year. Although the studies carried out so far are inconclusive on the likely changes in the frequency of cyclones with global warming, it is almost certain that an increase in sea surface temperature will be accompanied by a corresponding increase in cyclone intensity. This has been proved by the frequent occurrence of high intensity cyclones in Bay of Bengal during recent years resulting severe damage in all aspects. The impact of climate change and resulting floods would be on freshwater resources, industry, agriculture, fisheries, tourism and human settlements of the country.

The climate change over the globe may have its impact on increase in wavelength and amplitude of the ocean waves and also increase in sea water level resulting in the inundation of several low lying areas of the world as well as India. Sea level rise is likely to result in loss of land due to submergence of coastal areas, inland extension of sea water intrusion and groundwater contamination and may have wide economic, cultural and ecological repercussions over India.

The impacts of any increase in the intensity of cyclones due to global warming and consequent sea lea level rise could be disproportionately large not only in heavily populated coastal areas but also in terms of the paralysing devastation in low income rural areas as well as high income large coastal cities of India. The people of Andhra Pradesh, Tamilnadu, West Bengal and Orissa states are being severely affected by the storm risk and resulting floods. The coasts of the above states are flat and deltaic.

Area of Investigation

The state of Orissa extends from $17^0 49$'N to $22^0 34$'N latitude and $81^0 29$'E to $87^0 29$'E longitude in the eastern coast of India. It has an area of about 155,707sq.km and a coastline of 529km. It is bounded by West Bengal in the northeast, Bihar in the north, Chatisgarh in the west, Andhra Pradesh in the south and Bay of Bengal in the east. There are as many as 30 districts in the state. Physiographically, Orissa can be divided into three broad regions, namely, the costal plains, the middle mountainous country and the plateaus and rolling uplands. The costal plain Orissa is a gift of the six major rivers, which have brought silt from their catchments and have reclaimed this area from the depths of the Bay of Bengal. The costal plains can be termed as "a land of six deltas," but most of the rivers have failed to develop true delta characteristics (as far as shape is concerned) because of the strong offshore current, which moves from Chennai to Andhra Pradesh coast and passes through the Orissa coast. In order of the contribution of the rivers to the formation of the costal plains, the rivers may be arranged as the Mahanadi, the Brahamani, the Baitarani, the Subernarekha, the Rushikulya and the Burhabalanga. This arrangement is based upon the length of river channels, the area of their catchment basins, the velocity of water flow, the nature of terrain over which they flow, etc. Among all the rivers those, which cause flood in Orissa are Mahanadi, Brahmani, Baitarani, Subarnarekha, Salandi and Kopali . All the major rivers are draining into the Bay of Bengal after flowing for long distances in the State.

The Mahanadi Delta

The river Mahanadi that originates from Amarkantak plateau in Madhya Pradsh enters from the northwest and drains a very large part of western Orissa (Figure 1). After forming a gorge at Satkosia in the Eastern Ghats, it forms a delta at Cuttack and finally flows into the Bay of Bengal. The river Mahanadi has a total length of 852.8 km from its sources to delta head and 107 km from its upper point of the delta to the delta mouth. The Mahanadi delta has developed in seven stages. The stages of development have occurred at 26, 18, 15, 12, 9, 6 and 1.5 m contours. These individual contour lines touch the points of bifurcation of the distributaries. After the seventh stages of delta development there was an upliftment, as a result of which three sets of parallel sand dunes developed along the coast. These raised sand ridges brought changes in the drainage pattern as well as minor changes in the deltaic morphology. The entire coastline has become a pro-graded coast, as a consequence, except Paradeep, the river mouths have been chocked and this hinders the free discharge of the floodwater. Due to the upliftment of the mouth portions, the slope of the valley long-profiles have been reduced and free discharge of floodwater has been hindered. Several lagoons of varied sizes have been formed. The daily high and low tides have kept the mouths of the Devi and the Mahanadi open to form estuaries (Behera and Babu, 2004).

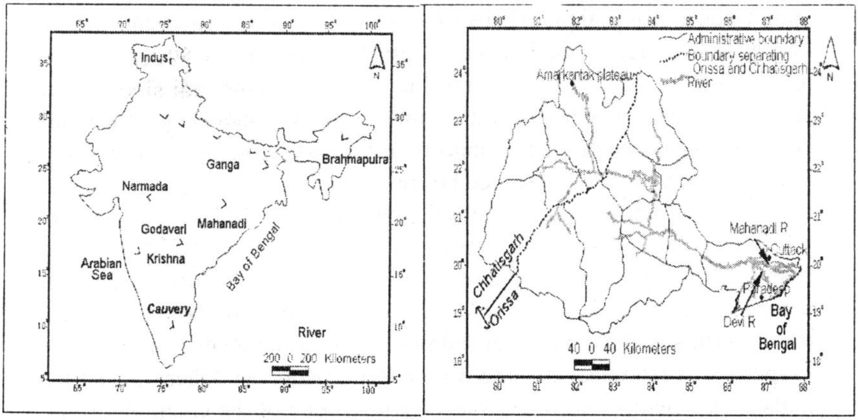

Figure- Map of Mahanadi River Basin of India

Floods vary not only according to periodicity of rainfall in the catchment but also according to capacity of the channels and the amount, size and shape of the basins, the distribution of tributaries, condition of the ground surface etc. The total channel capacity of all distributaries of the Mahanadi, at a distance of 40 km down-stream from Cuttack, is only half of what it is at the Mahanadi delta head. Thus free discharge of floodwater is obstructed and this leads to flooding. A cross section from the delta apex to the shore reveals that the lower reaches of the Mahanadi delta are of a higher level than those in its middle reaches. Thus the delta has assumed a basin shape because of the coastal sand ridges. The flood slope of the river is reduced and free and quick discharge of floodwater is hampered. The levees have helped breaching to a considerable degree and the sand banked mouths obstruct the free discharge of floodwater. Added to this is the fact that the tidal currents in the Bay of Bengal coupled with the high monsoon winds raise the level of the sea and thereby further influence the slope of the flood and obstruct free discharge of floodwater into the sea. Figure 2 shows the distribution of drainage channels in Mahanadi delta. Figure 3 presents the channels of Mahanadi delta about 8 km inside from the coast during monsoon and summer periods respectively.

History of Floods in Orissa

The flood data for the last 120 years (1868-1987) reveals that the maximum number of floods occurred in Mahanadi, followed by Baitarani and Brahmani. The common years of high floods in the above rivers occurred in 1896, 1960, 1961, and 1982 that affected the entire coastal plain when the coastal plain became a vast sheet of floodwater and caused immense damage. The Hirakud Dam has somewhat succeeded in preventing the occurrence of high floods in the Mahanadi since 1956. However, it has failed to prevent medium and low floods in the Mahanadi delta. A description of the four important floods that occurred in the post-independence India is presented below.

A devastating flood occurred in the year1982 causing its impact felt over 33,730 km^2. A total area of 7,270 km^2 of cultivable land got inundated, 6,308 million people were affected, 431,760 houses got damaged, 128 people died and 36,183 heads of cattle perished. The total loss to crops was estimated at 1.059 billion Indian Rupees and loss of property was estimated at 2.145 billion Indian Rupees (Behera and Babu, 2004). This heavy loss broke the very backbone of the economy of the state.

Similarly, there were two devastating cyclones in late 1999 in India. The first hit the coastal region of Orissa on October 18 and 19. With wind speeds of around 200 km/h and large tidal waves, it killed around 500 and affected two million people. A second and far more devastating cyclone 05B (dubbed the super cyclone) struck the Mahanadi delta on October 29. The cyclone with sustained wind speeds exceeding 260 km/h tore across 200 kilometres of the coast. At times, the wind speeds reached an astonishing 310 km/h. The cyclones affected approximately 15 million people and damaged three million homes (half of which were destroyed). Estimating that an average of five family members lives in each house, approximately 7.5 million people were made homeless. 2.3 million children were affected.

The flood crisis in Orissa in the year 2001 worsened the situation as rain poured incessantly over the state and many areas were submerged under water.

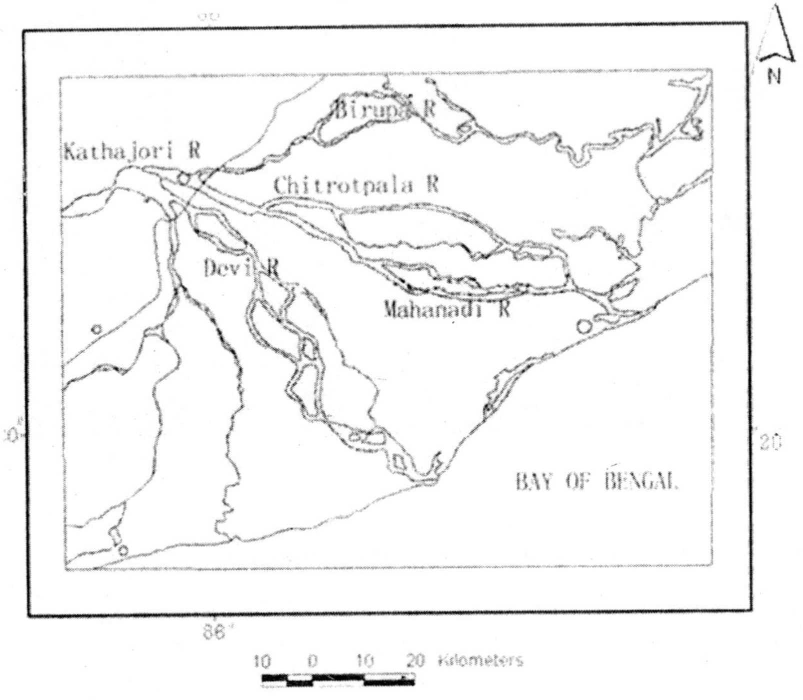

Figure 2: Drainage network of Mahanadi delta (after Mohanty & Swain, 2004)

(a)

(b)

Figure 3: Mahanadi delta during (a) monsoon months and (b) summer months respectively

Floods affected cities like Balasore, Cuttack, Jagasinghpur, Jajpur,Bhubaneswar, Puri, Rayagada etc. The state government on 16th July declared red alert for five districts: Puri, Cuttack, Kendrapada, Jajpur and Jagatsinghpur. On 18th July as many as 4 million people in 21 districts were reeling under the impact of flood. The river Mahanadi had nearly 50 per cent more water than normal. Jagatsinghpur was one of the worst flood affected districts of Orissa in the year 2001. Red alert was declared in Jagatsinghpur with the river systemc of Mahanadi in the delta flowing above the "danger" mark. The flood situation worsened further because the uninterrupted rain continued for few more days.

Due to depression over Bay of Bengal, there were heavy rains during 26th August to 4th September 2003 in many parts of the State of Orissa. As per the reports of the State Government 21 districts have been affected due the floods. The districts of Cuttack, Jagatsinghpur and Kendrapara were severely affected. 0.402 million hectares of cropped area were affected in 5836 villages. There were 584 breaches in embankments, a number of breaches could not be closed for quite some time. Figure 4(a) shows the breaches to the embankments during 2003 floods at Jagatsinghpur. Figure 4(b) gives a view of the elevated borewell used in Jagatsingpur to utilize the same during flood situation. Figure 4(c) presents a multi-purpose school Cum cyclone shelter constructed at several places along the Orissa coast. Figure 4(d) shows the extent of devastation due to floods for Paradeep port area. It is interesting to note that there are eighteen cities/towns located in Mahanadi delta including large cities like Bhunaneswar, Cuttack, Puri etc (Dept. of Water Resources, Govt. of Orissa, 2001). In addition, there exist 88 major and minor industries in the Mahanadi delta which are prone to periodic floods that occur in the region.

(a)

(b)

(c)

(d)

Figure 4: (a) Breaches to embankments during 2003 floods, (b) Elevated bore-well, (c) School cum cyclone shelter, and (d) Paradeep port under floods

Hydrological Model Study

Considering the frequent flood inundation of Mahanadi delta from cyclonic storms and surges, it has been proposed to carry out a study using a distributed hydrological model to assess the extent of flood inundation during some of the cyclonic storm and resulting flood periods and assess the socio-economic impacts on the population, buildings and transport network of the region due to such floods. The Distributed Hydrologic Model used in this study has been developed by the Institute of Industrial Science, University of Tokyo, Japan known as IISDHM (Dutta *et al.*, 2000). IISDHM is a physically based distributed hydrological model. It considers five major processes of hydrologic cycle like interception, evapotranspiration, river flow, over land flow, unsaturated and saturated zone flow. An implicit finite difference scheme is used to solve the governing equations of flow for river network. Though it has been proposed to carry out model simulation for a number of cyclonic storm events over the Mahanadi delta, due to data limitations, the simulation has been made for the 16-20 October, 1999 cyclonic storm. Depending on availability of data the model study will be further carried out for other major storm events.

Hydrological and Geo-spatial Datasets

The major temporal datasets collected includes hourly rainfall data, and monthly water level data. Rainfall data of hourly resolution for three rain-gauge stations, Bhubaneswar, Cuttack and Puri, was collected from Indian Meteorological Department, Pune. Hourly rainfall plot of these rain gauge stations for the severe cyclonic storm that occurred during 17th and 18th of October 1999 are illustrated in Figure 5. As already mentioned, Bhubaneswar, Cuttack and Puri are the major coastal cities located in the Mahanadi delta.

A severe cyclone crossed Gopalpur area adjoining Mahanadi delta during 16th to 20th of October 1999. On the analyses of hourly rainfall data it has been observed that the maximum precipitation occurred on 17th and 18th October, 1999 and hence this period is selected for model simulation. Maximum rainfall observed was at station Bhubaneswar for both the days. Water level at the outlet of Hirakud dam was considered as the upstream boundary condition and mean sea level was considered as the downstream boundary condition. Major spatial datasets required for model simulation includes land use, topography, soil and river cross-section. The above digital data sets have been secured from various

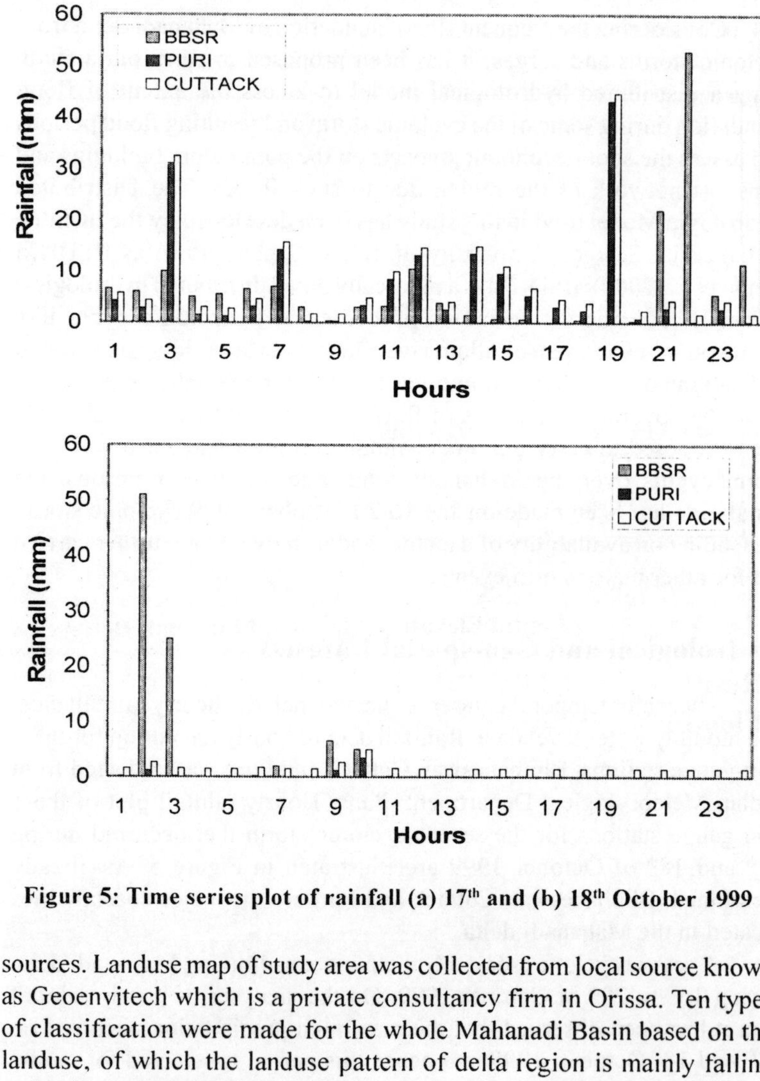

Figure 5: Time series plot of rainfall (a) 17th and (b) 18th October 1999

sources. Landuse map of study area was collected from local source known as Geoenvitech which is a private consultancy firm in Orissa. Ten types of classification were made for the whole Mahanadi Basin based on the landuse, of which the landuse pattern of delta region is mainly falling under Agricultural Land under Major Irrigation. Hydro1K DEM dataset of 1Km x 1Km horizontal resolution was used in this study. The data is available in the public domain. Figure 6 illustrates the DEM of the study area. It is observed that the maximum elevation of the delta is 4.7 meters and at the point where delta merges with the Bay of Bengal the elevation from m.s.l is about 0.1 meters. Soil map of Peninsular India produced by

Food and Agricultural Organization of the United Nations (Satyanarayana *et al.*, 2003) has been utilized in this study. In the delta region, the soil type is mainly ferric luvisol.

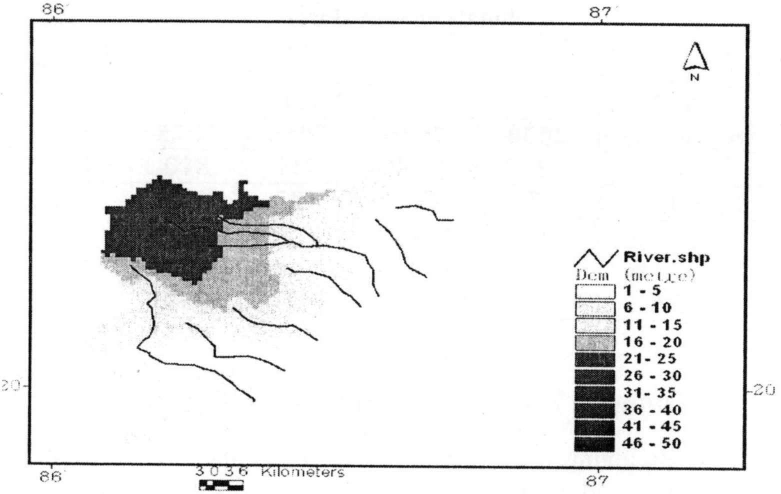

Figure 6: Digital Elevation Model of Mahanadi Delta

Result and Discussion

Flood Inundation

The simulation has been carried out for the present and future scenarios of climate change. Present scenario was generated for the year 2000, and future scenarios were considered for years 2025, 2050, 2075 and 2100 respectively. According to Inter Governmental Panel on Climate Change (IPCC), the A1 scenario indicated an increase in the mean sea level (msl) of 14, 32, 57 and 88 cm for the future years of 2025, 2050, 2075 and 2100 respectively. For the simulation, the observed sea level for year 2000 was considered as the basic, and the sea level was increased to 14, 32, 57 and, 88 cm from this basic msl to obtain future flood inundation profile in the Mahanadi delta. Figure 7 illustrates the flood inundation maps of different magnitudes generated as output of the model for various scenarios considered. Maximum flood inundation depth simulated was 5.68 m, for an increase in msl value of 88 cm. Table 1 indicate the simulated flood area for all the considered years for the situation of the severe cyclonic storm considered earlier. From the figure and table it can be observed that, in the year 2000 and 2025, inundation is caused mainly by flood depth of < 0.1 m. However, for the years from 2050 to 2100, the damage is caused mainly by inundation depth of > 3.5 m.

Table 1: Simulated flood inundation area for various scenarios of climate change (km^2)

Flood Depth (m)	Year				
	2000	2025	2050	2075	2100
< 0.1 m	1,319	1,430	811	810	811
0.1-0.6 m	174	27	382	380	376
0.6 – 1 m	--	--	--	--	--
1 – 3.5 m	--	--	740	747	787
> 3.5 m	151	149	583	590	612

Socio-Economic Impacts and Vulnerability

The preliminary information for assessing the socio-economic impact of the flood was obtained by conducting a questionnaire survey. The survey was aimed to develop Flood Impact Indices for various categories such as Buildings, Transportation and Population under different flood depth conditions. Tables 2, 3 and 4 comprehend the indices developed from the questionnaire survey.

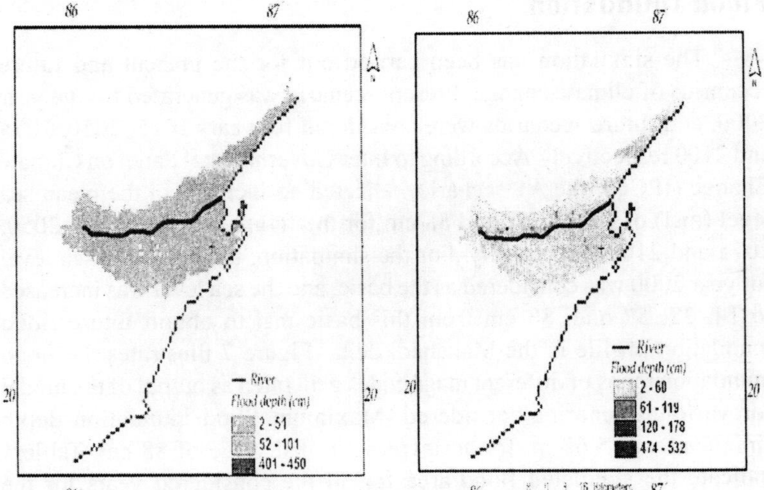

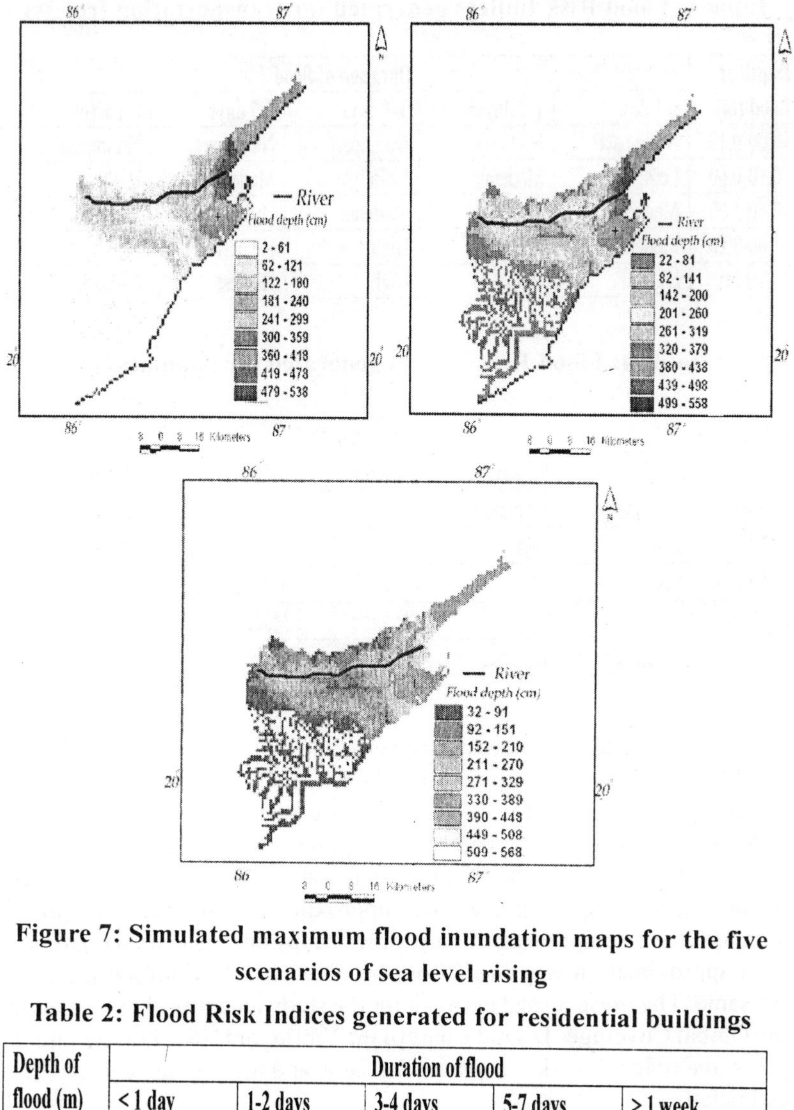

Figure 7: Simulated maximum flood inundation maps for the five scenarios of sea level rising

Table 2: Flood Risk Indices generated for residential buildings

Depth of flood (m)	Duration of flood				
	< 1 day	1-2 days	3-4 days	5-7 days	> 1 week
< 0.1 m	No Impact	No Impact	No Impact	No Impact	No Impact
0.10-0.60	Less	Less	Less	Moderate	High
0.60-1.00	Less	Less	Moderate	Moderate	High
1.00-3.50	Less	Moderate	Moderate	High	High
> 3.50	Moderate	High	High	High	Highest

Table 3: Flood Risk Indices generated for transportation (roads)

Depth of flood (m)	Duration of flood				
	< 1 day	1-2 days5	3-4 days	5-7 days	> 1 week
0.00-0.10	No Impact	No Impact	No Impact	No Impact	No Impact
0.10-0.60	Less	Moderate	Moderate	Moderate	High
0.60-1.00	Moderate	Moderate	Moderate	High	High
1.00-3.50	Moderate	Moderate	High	Highest	Highest
> 3.50	High	High	High	Highest	Highest

Table 4: Flood Risk Indices generated for population

Depth of flood (m)	Duration of flood				
	< 1 day	1-2 days	3-4 days	5-7 days	> 1 week
0.00-0.10	No Impact	No Impact	No Impact	No Impact	No Impact
0.10-0.60	Less	Less	Less	Less	Moderate
0.60-1.00	Less	Less	Less	Moderate	Moderate
1.00-3.50	Less	Less	Moderate	Moderate	High
> 3.50	Moderate	Moderate	Moderate	High	High

With the above indices as the basis, and by incorporating the flood depth values obtained as the output of the model, the Hazard classification for the three categories for the five years are developed. Tables 5, 6 and 7 depict the estimated damage for residential buildings, transportation and population categories from the present study. The number of residential buildings at present was approximated from the population data with the assumption that four people occupy one residential building. This approximation was carried out because of lack of information on the same. The present road network for the study area was obtained from the Global Coverage 'Digital Chart of the World' of ESRI. For the future years, the road network is retained the same and the damage caused was estimated.

The loss in this case is less compared to other categories because the length of road network falling in delta region is only to a lesser extent. Railway network was not considered in this study because the railway-line is passing adjoining the upper boundary of the delta and there is only a small railway line connecting Bhubanswar and Puri in the study area. Present population data was obtained from Primary Census Abstract of

Table 5: Estimated number of affected residential buildings

Impact Index	Year				
	2000	2025	2050	2075	2100
Less	7,422	6,484	519,737	780,995	831,289
Moderate	8,008	10,959	79,431	82,280	82,787

Table 6: Estimated number of affected road (in km)

Impact Index	Year				
	2000	2025	2050	2075	2100
Less	-	92	360	501	520
Moderate	-	-	35	110	92
High	-	-	-	82	67

Table 7: Estimated number of affected population

Impact Index	Year				
	2000	2025	2050	2075	2100
Less	29,688	25,936	2,078,949	3,123,980	3,325,159
Moderate	32,030	43,837	317,726	329,120	331,150

Orissa State (Census of India 2001). Future projection of population was estimated by following the percentage growth rate from 1991 to 2001.

Conclusion

The presented the preliminary outcomes of case study conducted in Mahanadi Delta to assess the socio-economic impacts of floods under sea level rising conditions. The simulated inundation results by the distributed hydrologic model show that magnitude of flood depths would increase significantly for similar cyclones if sea level rises by more than 30 cm. Subsequently the socio-economic impacts will be accelerated. The spatio-temporal resolutions of the modelling outputs can be enhanced for comprehensive understanding of socio-economic impacts at local level by improving the input temporal and spatial datasets for the model and model calibration and verification and detailed socio-economic factors.

References

- Behera, D.K. and P.V.Babu (2004). Flood situation in Jagatsingpur District of Orissa: An Analysis of Adaptation and Mitigation Strategies. RMRC Report 2004, 38p.
- Census of India (2001). Primary Census Abstract, Office of the Registrar General, India, New Delhi.
- Dept. of Ocean Development (DOD), Govt. of India (1993). Annual Report 1992-93: Chapter 5 Coastal Zone and Islands. Available at *http://dod.nic.in/ayr92-93/ann92-93.htm*
- Dept. of Water Resources, Govt. of Orissa (2001) . 3rd Spiral Study Report of Mahanadi Basin Plan, Volume I, July 2001, Bhubaneswar, 210p.
- Dutta, D., Herath, S. and Musiake, K. (2000). Flood inundation simulation in a river basin using a physically based distributed hydrologic model. *Hydrological Processes* 14: 497-519.
- IMD (1979) Track of storms and depressions over the Bay of Bengal and Arabian Sea, 1877 – 1970. Publication of India Meteorological Department, New Delhi, India 186p.
- Mc Bride, J. L. (1995). Tropical cyclone formation, In Elsberry R.L. (ed.) Global perspectives on tropical cyclones.
- Mohanti, M. and M.R.Swain (2004) Mahanadi river delta, east coast of India: An Overview on evolution and dynamic processes. Proceedings of First Annual Meeting of IGCP 475 Delta Map and APN Mega-Delta (Under press). Available at *http://www.megadelta.ecnu.edu.cn/main/upload/ mahanadi.pdf*
- Neumann, C. J. (1993). Global guide to tropical cyclone forecasting : Chapter 1 - Global Overview. WMO/TC-No.560, Report No. TCP-31, World Meteorological Organization, Geneva, Switzerland
- Satyanarayana, A., Rangaswamy, P., Muttaiah, R., Patil, M.S., Reddy, M.V., Naidu, N.V., Cherallu, C., Ramakrishna, A. and Chauhan, Y.S. (2003). Legumes in tropical rice-based cropping systems in Peninsular India. Avalilable at *http://www.icrisat.org/text/research/nrmp/cdlegumes/e-book/03.pdf*

4. Need to Estimate the Cost of Environmental Degradation for Developing Economies

Reeta Rautela

The contribution of the environment to the national economy can't be overlooked. The role of the environment and natural resource sector in economic development is taken for granted. Economic growth is usually brought about by technological innovation and positive external forces. Developing economies are highly populated as well as highly polluted. Pollution control and regulation are not cost-free activities. Environmental damage can have a negative effect on our quality of life and limits our sustainable rate of growth. The mechanisms of economic growth may impact the environment through many portals. Growing economies face environmental deterioration. Economic growth is accompanied with increase in industrial production, rise in income levels and greater exploitation of natural resources; but unhappily it also brings with it the curse of environmental degradation irrespective of whether the country is developed or developing. This paper suggests that there is a need of GDP growth that takes into account environmental cost.

Key words: Economic growth, Natural resource, Environment, Environmental degradation

Introduction

Economic growth has been defined as "a positive change in the level of production of goods and services by a country over a certain period of time." Technological progress can reduce ecological impacts of economic growth; it also can contribute to ecological problems. Some economists believe that economic growth provides a basis on which sustained environmental improvements can and often do occur. Fast growth of production and consumption can create negative externalities such as increased noise and air pollution and road congestion. High economic growth also leads to the degradation of land and unsustainable depletion of renewable and non renewable resources. **Sustainable Economic Development is the** *"development that meets the needs of the present without compromising the ability of future generations to meet their own needs."*-The **World Commission on Environment and Development** (1987)

The Relation between the Economic and Natural system

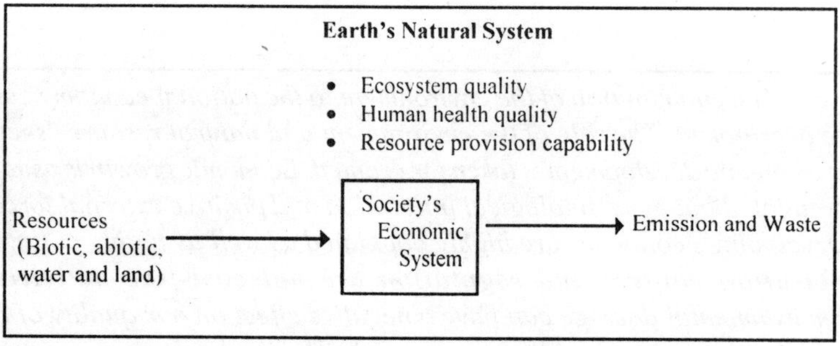

Impact of Economic Growth on Environment:

Is economic growth good for the environment? The relationship between economic growth and the environment is controversial. The **Kuznets Curve** suggests that the relationship between economic growth and the environment could be positive and hence growth is a prerequisite for environmental improvement. A Kuznets curve is the graphical representation of Simon Kuznet's hypothesis that economic inequality increases over time while a country is developing, and then after a certain average income is attained, inequality begins to decrease.[1]

This implies that the environmental impact indicator is an inverted U-shaped function of income per capita.[2]

Fig 1. The Environmental Kuznets Curve: A Development-Environment Relationship

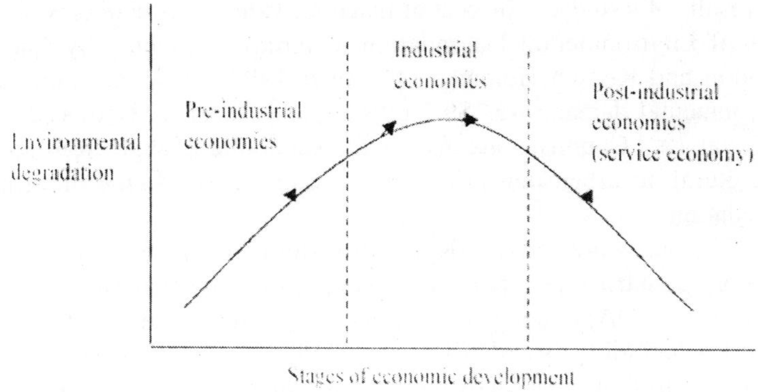

Source: Economic Survey of Europe, 2003 No. 2
http://www.unece.org/ead/pub/032/032_c2.pdf

Evaluation of Environmental Performance

One of the most comprehensive measures of a country's environmental performance is the Environmental Sustainability Index **(ESI).** Due to a shift in focus by the teams developing the ESI, in 2006, a new index was developed, the Environmental Performance Index **(EPI)** that uses outcome-oriented indicators, then working as a benchmark index that can be more easily used by policy makers, environmental scientists, advocates and the general public. The 2010 Environmental Performance Index (EPI) ranks 163 countries on 25 performance indicators tracked across ten well-established policy categories covering both environmental public health and ecosystem vitality. India lags behind other countries such as Brazil, Russia and China in environmental performance as indicated in the 2010 Environmental Performance Index (EPI). In spite of taking all out efforts, India is ranking at 123rd position due to increasing vehicular traffic, high rate of industrial and infrastructural growth, increasing urbanization and pressure of population. In a bid to bring environmental concerns into mainstream growth accounting, Indian government plans to release green GDP data from 2015.

Estimating the Cost of Environmental Degradation

Environmental degradation has kept pace with economic growth. The results of a study, "The cost of Inaction: Valuing the economy-wide costs of Environmental Degradation in India" conducted by Carter Brandon and Kristen Hommann (October 1995) show that the total environmental damage was $9.7 billion per year, or 4.5% of GDP in 1992 values. Countries are facing the economic change from rural agricultural, to urban industrial society, with increased environmental degradation.

The buzzword today is Kyoto Protocol of 1997 under which the signatory countries have agreed to accept Environmental Degradation Accounting (EDA) as a mandate for action. As of November 2007, 175 countries including Brazil, China and India have ratified the Kyoto Protocol signifying that EDA has now become an essential feature of environmental policies of most governments of the world.[3] The Gross Domestic Product (GDP) in India expanded 8.90 percent in the third quarter of 2010 over the previous quarter. According to govt. officials, "such figures fail to incorporate the toll on the economy of rising public health bills and other long-term costs linked to environmental damage, such as air and groundwater pollution. If we are reporting GDP growth of 8 percent a year, actual GDP growth that takes into account environmental costs may be 2.5 percentage points lower." Global environmental damage is estimated to cost $28 trillion by 2050.

Conclusion

No country in the world was better prepared for natural disasters. The simple fact is that the climate will continue to change with or without our help. It is a matter of justice that we seek sustainable clean environment for all of our fellow human beings. Let us each ask ourselves how our own materialism, over-consumption and nonrenewable ways of living have made possible and perpetuated such disasters?

Foot Notes

1. International Society for Ecological Economics, Internet Encyclopedia of Ecological Economics, the Environmental Kuznets Curve, David I. Stern, Department of Economics, Rensselaer Polytechnic Institute, Troy, NY 12180, USA. June 2003
2. *http://en.wikipedia.org/wiki/Kuznets curve*

3. GEO: Global Environment Outlook 3, past, present and future perspectives.

References

- Barker, Terry, 'The Economics of Avoiding Dangerous Climate Change', Cambridge: Tyndall Centre for Climate Change Research, Working Paper 117, June 2008
- Ekins, Paul, *Economic Growth and Environmental Sustainability: The Prospects for Green Growth,* London: Routledge, 2000
- Stokey, Nancy L. (1998). Are There Limits to Growth? *International Economic Review*, 39, pp. 1-31
- UNEP (2010)

Web links

- *http://www.worldbank.org.in/wbsite/external/countries/southasiaext/indiaextn/0,,contentmdk:2129*
- *http://en.wikipedia.org/wiki/Environmental_Sustainability_Index*
- http://www.indiaenvironmentportal.org.in/content/environmental-performance-index-2010-preliminary-results

5. Sustainable Development Through Good Environment

Jagadish Singh

Sustainable development is a requirement to our generation to manage the resource base that the average quality of life, we ensure ourselves can potentially be shared by all future generations. The sustainable development can be explained as the objective of fulfilling needs of present generation without damaging the resources for the future generation. So, the main goal of sustainable development is to be brings co-ordination among the population, resources, environment and development.

All development activities irrespective of their scale nature & magnitude, affect the environment. Energy is used for developmental activities and as a result- environment is degraded due to emission of greenhouse gases. India is a large developing country with nearly 700 million rural populations directly depending on climate – sensitive sectors (as-agriculture, forests & fisheries) and natural resources for their subsistence and livelihoods.

As we know that good environment is main requirement of the sustainable development. Thus, some steps to be taken for good environment as- Increase plantation & vegetation, increase bio- fuel production, use of renewable energy as alternative of thermal electricity, improved transport quality and low emissions technology etc.

Key words: Sustainable development, Environment, Deforestation, Climate change.

Introduction

Sustainable development is a requirement to our generation to manage the resource base that the average quality of life, we ensure ourselves can potentially be shared by all future generations (Asheim, 1991).Sustainable development is the concept of a relationship between economic growth and the environment (OECD2001). It is a term widely used by politicians all over the world; its main components are the economic, social, and environmental factors. An understanding of sustainable development involves disciplines such as ecology, biology, ethics, economics, chemistry, physics, statistics and engineering.

Economics has a crucial role to play in the societies, but critical for poor ones, which depend more heavily on natural resources such as soils, rivers, fisheries and forests than do the richer nations. It is argued that environmental problems in developing countries are predominantly driven by poverty. (Gillis.2005).

The Brundtland Commission which also called 'world commission on environment and development' in 1987 has brought the concept of 'Sustainable Development'. It has published a report of 'our common future' we can say: "sustainable development is a process in which the exploitation of resources, the direction of investment, the orientation of technological development and institutional change to meet the needs of the present generation without compromising the ability of future generation to meet their needs".

The sustainable development can be explained as the objective of fulfilling needs of present generation without damaging the resources for the future generation. So, the main goal of sustainable development is to bring co-ordination among the population, resources, environment and development.

According to Moore & Ryan(1995), "Sustainable development refers to creating a style of economic development which is sustainable with in the context of the planet's ecosystem and human society."

Objectives of Study

The overall objective of the present study is to know the sustainable development through good environment. The specific objectives are:

(1) To study the reasons of environmental crisis.

(2) To know the impact of climate change.

(3) To study the measures of good environment.

Reasons of Environmental Crisis

In our surrounding we have social, biological and physical or chemical factors which constitute a resource. Men utilize this resource in different ways. In the primary stage man was in equilibrium with the environment. But in the quest of its development, man broke this equilibrium. All development activities irrespective of their scale, nature and magnitude, affect the environment. Energy is used for developmental activities and as a result, environment is degraded due to emission of greenhouse gases.

In 2002 about 40% of US CO_2 emission stem from the burning of fossil fuels for the purpose of electricity generation. Coal accounts for 93% of the emission from the electric utility industry. About 20% of US, CO_2 emission comes from the burning of gasoline in internal- combustion engines of cars and light trucks. The US is the largest consumer of oil, using 2.4 million barrels per day. Methane is second most important of greenhouse gas. According to the IPCC, methane is more thanes 20 times as effective as CO_2 at trapping heat in the atmosphere.

Deforestation is the second principle causes of atmospheric CO_2. It is responsible for 25% of all carbon emission entering the atmosphere by the burning and cutting of 34 million acres of trees each year. During the past 20 years, about three quarters of human made CO_2 emission were from burning fossil fuels. Rising temperatures may produce changes in weather, sea level and land use pattern, commonly referred to as "climate change".

Thus, we can say that environmental crisis derives from imbalance used natural resources by human.

Impact of Climate Changes

The modern life is very complex and complicated due to the climate change. The immediate and most essential need of the time is to control the bad effects of climate change; otherwise the entire life will be vanished. Climate change, to be put very simply is the change in earth's weather pattern. As it was thought to be a warming of the glob that is happening but this warming is also causing other phenomenon. Therefore, in my view, it can be said that climate change has been caused by global warming which in turn is a result of the green house effect.

According to Kin & Keller in 1982 sunlight passes through the glass of green house. The glass of green house and the CO_2 layer does not allow the heat to be warm up the entire atmosphere. The higher is the

CO_2 concentration in earth's atmosphere, the greater will be the amount of infra-red radiation absorption and consequently the warmer will be the atmosphere carbon dioxide layer acts like a blanket and traps the infra-red rays. Carbon dioxide is the main offender in the global warming, according to the consensus of 1990.In the present time, many impacts have been seeing of the climate change. They are:-

(1) Increasing environmental temperature.
(2) Increasing sea level.
(3) Impact on water resources and natural disaster.
(4) Bad impact on agricultural production.
(5) Bad impact on human health.
(6) Bad impact on animals & plants.
(7) Change in pattern and amount of rainfall.
(8) Extreme weather events etc.

Developing countries are faced with immediate concerns that relate to forest and land degradation, freshwater shortages, food security, air & water pollution. Increasing global temperatures will result in rising sea levels.

The question why India is so much concerned about climate change? Because first, India is a part of globalize world and anything that happens in the world will affect India also. Secondly, India is a large developing country with nearly 700 million rural populations directly depending on climate- sensitive sectors (such as – water, biodiversity, mangroves, coastal zones, grasslands etc.) for their subsistence and livelihoods.

Measures of Good Environment

As we know that for good production & productivity and good human health need to be compulsory of good environment. India is well aware of the problems relating global warming or climate change and the need for reducing green house gases. There includes-

(1) Ozone monitoring and development of CFC substitutes.
(2) Green house molecules monitoring.
(3) Sea- level monitoring and modeling.
(4) Air- sea interactions.
(5) Effect of increased UV-B and climatic changes on eco-system, aquatic systems and industrial materials.
(6) Environment oriented studies.
(7) Health hazards etc.

Their wide varieties of programmes are carried out with varying techniques- (a) At the ground level, (b) Instruments carried by balloons, &(c) Apparatus and other sophisticated instruments in rockets. India has very important role to play in the global negotiations, as a major developing country.

Suggestions for Good Environment

Some steps to be taken for good environment. They are given below-

(1) Increase plantation & vegetation.
(2) Increase bio-fuel production and use in automobiles.
(3) Use of renewable energy as alternative of thermal electricity.
(4) Appropriate technology change can be reducing increasing level of green house gases effectively.
(5) Improved transport quality, As-low emission CO_2 etc.

Conclusion

The sustainable development can be explained as the objective of fulfilling needs of present generation without damaging the resources for the future generation. So, the main goal of sustainable development is to be bringing co-ordination among the population, resources, environment and development.

All development activities irrespective of their scale nature & magnitude, affect the environment. Energy is used for developmental activities and as a result environment is degraded due to emission of greenhouse gases. India is a large developing country with nearly 700 million rural populations directly depending on climate – sensitive sectors (as-agriculture, forests & fisheries) and natural resources for their subsistence and livelihoods.

As we know that good environment is a main requirement of the sustainable development. Thus, some steps to be taken for good environment as- Increase plantation & vegetation, increase bio-fuel production, use of renewable energy as alternative of thermal electricity, improved transport quality and low emissions technology etc.

References

- Balasubramanian, Arun (1984),Ttowards a Philosophy of Environmental Education. Regional Institute of Higher Education and Development, Singapore.
- Climate change: Perspective of India, UNDP, India-November-2009.
- Pattanik,B. K.(2010), Sustainable agriculture & food security. Kurukshetra: A Journal of Rural Development, March, p-27-33.
- Kumar, Velu Suresh (2010), Climate Change: It's Impact on India. Southern Economist Journal, May 15 P-5-8.
- Catherine Mcmullen & Thomas Hayden (ed.), UNDP Year Book 2009 Nairobi, P-21.
- The Brundtland Report, Our Common Future, 1987.

6. Global Environmental Issues, Challenges and Strategies to achieve Sustainable Development

Mridul Kumar Verma

The economic development that we have achieved so far has come at a very heavy price-at the cost of environmental quality. As we step into an era of globalization that promises higher economic growth, we have to bear in mind the adverse consequences of the past development path on our environment and consciously choose a path of sustainable development. To understand the unsustainable path of development, the present paper tries to attempt to visualize the significance and contribution of environment to economic development. Keeping this objective in the mind, this paper deals with the role of Environment, Global Environmental Issues and challenges, state of India's environment and finally the strategies to achieve sustainable development.

Key Words: Environment, Global warming, Resources, Degradation, Sustainable development.

Introduction

Environment is defined as the total planetary inheritance and the totality of all resources. It includes all the biotic and abiotic factors that influence each other. While all living elements-the birds, animals and plants, forests, fisheries etc. are biotic elements, abiotic elements include air, water, land etc. Rocks and sun lights are examples of abiotic elements of environment. A study of the inter-relationship between the biotic and abiotic components of the environment.

Role of Environment

- **Environment Supplies Resources:** Resources include both renewable and non-renewable resources. Renewable resources are those which can be used without the possibility of the resource becoming depleted or exhausted. In other words, a continuous supply of the resource remains available. They are trees in the forest and fish in the ocean. Non-renewable resources are those which get exhausted with extraction and use. For example, fossil fuels.
- **Environment Sustain Life:** Environment includes sun, soil, water and air which are essential ingredients for the sustenance of human life. The carrying capacity of the environment implies that the resources extraction is not above the rate of regeneration of the resources and waste generated are within the assimilating capacity of the environment. Carrying capacity of the environment helps to sustain life. Absence of carrying capacity of environment means the absence of life.
- **Environment Assimilates Waste:** Production and consumption activities generate waste. This occurs mostly in the form of garbage. Environment absorbs garbage.
- **Environment Enhances Quality of Life:** Environment includes oceans, mountains, deserts, etc. Man enjoys these surroundings, adding to the quality of life.

Global Environmental Issues

1. **Global Warming:** Global warming is a gradual increase in the average temperature of the earth's lower atmosphere as a result of the increase in greenhouse gases due to industrialization in recent times. Most of the global warming has been caused by man-made increases in carbon dioxide and other green house gases through the burning of fossil fuels and deforestation. It is mainly caused by burning of coal and petroleum products, deforestation, and release of methane gas from animal waste and increased cattle production. In 1992, the Earth Summit had a frame work convention on climate change; few warnings have been seriously ignored. Climate changes pose threat to human development. While

outcomes of global warming will vary across regions and within countries, some long-term results of global warming, which can be predicated are:

- Melting of polar ice caps with a resulting rise in the sea level and coastal flooding.
- Disruption of drinking water supplies as snow melts.
- Extinction of spices.
- Frequent tropical storms and tropical diseases.
- Agricultural and rural development will bear the brunt of climate risk.
- Extreme poverty and malnutrition will increase as water insecurity rises.

2. **Ozone Depletion:** Ozone depletion refers to the phenomenon of reduction in the amount of ozone in the stratosphere. Ozone layer protects the earth from the ultraviolet rays of the Sun. The depletion of the ozone layer has been caused by high levels of chlorine and bromine compound in the stratosphere. The origin of these compounds is chlorofluorocarbons (CFC). CFC is used as cooling agents in air conditioners or refrigerators, or as aerosol propellants, and bromoflurocarbons (halons) is used as fire extinguishers. Long run effects of ozone depletion are skin cancer; it lowers the production of phytoplankton, other aquatic organisms and also lowers the growth of terrestrial plants. Montreal protocol is an international treaty designed to protect the ozone layer by phasing out the production of a number of substances believed to be responsible for ozone depletion. It banned the use of compounds containing ozone depleting chemicals.

3. **Environmental Crisis:** Today the world is facing environmental crisis. The rising population of the developing countries and the affluent consumption and production standards of the developed world have put a great stress on the environment in terms of its functions of supplying resources and assimilating waste. Many resources have become extinct and the waste generated is beyond the absorptive capacity of the environment to absorb degradation. The result is that today we are at the threshold of environmental crisis.

4. **Rise in Opportunity Cost:** Opportunity cost is the cost of alternative opportunity given up. The country has to pay huge

amount for damages done to human health. The health cost due to degraded environmental quality have resulted in diseases like asthma, cholera etc. Huge expenditure is incurred in treating these diseases.

5. **Supply demand reversal of environmental resources:** The demand for resources for both production and consumption has gone beyond the rate of regeneration of the resources increasing the pressure on the absorptive capacity of the environment. This reversal of the supply demand relationship with demand for resources exceeding the supply has led to degeneration of the environment.

6. **Massive overuse and misuse of environmental resources:** There is massive overuse and misuse of environmental resources. Examples of overuse of environmental resources are deforestation and land degradation. Examples of misuse of environmental resources are ozone depletion and global warming.

State of India's Environment

India has rich quality of natural resources in plenty. India has rich quality of soil, hundreds of soil, hundreds of rivers and tributaries, lush green forest, abundant mineral deposits under the land surface, vast stretch of the Indian Ocean, rangers etc. The black soil of the Deccan Plateau is particularly suitable for cultivation of cotton. It has lead to concentration of textile industries in this region. The Indo-Gangetic plains-spread from Arabian Sea to the Bay of Bengal-are one of the most fertile, intensively cultivated and densely populated regions in the world. India's forests provide green cover for the majority of its population and natural cover for its wild life. Large deposits or iron-ore, and natural gas are found in the country. India alone accounts for nearly 20 percent of the world's total iron-ore reserves. Bauxite, copper, chromate, diamonds, gold, lead, lignite, manganese, zinc, uranium, etc are also available in different parts of the country. However, the developmental activities in India have resulted in pressure on its finite natural resources, besides creating impacts on human health and well being. The threat to India's environment poses a dichotomy-threat of poverty-induced environmental degradation and at the same time, threat of pollution from affluence and a rapidly growing industrial sector. Air pollution, water contamination, soil erosion, deforestation and wild life extinction are some of the most pressing environmental concerns of India.

Environmental Challenges: Major challenges to India's environment are :

- **Land Degradation:** Land in India suffers from varying degree and types of degradation arising from unstable use and inappropriate management practices. Following are the factors responsible for land degradation.

  a. Loss of vegetation occurring due to deforestation
  b. Unsustainable fuel wood and fodder extraction
  c. Shifting cultivation
  d. Encroachment in to forest lands
  e. Forest fires and over grazing
  f. Non-adoption of adequate soil conservation measures
  g. Improper crop rotation
  h. Indiscriminate use of agro-chemicals such as fertilizers and pesticides
  i. Improper planning and management of irrigation systems
  j. Extraction of ground water in excess of the recharge capacity
  k. Open access resource
  l. Poverty of the agriculture-dependent people

- **Biodiversity Loss:** India has approximately 20% of live stock population on a mere 2.5 percent of the world's geographical area. The high density of population and live stock and the competing uses of land for forestry, agriculture, posture, human settlements and industries exert an enormous pressure on the country's finite land resources. The per capita forestland in the country is only 0.08 hectares. There is an excess felling of about 15 million cubic meter forests over the permissible limit. Indiscriminate feeling towards trees has led to destruction of forest cover. Trees are natural consumers of carbon dioxide-one of the green house gases, which is build up in the atmosphere contributes to global warming. Destruction of trees throws into the atmosphere even more carbon dioxide, along with methane, another major green house gas. Once forests have been cut down, essential nutrients are washed out of the soil all-together. This leads to soil erosion. This makes the land unusable and can in turn lead to disastrous flooding since there is no soil to soak up the rain.

- **Air Pollution:** In India, air pollution is widespread in urban areas where vehicles are the major contributors. Vehicular emissions are of particular concern since these are ground level sources and, thus, have maximum impact on the general population. The number of motor vehicles has increased substantially over a period. India is one of the ten most industrialized nations of the world. It has led to unplanned urbanization, pollution and the risk of accidents. The Central Pollution Control Board (CPCB) has identified 17 categories of industries which are significant polluters.

- **Management of Fresh Water and Solid Waste:** Many states in India are on the edge of famine. Whatever water available, it is polluted or contaminated. It causes diseases like diarrhea and hepatitis. Also, there is lack of planned management of solid waste compost uncollected garbage not only causes diseases but it is also visibly ugly, dumping garbage into landfills is not an attractive option.

Environmental Degradation –The following contributors for environmental degradation:

- **Increasing Population Density:** The high rate of growth of population adversely affects the environment. It certainly leads to soil and water pollution.

- **Urbanization:** Whenever there is large migration of population from rural to urban areas, it leads to fast growth of slum areas. There is excess of load on the existing infrastructural facilities. It causes environmental degradation and ill health.

- **Use of High Yielding Variety of Seeds:** Farmer and workers suffer health problems due to increased use of poisonous insecticides, pesticides and fertilizers. Also, consumers' food grains which have too much undesirable chemical element as compared to international standard. Overuse of fertilizers, insecticides, etc. lead to ill health.

- **Rise in Economic Activity:** With rise in national income or economic activity, there is rise in industrialization and urbanization. This raises pollution of air, water and noise. There are accidents, shortage of water, housing problems, etc. In other words, with rise in national income there is ecological degradation which reduces welfare of the people.

Sustainable Development: Sustainable development implies meeting the basic needs of everyone and extending to all the opportunity to satisfy their aspirations for better life, without compromising on the needs of future. The strategies for sustainable development imply the use of non-conventional sources of energy to minimize the adverse environmental impacts. According to Herman Daly, a leading environmental economist, to achieve sustainable development the following needs to be done.

- Limiting the human population to a level within the carrying capacity of the environment. The carrying capacity of the environment is like a "plimsoll line" of the ship which is its load limit mark. In the absence of the plimsoll line for the economy, human scale grows beyond carrying capacity of the earth and deviates from sustainable development.
- Technological progress should be input efficient and not input consuming.
- Renewable resources should be extracted on a sustainable basis, that is, rate of extraction should not exceed rate of regeneration.
- For non-renewable resources rate of depletion should not exceed the rate of creation of renewable substitutes.
- Inefficiencies arising from pollution should be corrected.

Strategies to achieve Sustainable Development

1. **Use of Non-conventional Source of Energy:** India is hugely dependent on the thermal and hydro power plants to meet its power needs. Both of these have adverse environmental impacts. Thermal power plants emit large quantities of carbon dioxide which is a greenhouse gas. It is also products fly ash which, if not used properly, can cause pollution of water, land and environment. Hydroelectric projects inundate forests and interfere with natural flow of water in catchment areas and the river basins. Wind power and solar rays are good examples of conventional but cleaner and

greener technologies which can be effectively tapped to replace thermal and hydro-powder.

2. **LPG, Gobar Gas in Rural Areas:** Households in rural areas generally use wood, dung cake or other biomass as fuel. This practice has adverse effect like deforestation, reduction in green cover, wastage of cattle dung and air pollution. To rectify the situation, subsidized LPG is being provided. Also, Gobar Gas Plants are being provided through easy loans and subsidy for the gobar gas plant to function, cattle dung is fed to the plant and gas is produced which is left over is a very good organic fertilizer and soil conditioner. Liquid Petroleum Gas (LPG) is a clean fuel. It reduces household pollution to a large extent.
3. **CNG in Urban Areas :** In Delhi, the use of Compressed Natural Gas (CNG) as fuel in public transport system has significantly lowered air pollution and the air has become cleaner in the last few years.
4. **Wind Power :** In the areas where speed of wind is usually high, wind mills can provide electricity without any adverse impact on the environment. Wind turbines move with the wind and electricity is generated. Although initial cost of high yet the benefits are more than cost.
5. **Solar Power through Photovoltaic Cells :** India is naturally endowed with a large quantity of solar energy in the form of sunlight. With the help of photovoltaic cells, solar energy can be converted into electricity. These cells use special kind of materials to capture solar energy and then covert the energy into electricity. This technology is extremely useful for remote areas and for places where supply of power through grid of power lines is either not possible or proves very costly. This technology is totally free from pollution.
6. **Mini-Hydel Plants:** In mountainous regions, streams can be found almost everywhere. Mnin-hydel plants use the energy of such streams to move small turbines. The turbines generate electricity which can be used locally. Such power plants are environment-friendly as they do not change the land use pattern in areas where they are located. Also, they generate enough power to meet local demands.
7. **Traditional Knowledge and Practices:** Traditionally, Indian people are close to their environment. In earlier days, agriculture system, health care system, housing, transport etc., were environmental friendly. With modernization, we have gone far

away from the traditional systems. It has caused massive damage to the environment and to our rural heritage. We should go back to our traditional knowledge and practices. India has about 15,000 species of plants which have medical properties. About 8000 of these are in regular use in various systems of treatment like Ayurveda, Unani, Tibetan and Folk systems. The health care systems are in great demand for treating chronic health problems. Cosmetic products like hair oil, toothpaste, body lotion, face cream, etc. are herbal in composition. These products are environment friendly and have no side effects.

8. **Bio-composting:** The use of chemical fertilizers has resulted in large tracks of unproductive land, chemical contamination of water bodies including ground water system. Farmers, in large numbers of all over the country, have again started using compost made from organic waste of different types. In certain parts of the country, cattle are maintained only because they produce dung which is an important fertilizer and soil conditioner. Earthworms can convert organic matter into compost faster than the normal composing process. This process is now being widely used.
9. **Bio-pest Control:** With Green Revolution, states started using more and more of chemical pesticides for higher yield. Adverse impacts started showing in food products which were contaminated. Soil, water bodies and even ground water were polluted with pesticides. Milk, meat and fishes were also found to be contaminated. Efforts are made to have safer methods of pest control. Neem based pesticides are environment friendly and free from side effects.

Conclusion

Economic development, which aimed at increasing the production of goods and services to meet the needs of a rising population, puts greater pressure on the environment. In the initial stages of development, the demand for environmental resources was less than that of supply. Now world is faced with increased demand for environmental resources but their supply is limited due to over use and misuse. Sustainable development aims at promoting the kind of development that minimizes environmental problems and meets the needs of the present generation without compromising the ability of the future generation to meet their own needs.

References

- Agarwal, Anil and Sunita Narain. (1996): Global Warming in an Unequal World, Centre for Science and Environment, Reprint Edition, New Delhi.
- Bharucha. E. (2005), Textbook of Environmental Studies for Undergraduate courses, University Press (India) Pvt. Ltd.
- Karpagam.M. (2001), Environmental Studies: Crisis to cure, Oxford University Press.
- NCERT – Indian Economic Development (2006) pp-163-175
- Indian Economic Development – Sarawati Publication (2009), New Delhi, pp-10.1 to 10.8
- Srivastava. P. (1995) "Ecocentric management for a risk society", Academy of Management Review, Vol.20, no.1,pp 118-137.
- Sanjay J.Bhyani (2010), "Environmental Social Responsibility: A Comparative Analysisof Perceptions of Indian Managers", Journal of Accounting and Finance, Vol. 23, No. 2, April-Sept 2009, pp. 15-33
- Klassen, R (2000), "Just-in-time manufacturing & Pollution prevention generate mutual benefits in furniture industry", Interfaces 30 (3), 95.
- Guide, V.D.R. Srivastava, R and Spencer, M.S., (1996), "Are production system ready for green revolution?" Production & Inventory Management Journal, Vol. 37, No4, pp34-51.
- Ottman, J and Terry,V, (1998), "Strategic marketing of Greener Products," The Journal Sustainable Product Design, Vol. 5.

7. Geological History of Earth and Climate Transformation Precambrian to Cenozoic

Vikram Singh
Rajesh Kumari

The present-day configuration of the continents and ocean basins is merely a snapshot in time. As the plates move about, the location of continents and ocean basins constantly changes. One of the goals of historical geology is to provide paleogeographic reconstructions of the world during the geologic past. By synthesizing all the pertinent paleoclimatic, paleomagnetic, paleontologic, sedimentologic, stratigraphic, and tectonic data available, geologists can construct paleogeographic maps. Such maps are simply interpretations of the geography of an area for a particular time in the geologic past. The majority of paleogeographic maps show the distribution of land and sea, possible climatic regimes, and such geographic features as mountain ranges, swamps, and glaciers. Geological history also tells us about the climate change of the earth and its impact on the flora and fauna.

Key Words: Archean, Proterozoic, Phanerozoic, Paleozoic, Mesozoic, Cenozoic.

Introduction

The history of the earth is 4.5 billion year old. In the movie we would see a planet under gang remarkable change as continents moved about its surface. As a result of these movement ocean basins would open and close and mountains ranges would from along continental margin or where continents collided with each other. Ocean and atmospheric circulation Pattern would shift is response to the moving continents, sometimes causing massive ice sheets to form and then melt away at other times, extensive swamps or vast interior deserts would appear. Earth's very earliest atmosphere was probably composed of hydrogen and helium but it would have quickly been lost into space for two season, first earth's gravitational attraction in insufficient to retain gases with such low molecular weights. And second before earth differentiated if had no core or magnetic field. And lacked of magnetosphere so a strong solar wind, an out flow of ions from the sun, would have swept away any atmospheric gases. One earth had differentiated and a magnetosphere was present thought and atmosphere began accumulating as a result of our gassing inviting the release of gases from earth's interior during volcanism. The earth history has been divided into Eon, Era, Period and epoch. Each Eon, Era, Period and epoch have their own impact on the evolution of life adaptation of environment, extinction of life and after long time formation particular species and their society. In this article we simplified the complicated process of evolution life and their adaptation to climate and their extinction on the geological time scale.

Objectives

1) To study the evolution of the earth.
2) To study the dynamic pattern of the earth.
3) To study the climate transformation of the earth from Precambrian to Cenozoic.
4) To study the evolution of flora and fauna and their extinction.

Research Methodology

1. The study based on the geological past History of the earth and their continuous changing nature of the earth.
2. The study based on the plate tectonic motion of earth.
3. The study based of changing flora and fauna pattern of earth.
4. The study based in the Paleoclimate, Paleosoil, Paleofossil of and Paleomagnetism of earth.
5. The study based on the post igneous activity, sedimentation and metamorphism
6. The study based on the past minerals formation.

7. The study based on the factor like, Milankovitch Cycle, Sunspot activity, Earth Rotation, Carbon Isotopes.

Precambrian Earth and Life History The Archean Eon

The Precambrian, which lasted for more than 4 billion years. Earth has existed for 4.6 billion years. Suppose that a 24-hours clock represented 4.6 billion years. In this case, the Precambrian alone would be more than 21 hours long and constitute about 88% of all geologic item. Precambrian is a widely used term that refers to both time and rocks. Unfortunately, no rocks are known for the first 640 million years of geologic time, so our geologic record actually begins 3.96 billion years ago with the oldest known rocks on Earth other than meteorites. The geologic record we do have for the older part of the Precambrian is difficult to interpret, particularly for the older part of the Precambrian, because many of the rocks have been metamorphosed and complexly deformed, and in many areas they lie deeply buried beneath younger rocks. The formal subdivisions of the Precambrian is the *Archean and* the *Proterozoic.* The oldest known rocks on Earth are 3.96 billion years old, so we have no geologic record for most of the Eoarchean. Nevertheless, we know of some events that took place during this time. For one thing, it was during the Eoarchean that Earth accreted form planetesimals and differentiated into a core and mantle, and at least some crust was present. Like the other terrestrial planets and the Moon, Earth was bombarded by comets and meteorites, and volcanic activity was ubiquitous. A atmosphere formed, but one that was quite different from the oxygen-rich one we have now, and once the planet cooled sufficiently, the surface water began to accumulate. The oldest known rocks on Earth, the 3.96-billion year. Eath's rate of rotation is slowing because of tidal effects. That is friction caused by the moon on Earth's oceanic waters, as well as its landmasses, causes its rate of rotation to slow very slightly every years. When earth first formed, it probably rotated in as little as 10 hours. The first crust that formed was probably thin and made up of ultramafic rock-that is, igneous rock with less than 45 Percent silica, Upwelling mantle currents of mafic magma disrupted this early crust, and numerous subduction zones developed to form the first island arcs. Collisions between island arcs eventually formed a few continental cores as silica-rich materials were metamorphosed (Figure 8.4b). Larger groups of merged island arcs, or proto continents, grew faster by accretion the first continental nuclei or cartons formed. Earth did not fully differentiate into core, mantle, and crust until millions of years after it formed. Outgassing was probably responsible for the early atmosphere and the hydrosphere. However, the atmosphere so formed lacked free oxygen but contained abundant carbon dioxide and water vapour. Models for the origin of life by natural processes require an oxygen-deficient atmosphere, the necessary elements for organic molecules, and energy to promote the synthesis of organic molecules. The first naturally formed organic molecules were probably monomers, such as amino acids, that linked together to form more complex polymers, including nucleic acids and proteins. RNA molecules may have been the first molecules capable of self-replication. However, the method whereby a reproductive system formed is not known. The only known Archean fossils are of single-celled, prokaryotic bacteria such as blue-green algae, but chemical compounds in some Archean rocks may indicate the presence of archean. Stromatolites that formed by the activities of photosynthesizing bacteria are found in rocks as much as 3.5 billion years old. Archean mineral resources include gold, chrome, zinc, copper, and nickel.

Precambrian Earth and Life History The Proterozoic Eon

At 1.958 billion years long, the Proterozoic Eon along accounts for 42.5% of all geological time. Some Proterozoic rocks are also metamorphic but there are also widespread exposure that show little or no metamorphic effects but the contain few fossils that are useful in stratigraphy. Proterozoic Eon is subdivided into three eras with the prefixes Paleo, Meso and New, each era is further subdivided into periods. Geologists have place the Archean-Proterozoic boundary at 2.5 billion years ago because that marks the approximate time of changes in the style of crustal evolution. Many Archean rocks have been metamorphosed, although their degree of metamorphism varies and some are completely unaltered. In contrast, vast exposure of Proterozoic rocks show little or no effects of metamorphism and in many areas they are separated from Archean rocks unconformity. And finally, the Proterozoic is characterized by widespread sedimentary rock assemblages that are rare or absent in the Archean. Proterozoic was also an important time in the evolution of the atmosphere and biosphere, as well as the origin of some important mineral resources. It was during the Proterozoic that oxygen-dependent organisms appeared in the fossil records, and the kind of cells that make up most organisms today made their appearance. The appearance of eukaryotic cells that is, cells with a nucleus and internal organelles-marks a milestone in evolution. Most eukaryotes are multicelled and reproduce sexually, and nearly all are aerobic. The crust-forming process the yielded

Archean granite-gneiss complexes and greenstone belts continued into the Proterozoic but at a considerably reduced rate. Paleoproterozoic collisions between Archean cartons formed larger cartons that served as nuclei, around which crust accreted. One large landmass so formed was Laurentia. Paleoproterozoic amalgamation of cartons, followed by Mesoproterozoic igneous activity. Ophiolite sequences marking convergent plate boundaries are first well documented from the Neoarchean and Paleoproterozoic, indicating that a plate tectonic style similar to that operating now had become established. The supercontinent Rodinia assembled between 1.3 and 1.0 billion years ago, fragmented, and then reassembled to form Pannotia about 650 million years ago, which began fragmenting about 550 million years ago.

		(Pannotia)	
Rodina →	Fragment →	Reassembled →	Fragment
(1.3 billion years)	(750 million years)	(650 million years)	(550 million years)
		(Late Proterozoic Eon)	(Onset of Phanerozoic Eon)

Glaciers were widespread during both the Paleoproteroic and the Neoproterozoic. Photosynthesis continued to release free oxygen ito the atmosphere, which became increasingly rich in oxygen through the Proterozoic. Fully 92% of Earth's iron or deposits in the form of banded iron formations were deposited between 2.5 and 2.0 billion years ago. Widespread continental red beds dating from 1.8 billion years ago indicate that Earth's atmosphere had enough free oxygen for oxidation of iron compounds. Most of the known Proterozoic organisms are single-celled prokaryotes (bacteria). When eukaryotic cells first appeared is uncertain, but they may have been present by 2.1 billion years ago. Endosymbiosis is a widely accepted theory for their origin. The oldest known multicelled organisms are probably algae, some of which might date back to the Paleoproterozoic. Well-documented multicelled animals are found in several Neoproterozoic localities. Animals were widespread at this time but because all lacked durable skeletons their fossils are not common. Most of the world's iron ore production is from Proterozoic banded iron formations. Other important resources include nickel and platinum.

Early and Late Paleozoic Earth History

The Paleozoic history of most continents involves major mountain-building activity along their margins and numerous shallow-water marine transgressions and regressions over their interiors. These transgressions and regressions were caused by global changes in sea level that most probably were related to plate activity and glaciations. In the beginning of Paleozoic Era six major continents were present. Baltica (Russia west of the Ural Mountains and the major part of northern Europe), China (China, Indochina, and the Malay Peninsula), Gondwana (Africa, Antarctica, Australia, India, Madagascar and parts of the Middles East and southern Europe. Kazakhstania, Laurentia (America, Greenland, Ireland, and Scotland), and Siberia (East of Ural mountain). During the Early Paleozoic (Cambrian-Silurian), Laurentia was moving northward and Gondwana moved to a south polar location, as indicated by Tillie deposits. Most continents consist of two major components; a relatively stable craton over which eperic seas transgressed and regressed, surrounded by mobile belts in which mountain building took place. Early Paleozoic-aged rocks contain a variety of mineral resources, including building stone, milestone for cement, silica sand, hydrocarbons, evaporates, and iron ore.

The late Paleozoic was a time marked by continental collisions, mountain building, fluctuating sea levels, and varied climates, coals, evaporites, and tillites testify to the variety of climatic conditions experienced by the different continents during the Late Poleozoic. Major glacial and interglacial episodes took place. During the Late Paleozoic, Baltica and Laurentia collided, forming Laurasia, Siberia and Kazakhstania collided and finally were sutured to Laurasia. Gondwana moved over the South Pole and experienced several glacial interglacial periods, resulting in global sea level changes and transgressions and regressions along the low-lying craton margins. Laurasia and Gondwana underwent a series of collisions beginning in the Carboniferous. During the Permian, the formation of Pangaea was completed. Surrounding the supercontinent was the global ocean Panthalasa.

Life – Scientist have found the fossil in 3.3 to 3.5 billion year old Archean rock, and chemical evidence in 3.85 billion year old rocks in Greenland. But Archean was biologically. Impoverished. Member of archean are single cell in fact they look much like bacteria but differ from them in their genetics and chemical make up.

Poleozoic Life History Invertebrates – Invertebrates with hard parts suddenly appeared during the early Cambrian in what called the cambrian explosion. Skeleton provided such advantages as protection against predators and support for muscles.

The Cambrian invertebrate community was dominated by three major groups the trilobites made up about help of the total fauna.

Trilobites – Trilobites were benthic mobile, sediment deposit feeders. They first appeared in the early Cambrian, rapidly diversified, reach their maximum diversity in the Late Cambrian and then suffered mass extinction near the end of the Cambrian. Cambrian brachiopods were mostly types called in articulate have a tooth were also present but did not abundant until the Ordovician period. The major group to Cambrian organism were the archaeocythids. These organism were benthic, sessile, presumably suspension feeder. It became extinct at the end of Cambrian.

Brachiopods – Brachiopods were inarticulate. Inarticulate brachiopods also lacked a tooth-and-socket arrangement and were also present but did not abundant until the Ordovician period.

Archaeocyathids – These organisms were benthic, sessile, presumably suspension feeders that constructed reeflike structures beginning in the Early Cambrian. Archaeocyathids went extinct at the end of the Cambrian.

Ordovician Marine Community – The Ordovician marine invertebrate community marked the beginning of the dominance by the shelly fauna and the start of the large scale reef building. Three Ordovician fossil groups have proved particularly useful for biostratigraphy correlation. The articulate brachiopods, graptolites and condones.

Articulate Brachiopods – Present since the Cambrian, began a period of major diversification in the shallow water marine environment during the Ordovician.

Graptolites – Were Plank tonic animal. They were specially abundant during the Ordovician and Silurian periods.

Conodonts – Are a group of well known, small tooth like fossils composed of marine apatite. Studies of specimens indicate the conodonts animals was probably an elongate swimming organisms.

The end of the Ordovician period marked by the mass extinction for many invertebrate phyla. Many geologist think this extinction result of extensive glaciations in Gondwana at the end of Ordovician.

Silurian and Devonian Marine Communities – The Silurian and Devonian periods were times of diverse faunas dominated by the reef building animals.

Carboniferous and Permian Marine Communities – Carboniferous and Permian periods saw the great decline in invertebrate diversity. A major extinction occurred at the end of the Paleozioc Era, attesting the invertebrate as well as the vertebrates.

Paleozoic Life History Vertebrates and Plants

Vertebrate are animals with back bone, fish are the earliest known vertebrate with their first fossil occurrence in upper Cambrian rocks. The oldest and most primitive are ostracoderms (bony skin). There are jawless and reached their zenith during the Silurian and Devonian and then extinct. The evolution of the jaw may thus have been related to respiration rather than to feeding. The fossils remain of the first jawed fish are found in lower Silurian rocks. The jaw fish most abundant during the Devonian and declined in carboniferous and became extinct during the Permian. The other jawed fish placoderm cartilaginous fish evolved during the Devonian so that Devonian informally called age of fish became all major fish group were present during this period.

Amphibians – Amphibian were the first vertebrates to live on land. The transition from water to land required animals to surmount several barriers. The most critical were decocting reproduction, the effects of gravity and the extraction of oxygen form the atmosphere by lungs rather than form water by gills. The oldest amphibian fossils found in the upper Devonian and belonging to genera Ichtheorteger, had skeletal feature that allowed it to spend life on land amphibian became abundant during the Carboniferous and early Permian. In 2006 discovery of 1.2-2.8m long 375 million year old fishpod on Ellesmere Island Canada Tiktaalik roseae was an intermediary between the finned fish and the earliest terapod (Acanthostega). Tiktaalik roseae truly a fishpod in that if has mixture of both fish and terapod characteristic. Acanthostega, a Late Devonian (365 million year ago) terapod to be the perfect intermediary between fish and true land dwelling terapods. However it limbs could not supports its weights on land and thus it was an aquatic animals using it limbs to navigate in water rather than walk on land.

(Upper Devonian) Crossopterygian → (Late Devonian) Rhipidistian → (Late Devonian) Tiktaalik roseae → (Late Devonian) Acanthostega → (Late Devonian) Ichtheostega → (Carboniferous-Permian) Labyrinthodonts

(Evolutionary Development of Amphibian)

Evolution of Reptiles – Amphibians were limited in colonizing the land because they had to return to water to lay this egg. The evolution of amniotes egg freed reptiles from this constraint. In amniotes egg, the embay is surrounded by liquid sac and provided with a food source and waste sac. During the Permian period reptiles diversified and began displacing may amphibians.

(Permian)		(Permian)		(Permian-Triassic)
Protorothyrids	→	Pleycosaurs	→	Therapsids

(Evolutionary Relationship Among the Paleozoic Reptiles)

The reptiles succeeded partly because of then advanced method of reproduction and advanced Jaw, Teeth the Pelycosaurs evolved during the Pennsylvanian and were dominant reptile group the early Permian. The Pelycosaurs succeeded by therapsids. As the Poleozoic Era came to an end the therapsids constituted about 90% of all marine invertebrate species were extinct.

Plant Evolution – Most experts agree that the ancestors of land plants first evolved in a marine environment and then moved into a freshwater environment before finally moving onto land. In this way, the differences in osmotic pressures between salt and freshwater were overcome while the plant was still in the water. The high land plants are divided into two major groups. Nonvascular and vascular. Most land plants are **vascular,** meaning they have a tissue system of specialized cells for the movement of water. The earliest land plants form the Middle to Late Ordovician were probably small and bryophyte-like in their overall organization. The ancestor of terrestrial vascular plants was probably some type of green algae. The problem of desiccation was circumvented by the evolution of *Cutin,* an organic compound found in the outer-well layers of plants. Cutin also provides additional resistance to oxidation, the effects of ultraviolet light, and the entry of parasites.

Eon	Era	Period		Epoch	Year
Phanerozoic	Cenozoic	Neogene	Quaternary	Recent or Holocene	10,000
				Pleistocene	1.8
				Pliocene	5
				Miocene	23
				Oligocene	34
				Eocene	56
				Paleocene	66
	Mesozoic	Cretaceous			146
		Jurassic			200
		Triassic			251
	Paleozoic	Permian			299
		Carboniferous	Pennsylvanian		318
			Mississippian		359
		Devonian			416
		Silurian			444
		Ordovician			488
		Cambrian			542
Precambrian	Proterozoic				2500
	Archean				4600

The Geologic Time Scale the number to the right of the columns are the ages in millions of years before the present

Silurian and Devonian Floras – During the Devonian the number of plant genera remained about the same, yet the composition of the flora changed. Whereas the Early Devonian landscape was dominated by relatively small, low-growing, bog-dwelling types of plants, the Late Devonian witnessed forests of large, tree-sized plants up to 10m tall. The evolution of the seed at this time liberated land plants form their dependence on most conditions and allowed them to spread over all parts of the land.

Late Carboniferous and Permian Floras – The Carboniferous period was a time at vast coal swaps where condition were ideal for the seedless vascular plants. With the onset of more arid condition during the Permian the gymnosperms became the dominant element at the world flora.

Mesozoic Earth History – The Mesozoic Era (251 to 66 million year ago) was an important time in Earth history. The major geologic event was the breakup of Pangaea, which affected oceanic and climatic circulation patterns and influenced the evolution of the terrestrial and marine biotas. Pangaea's breakup began with rifting between Laurasia and Gondwana during the Triassic. During the Late Triassic and Jurassic periods, Antarctica and Australia, which remained sutured together, began separating from South America and Africa. Also during this time, India began rifting from the Gondwana continent. South America and Africa began rifting apart during the Jurassic. By the and of the Cretaceous, Australia and Antarctica had detached from each other, and India had moved into the low southern latitudes and was nearly to the equator. Greenland was essentially an independent landmass with only a shallow sea between it and North America and Europe. A global rise in sea level during the Cretaceous resulted in worldwide transgressions onto the continents. By the Middle Cretaceous, sea level was probably as high as at any time since the Ordovician, and about one-third of the present land area was inundated by epeiric seas. The final stage in Pangaes's breakup occurred during the Cenozoic. During this time, Australia continued moving northward, and Greenland was completely separated and Europe and North America and formed a separate landmass.

Life of the Mesozoic Era

Marine Invertebrates – Following the Paleozoic mass extinctions, the Mesozoic was a time when marine invertebrates repopulated the seas. The Early Triassic invertebrate fauna was not very diverse, but by the Late Triassic the seas were once again swarming with invertebrates form planktonic foraminifera to a wide variety of cephalopods. Mollusks such as cephalopods, bivalves, and gastropods were the most important elements of the Mesozoic marine invertebrate fauna. And important reef builder throughout the Mesozoic was a group of bivalves known as rudists. Rudists are important because they displaced corals as the main reef-builders during the later Mesozoic and are excellent guide fossils for the Late Jurassic and Cretaceous. A new and familiar type of coral also first appeared during the Triassic : the scleractinians. Another invertebrate group the prospered during the Mesozoic was the echinoids. The foraminifera diversified rapidly during the Jurassic and Cretaceous and continue to be diverse and abundant to the present.

A major difference between the paleozoic and Mesozoic marine invertebrate faunas was the increased abundance and diversity of burrowing organism.

The primary producers in the Mesozoic seas included the microscopic coccolithophores, diatoms, and dinoflagellates. In genera terms we can think of the Mesozoic as a time of increasing complexity among the marine invertebrate fauna. At eh beginning of the Triassic, diversity was low and food chains were short. Near the end of the Cretaceous, though, the marine invertebrate fauna was highly complex, with interrelated food chains. This evolutionary history reflects changing geologic conditions influenced by plate tectonic activity.

Aquatic and Semi-aquatic – Sharks and the other cartilaginous fishes became more abundant during the Mesozoic. Few species of lungfishes and crossopterygians existed during the Mesozoic, and the latter declined and were nearly extinct by the end of the era. Only one crossopterygian species exists now and the group has no known Cenozoic fossil record. The labyrinthodont amphibians were common during the latter part of the Paleozoic, but the few surviving Mesozoic species died out by the end of the Triassic, amphibians have made up only a small part of the total vertebrate fauna. Frogs and salamanders evolved during the Mesozoic, but both have poor fossil records.

Plants – Just as during the Late Paleozoic, seedless vascular plants and gymnosperms dominated Triassic and Jurassic land plant communities, and, in fact, representatives of both groups are still common. Among by gymnosperms, the large seed ferns became extinct by the end of the Triassic, but ginkgos remained abundant and still exist in isolated regions, and conifers continued to diversify and are now widespread, particularly at high elevations and high latitudes. A new group of gymnosperms known as cycads made their appearance during the Triassic. They became widespread and now exist in tropical and semitropical areas.

The long dominance of deedless plants and gymnosperms ended during the Early Cretaceous, perhaps the Late Jurassic, when many were replaced by **angiosperms,** or flowering plants. Since they first evolved, angiosperms have adapted to nearly every terrestrial habitat from mountains to deserts, and some have even adapted to shallow coastal waters. Several factors account for their phenomenal success, but chief among them were the evolution of flowers, which attract animal pollinators, especially incest, and the evolution of enclosed seeds.

Archosaurs and the Origin of Dinosaurs – Crocodiles, flying reptiles, dinosaurs, and birds are all **archosaurs** (archo, meaning "ruling", and sauros, meaning "lizard"). Including such diverse animals in a single group implies that they share a common ancestor and in fact, they possess several characteristics that unite them and distinguish them from other reptiles. All dinosaurs recognize two distinct orders: The **Saurischia** and **Ornithischia**. Saurischian dinosaurs have a lizardliek pelvis and accordingly are called lizard-hipped dinosaurs, whereas a birdlike pelvis is present in the ornithiscians, so they are called bird-happed dinosaurs. Palaeontologists are convinced that both orders of dinosaurs share a common ancestor.

Flying Reptiles & Marine Reptiles – Paleozoic insects were the first animals to achieve flight, but the first among vertebrates were **pterosaurs,** or flying reptiles, which were common in the skies form the Late Triassic until their extinction at the end of the Cretaceous. Some were no bigger than today's sparrows, robins, and crows. Even the very largest species probably weighed no more than a few tens of kilograms. Several types of Mesozoic reptiles adapted to a marine environment. Including turtles and some crocodiles. As well as the Triassic mollusc-crushing placodonts.

Crocodiles, Turtles, Lizards and Snakes – By Jurassic time, crocodiles had become the most common freshwater predators. All crocodiles are amphibious, spending much of their time in water, but they are well equipped for walking on land. Turtles, appearance during the Triassic. Turtles are more thoroughly armoured than any other vertebrate animal, living or fossils. Turtle ancestry is uncertain. Lizards and snakes are closely related and lizards were in fact ancestral to snakes. The limbless conditions in snakes and skull modifications that allow snakes to open their mouths very wide are the main difference between these two groups. Lizards are known form Upper Permian strata, but they were not abundant until the Late Cretaceous, Snakes first appear during the Cretaceous.

Birds and Mammals – Jurassic-ages Archaeopteryx, the oldest known bird, possesses so many theropod characteristics that it has convinced most palaeontologists the two are closely related. Mammals evolved by the Late Triassic, but they differed little from their ancestors, the cynodonts. Minor differences in the lower jaw, teeth, and middle ear differentiate one group of fossils from the other. Several types of Mesozoic mammals existed, but most were small, and their diversity was low. Both marsupial and placental mammals evolved during the Cretaceous form a group known as eupantotheres.

Among the victims of the Mesozoic mass extinction were dinosaurs, flying reptiles, marine reptiles, and several groups of marine invertebrates. A huge meteorite impact may have caused these extinctions, but some palaeontologists think other factors were important, too.

Cenozoic Geologic History – At 66 million years long, the Cenozoic Era is only 1.5% of all geologic time, or just 20 minutes on our hypothetical 24-hours clock for geologic time. The cenozoic era is generally divided into three periods Palaeogene (66 to 23 million years ago) include Palaeocene, Eocene and Oligocene epochs, and Neogene Period (23 million to 2 million year ago) include Miocene, Pliocene and quaternary (form 2 million year to present) period divided into Pleistocene and Holocene. The progressive fragmentation of Pangaea, the supercontinent that existed at the end of the Paleozoic account s for the present distribution of Earth's landmasses. The Americas separated from Europe the Arfica, the Atlantic Ocean basin opened, first in the south and later in the north. Spreading ridges such as the Mid-Atlantic Ridge and East Pacific Rise were established, along which new oceanic crust formed and continues to form. Another important plate tectonic event was the north-ward movement of the Indian plate and its eventual collision with Asia. Simultaneous northward movement of the African plate caused the closure of the Tethys Sea and intiated the tectonic activity that currently takes place throughout the tectonic activity that currently takes place throughout an east-west zone form the Mediterranean through northern India. Erupting volcanoes in Italy and Greece as well as seismic activity in Italy, plate interactions in this part of the world. Neogene rifting began in East Africa, the Red Sea, and the Gulf of Aden. In the meantime, the North and South American plates continued their westerly movement as the Atlantic Ocean basin widened. Subduction zones bounded both continent on their western margins, but the situation changed in the East Pacific Rise and it now has a transform plate boundary.

Pleistocene is best known for widespread glaciers, it was also a time of volcanism and tectonism. Pleistocene glaciers covered about 30%

of the land surface, and were most widespread on the Northern Hemisphere continents. At least four intervals of extensive Pleistocene glaciation took place in North America, each separated by interglacial stages. Fossils and oxygen isotope data indicate about 20 warm-cold cycles occurred during the Pleistocene. Areas far beyond the ice were affected by Pleistocene glaciers : Climate belts were compressed towards the equator, large pluvial lakes existed in what are now arid regions, and sea level was as much as 130 m lower than now. Moraines, striations, outwash, and various other glacial landforms are found throughout Canada, in the northern tier of states, and in may mountain rages where valley glaciers were present. The tremendous weight of Pleistocene glaciers caused isostatic subsidence of Earth's curst. When the glaciers melted, isostatic rebound began and continues even now in some areas. Major glacial episodes separated by tens or hundreds of millions of years probably stem form changing positions of plates, which in turn profoundly affects oceanic and atmospheric circulation patterns. According to the milankovitch theory, minor changes in Earth's rotation and orbit bring about climatic changes that produce glacial-interglacial intervals. The causes of short-term climatic changes such as occurred during the Little Ice Age are unknown; tow proposed causes are variations in the amount of solar energy and volcanism.

Life of the Cenozoic Era – The Cenozoic Era constitutes only 1.4% of all geologic time, this comparatively brief 66 million years of Earth and life history was short only in the context of geologic time. The marine invertebrate groups that survived the Mesozoic extinctions diversified throughout the Cenozoic, Bivalves, gastropods, corals, and several kinds of phytoplankton such as foraminifera proliferated. Birds belonging to the living orders and families evolved during the Paleogene Period. Large, flightless predatory birds of the Paleogene were eventually replaced by mammalian predators. Evolution history is better known for mammal than for other cases of vertebrates, because mammals have a good fossil record and their teeth are so distinctive. Egg-lying mammals (monotremes) and marsupials exist mostly in the Australian region. The placental mammals-by far the most common mammals-owe their success to their method of reproduction. All placental and marsupial mammals descended from shrew like ancestors that existed from Late Cretaceous to Paleogene time. Small mammals such as insectivores, rodents, and rabbits occupy the microhabitats unavailable to larger mammals. Bats, the only flying mammals, have forelimbs modified into wings but otherwise differ little from their ancestors. Most carnivorous mammals have well-developed canines and specialized shearing teeth, although some aquatic carnivores such as seals have peglike teeth. The most common angulates are the even-toed hoofed mammals (artiodactyls) and odd-toed hoofed mammals (perissodactyls), both of which evolved during the Eocene. Many ungulates sow evolutionary trends such as moralization of the premolars as well as lengthening of the legs for speed. During the Paleogene, perissodactyls were more common than artiodactyls but now their 16 living species constitute only about 10% of the world's hoofed mammals fauna. Even though horses, rhinoceroses, and tapirs as well as the extinct titanotheres and chalicotheres do not closely resemble one another, fossils show they diverged from a common ancestor during the Eocene. The fossil record for whales is now complete enough to verify that they evolved from land-dwelling ancestors. Elephants evolved form rather small ancertors, became quite diverse and abundant, especially on the Northern Hemisphere continents, and then dwindled to only two living species. Horses, camels, elephants and other mammals spread across the northern continents during the Cenozoic because land connections existed between those land-masses at various times. During most of the Cenozoic, South America was isolated, and its mammals fauna was unique. A land connection was established between the Americas during the Late Cenozoic and migrations in both directions took place. One important evolutionary trend in Pleistocene mammals and some birds was towards giantism. Many of these large species died out, beginning about 40,000 years ago. Changes in habitat and prehistoric overkill are the two hypotheses explaining Pleistocene extinctions.

Primate and Human Evolution – Primates are difficult to characterize as an order became they lack the strong specializations found in most other mammalian orders. We can however, point to several trends in their evolution that help define primate are related to their arboreal or tree dwelling ancestry. Other change include, change in skeleton and mode of locomotion, an increase in brain size, less specialized teeth, opposable thumb. Not all these trends took place in every primate group, nor did they evolve at the same rate in each group. In fact some primate have retained certain primitive have retained certain primitive features, whereas others show all or most of these trends.

The primate other is divided into two suborders first is the prosimians (before ape) or lower primate include lemurs, lorries, tree Shrews second is anthropoids on higher primates include monkey aps and humans.

Prosimians – Prosimians are generally small, they or arboreal. Most are nocturnal. They are omnivorous. During the Eocene prosimians were abundant diversified and widespread in north America, Europe Area. As the continents move northwards darning the Cenozoic and the Climate change form ward tropical to cooler mid Latitude conditions, the prosimians population decreased in both abundance and diversity. By the digocene hardly any prosimians left in the northern continent as the once, widespread Eocene population migrated south to the warmer latitudes of Africa, Area and South East Asia. Presently prosimians are found only in the tropical regions of Area, India Africa and Madagascar.

Anthropoids – Anthropoids evolved from a prosimians lineage sometime during the Late Eocene, and by the Oligocene they were well established. Much of our knowledge about the early evolutionary history of anthropoids, comes from fossils found in the Fayum district, a small desert area southwest of Cairo Egypt. During the Late Eocene and Oligocene, this region of Africa was a lush, tropical rain forest that supported a diverse and abundant fauna and flora. One of the earliest anthropoids was Aegyptopithecus, as small, Late Eocene-fruit-eating, arboreal primate. Anthropoids are divided into three super families: Old World monkeys, New World monkeys, and hominoids.

Old World Monkeys are characterized by close-set, downward-directed nostrils, grasping hands, and a no prehensile tail. They include the macaque, baboon, and proboscis monkey. Present-day Old World Monkeys are distributed in the tropical regions of Africa and Asia and are thought to have evolved from a primitive anthropoid ancestor, like *Aegyptopithecus*, sometime during the Oligocene.

New World Monkeys are found only in Central and South America. They probably evolved from African monkeys that migrated across the widening Atlantic sometime during the Early Oligocene, and they have continued evolving in isolation to this day. New World monkeys are characterized by a prehensile tail, flattish face, and widely separated nostrils, and include the howler, spider, and squirrel monkeys.

Hominoids consist of three families: the great apes, which include chimpanzees, orang-utans, and gorillas the lesser apes, which are gibbons and siamangs; and the hominids which are humans and their extinct ancestors. The hominoid lineage diverged form Old World monkeys sometime before the Miocene.

In the Late Eocene the northward movement of the continents resulted in pronounced climatic shifts. In Africa, Europe, Asia, and elsewhere, a major cooling trend began, and the tropical and subtropical rain forests slowly began to change to a variety of mixed forests separated by savannas and open grasslands as temperatures and rainfall decreased. As the climate changed, the primate populations also changed. Prosimians and monkeys became rare, whereas hominoids diversified in the newly forming environment and came abundant. Two apelike groups evolved during the Miocene that ultimately gave rise to present-day hominoids.

The first group, the ***dryopithecines,*** evolved in Africa during the Miocene and subsequently spread to Eurasia following the collision between the two continents.

The second group, the ***sivapithecids,*** evolved in Africa during the Miocene and then spread throughout Eurasia. The fossil remains of sivapithecids are plentiful and consist mostly of skulls, jaws, and isolated teeth. Body or limb bones are rare, limiting our knowledge about what they looked like and how they moved around. It is clear from the fossils evidence that sivapithecids were not involved in the evolutionary branch leading to humans but they were probably the ancestral stock from which present-day orang-utans evolved.

Hominids – The hominids, the primate family that includes present-day humans and their extinct ancestors, have a fossil record extending back to almost 7 million years. Several features distinguish them from other hominoids. Hominids are bepedal; that is, they have an upright posture, which is indicated by several modifications in their skeleton. They show a trend toward a large and internally reorganized brain. Other features include a reduced face and reduced canine teeth, omnivorous feeding. At present, there is no clear consensus on the evolutionary history of the hominid lineage. This is due, in part, to the incomplete fossil record of hominids.

Australopithecines – ***Australopithecine*** is a collective term for all members of the genus *Australopithecus*. Currently, five species are recognized : A. *anamensis,* A. *afarensis,* A. *africanus*, A. *Robustus*, and A. *boisei*. The oldest known australopithecine is *Australopithecus anamensis*, is more primitive in other characteristics, such as its teeth and skull. *Australopithecus afarensis*, who lived 3.9-3.0 million years ago, was fully bipedal and exhibited great variability in size and weight. The skull of A. *afarensis* retained may apelike features, but its teeth were intermediate between those of apes and humans. A. *afarensis* was strati graphically succeeded by *Australopithecus africanus*, who lived 3.0-2.3 million years ago. The differences between the two species are relatively minor. They were both about the same size and weight, but A. *africanus* had a flatter face and some-what larger brain. A. *Boisei*, A. *robustus* : A. *Boisei* was had a powerful upper body, a distinctive bony crest on the top of its skull, a flat face, and the larges molars of any hominid. A. *robustus* had a lighter and flat face. Its broad flat molars indicated A. *robustus* was a vegetarian.

The Human Lineage – Homo habilis The earliest member of our own genus, Homo, is Homo *habilis*, who lived 2.5-1.6 million years ago. Its remains were first found at Olduvai Gorge in Tanzania, but it is also known form Kenya, Ethiopia, and South Africa. H. *habilis* evolved from the A. *afarensis* and A. *africanus* lineage and coexisted with A. *africanus* for about 200,000 years. H. *habilits* and a larger brain (700 cc average). It was about 1.2-1.3 m tall and only. Weighed 32-37 kg. **Homo erectus** In contrast to the australopithecines and *H. habilis*, which are unknown outside Africa, *Homo erectus* was a widely distributed species, having migrated from Africa during the Pleistocene. Specimens have been found not only in Africa abut also in Europe, India, China ("Peking Man"), and Indonesia ("Java Man"). *H. erectus* evolved in Africa 1.8 million years ago and eastern Asia, where it survived until about 100,000 years ago. Although *H. erectus* developed regional variations in form, the species differed from modern humans in several ways. Its brain size of 800-1300 cc, although much larger than that of *H. habilis*, was still less than the average for *Homo sapiens* (1350 cc). It had prominent brow ridges, and its teeth were slightly larger than those of present-day humans. *H. erectus* was comparable in size to modern humans, standing between 1.6 and 1.8 m tall and weighing between 53 and 53 kg our species, H. sapiens, most certainly evolved from *H. erectus*.

Neanderthals – Perhaps the most famous of all fossil humans are the ***Neanderthals***, who inhabited Europe and the Near East form about 200,000 to 30,000 years ago. Their name comes from the first specimens found in 1856 in the Neander Valley near Dusseldorf, Germany. The most notable difference between Neanderthals and present-day humans is in the skull. Neanderthal skulls were long and low with heavy brow ridges, a projecting mouth, and a weak, receding chin. Their brain was slightly larger on average than our own and somewhat differently shaped. The Neanderthal body was more massive and heavily muscled than ours, with rather short lower limbs, much like those of other cold-adapted people o today. **Cro-Magnon** the name given to the successors of the Neanderthals is France, lived form about 35,000 to 10,000 years ago. During this period the development of art and technology far exceeded anything the world had seen before. Highly skilled nomadic hunters, Cro-Magnons followed the herds in their seasonal migrations. Cro-Magnons were cave painters. Cro-Magnon people painted hundreds of scenes on the ceilings and walls of caves in France and Spain.. With the appearance of Cro-Magnons, human evolution has become almost entirely cultural rather than biological. Since the evolution of the Neanderthals about 200,000 years ago, humans have gone form a stone culture to a technology that has allowed us to visit other planets with to be seen how we will continue as a species, evolve into another species, or become extinct as many groups have before us.

Climate During Geological Past – The early Proterozoic was characterised by Huronian glaciations which occurred around 2700 to 1800 million years ago. This was followed by a warm period until about 950 million years ago. During the late Precambrian Period, the glacial epochs-the Gnesjo (940 million years ago), the Strutian (770 million years ago), and the Varangian (615 million year ago)-occured each lasting for about 100 million years.

The Phanerozoic Eon – The Phanerozoic Eon includes the Palaeozoic, the Mesozoic and the Cenozoic Eras. Palaeozoic climates (570 to 225 million years ago) were warmer than the Cambrian and Ordovician Periods, interspersed as the latter were by brief glacial events. The Silurian Period was warm and with the advent of the Devonian Period, North America, Europe, China and Australia experienced tropical climate. Much cooler climate prevailed during the Carboniferous Period and by the Permian Period the great deserts started expanding. During the Mesozoic Era, climate was warm, and temperature fluctuated form 10°C to 20°C at the Poles and from 25°C to 30°C at the Equator. In the middle of the Triassic Period, the climate was hotter than today. The Jurassic Period witnessed a slight drop in temperature. For analysing the Cretaceous Period, oxygen isotope analysis of deep sea cores using marine fauna is important. The end of the Cretaceous Period is characterised by global cooling. The Cenozoic Era was a time of continuous cooling. At the start of the Eocene Epoch (55 million years ago), there was a long-term cooling. The early Tertiary Period was much warmer than today. There was a great advance of massive glaciers during the Quaternary period of the Cenozoic Era. The Quaternary Period is divided into the Pleistocene Epoch and the Holocene Epoch. The Pleistocene is characterised by ice ages. (The present ice sheets in different parts of the world are remnants o the Pleistocene glaciations). There were ice advances and retreats.

The Cenozoic Era had four well-marked short-term cooling episodes:

(i) The middle Palaeocene Epoch about 60 million year ago;
(ii) The middle Eocene Epoch about 45 million years ago;
(iii) The Eocene-Oligocene Epoch transition, 38 million years ago;
(iv) The mid-Oligocene Epoch around 28 million years ago.

The transition of the Eocene and Oligocene Epochs witnessed the most severe short-term cooling episode. Within a span of 100,000 to 1,000,00 years, the entire ocean water column had cooled by 3°C to 5°C in high latitudes while in low latitudinal regions the cooling took place at depth alone. The Antarctica ice sheet probably formed between about 25 million to 15 million years ago, and by the late Miocene the Antarctica ice sheet assumed the present status. During the past two million years, three distinct phases in the world climate occurred-ice age, warm (as today) and interstadial (intermediate). The final (Eemian Epoch) interglacial took place about 120,000 years ago, and it lasted about 10,000 years. The last major glaciations in the northern hemisphere took place around 18,000 years ago the world mean summer ocean temperature became 2.3^{0}C lower than present.

Conclusion

Today much debate on the climate change which is related to only anthropogenic factor and we neglected the other important factor of past geological transformation of the earth. We only focus on the global warming and ignore the global cooling which is more important than global warming. If we see from the earth origin we found five major ice age. The earliest called formed around 2.4 to 2.1 billion year ago during the early **Proterozoic** Eon. The second ice aged formed which is most severe occurred from 850 to 630 million year ago (The cryogenian Period) in Devonian Era and glacier ice sheet reach the equator. Third minor ice age occurred from 460 to 430 million year ago, during the Late Ordovician and Silurian period. Fourth occurred at intervals to 350 to 260 million year ago during carboniferous and early Permian period fifth ice age, occurred in the Pliocene-quaternary glaciations started about 2.58 million year ago. The causes of the ice age is not fully understood but consensus on several factor like, plate tectonic, milankovitch, Sun Spot Cycle, Asmosphere Composition. All these factor have been operated present also. So we consider all these factor when talk about climate change. I my view rather than global warming ice age have been come in future but when we only conjecture about the time but it will come sure and we all preparing our shelf to them.

References

- Reed wicander, James, S. Monroe, (2007) "Historical Geology" (Fifth edition Thomson Books/Cole, Belmont (USA)
- Stanley Chernicoff, "An introduction to physical Geology" (second edition) Houghton Mifflin company, Boston.
- Betz., F. Ir. (1975) : "Environmental Geology" Halsted, Stroudsburg.
- Coates, D.r. (1981) : "Environmental Geology" John Wiley New York.

8. Climate Change and Indian Strategy

Ranju Narang

The Climate Change is a global phenomenon that involves all the aspects of this earthen system. The adverse impact of climate change in the form of higher average temperatures near earth's surface and changed rainfall patterns would create problems for man and nature e.g. De-glaciations and sea level changes, ecological productivity, flooding threat to deltas ,loss of coastal mangroves, reduction in the crop yield per unit area due to high temperature and many more. The recent global climate change treaty, also known as Copenhagen accord was signed on Dec, 18th 2009. Delegates from 194 countries and numerous NGOs covered in the 15th Conference of Parties (COP) and more than 100 heads of states participated during the final days of conference. Under prevailing chaos and confusion, the accord was signed by Brazil, South Africa, India, and China along with United States outside the UNFCCC process.

India is emerging as one of the top ten industrialized countries of the world. While it signals, India's prosperity in industrial production, it also sounds alarming bells for environment pollution. India has rich environmental laws. Indian constitution is one among the few constitutions in the world that has provisions on environmental protection. We have Water Act, Air Act, Forest Conservation Act and many more principles like Pollution Pays Principle, Precautionary Principle etc for environment sustainability. But mere enumeration of the statutes and various constitutional provisions will not work. There is a need to find a new strategy that manages all natural resources and human resources as well as financial and physical assets for increasing long term wealth and well being so that we can leave behind us a greener planet for our future generations. The present paper is an attempt to describe the origin of climate change in India and Indian strategy to solve the problem and to explore some new policy measures which can help our economy to be a low carbon economy so that we may on the top of the world with our culture of worshiping mother nature.

Key words: Climate Change, Environmental protection, Ecosystem, Low carbon economy, Global warming.

Introduction

Man's March from primitive age to the modern age is often described as the process of growth and development. In the primitive age 'quest for survival was his principal thought and 'hunting for food was his principal activity. Today has discovered thousands of varieties of food, which are mechanically produced on the farms and hygienically processed in the industries. He has dived deep into the sea, flown in to the air and breaking the shackles of gravity, is all set to venture into space tourism. But this growth is achieved at the cost of our natural resources. This has resulted in environmental pollution and that way 'Climate Change'.

United Nations Framework Convention on Climate Change (UNFCCC) defines "Climate Change as a change that is attributed directly or indirectly to human activity that alters the composition of global atmosphere and that is in addition to natural climate variability observed over a comparable time periods" given the fact that human civilization and industrialization have amplified the emission of Green House Gasses (G.H.Gs) there is a need to develop a unique strategy that manages all natural resources and human resources as well as financial and physical assets for increasing long term wealth and well-being i.e. strategy of sustainable development.

The Indian Constitution is one among the few constitutions in the world that has provisions on environmental protection. Articles 48 and 51A (g) are inserted by the 42nd Constitutional Amendment. Act in 1976 imposes the duty of environmental protection on both the state and its citizens. In response to the 1972 United Nations Conference on the Human Environment at Stock Holm, India enacted "The environment (protection) Act 1986" and the corresponding Environment protection rules 1986. This is a general enactment that empowers the central government to prevent, control and abate environment pollution. For the effective implementation of the act, the various provisions of this Act have been amended from time to time.

Climate Change presents many challenges at the legal level. The present paper is an attempt to evaluate Indian strategy to handle the problem of the climate change and to explore some new policy measures which can make our economy a low carbon economy.

The paper is civilized into three parts.

Part I Describes meaning, origin, causes of climate change and its impact on ecosystem.

Part II Explains strategic approach of India towards climate change and enumerates Indian Environmental Law as well as strategies adopted in XIth FYP for environment protection.

Part III Tries to explore some more suggestive measures for low carbon economy.

PART –I

In the present context of environmental awareness, global warming is a buzz word that reverberates in every corner of our planet. The Chaotic Climate behaviour perceptible on global scale during the past decade arising out of rise in average temperature of the earth due to large scale emission of anthropogenic green house gases is a matter of grave concern threatening ecosystem of the entire planet. The survival of life on earth depends on the absorption of in coming solar radiation which warms the surface of the planet. This incoming energy is reflected back in to space as low intensity infra-red radiation. However the atmosphere of the earth contains small quantities of carbon dioxide and some other gases (collectively called Green House Gases or GHGs) which absorb some of the out going infra-red radiation and reflect it back to earth and thus increasing the warming of the surface. This "secondary warming" in turn, leads to feather emission of infra-red radiation into space. But any given level of concentration of green house gases, the average temperature of the earth settles at a level at which emerges that comes in solar radiation is balanced by the energy radiated out. The global warming is increase in the average temperature of the Earth's near-surface air and ocean.

The consensus of the world's climate scientists is that human induced green house gas emissions are crating significant change in global climate system. Higher temperatures rising sea levels and more extreme storms will fundamentally change where and how we live. It is now well established that human induced act activities are speeding up global warming which in turn is creating numerous and varied negative impacts on environment. The Inter-Government Panel on Climate Change (IPCC) defines climate change as 'a change in the state of climate that can be identified (e.g. using statistics) by change in the mean and variability or as a result of human activity' According to United Nations Industrial Development Organization (UNIDO) climate change is likely to have a greater impact on India compared to other countries similarly positioned, on account of the unique combination of its geography, diverse population characteristics and extremely higher carbon related energy dependence.

India maintains that climate change is not taking place due to the current level of GHG emissions but is the cumulative impact of the accumulated stock of emission in the atmosphere. The stock is the results of carbon based industrial activity of developed countries over the last two hundred odd years. This historical responsibility mandates that the industrialized countries take the lead in reducing emission and enable developing countries to make a shift to a low carbon development path through finance and technology. The following table supports this fact.

Carbon Dioxide Emission from Fossil Fuels burning in Top Ten countries 2006		
Country	Emission (Millions Tones of Carbon)	Share of Global Total (In Percent)
United States	1,656	19.8
China	1,480	17.7
Russia	437	5.2
India	**391**	**4.7**
Japan	342	4.1
Germany	221	2.6
Canada	177	2.1
United Kingdom	171	2.0
South Korea	130	1.6
Mexico	123	1.5
All other Countries	3,249	38.8
Global Total	**8,379**	**100.0**

Source: *Compiled by Earth Policy Institute from G. Marland, T.A Border and R.J. Andres, 'Global Regional and National CO2 Emissions "Trends: A Compendium of Data on Global Change (Oak Rifge, TN: Carbon Dioxide Information Analysis Center, (2007); BP, Statistical Review of World Energy (London: 2007).*

Need of the hour is to concentrate on the main causes of global warming in India and then to work together for reducing the carbon foot print of industries so that we can leave behind us a greener planet for the future generation. The fundamental causes of environmental degradation are:

- Unsustainably high rates of human population growth and natural resources consumption. Pressure of population on land has tremendously increased and consequently land has been ruthlessly exploited.
- A large section of Indian population is absolutely poor. These people cut trees to sell fuel wood in order to earn their livelihood, and thereby exploit natural capital.
- Inequality in ownership and access to natural resources, including the benefits from use and conservation of biodiversity.
- Increasing use of insecticides, pesticides and chemical fertilizers has also added to environmental pollution.
- India is emerging as top 10 industrialized countries of the world, while it signals India's prosperity in industrial production; it also sounds alarming bells for environmental pollution. Industrial smoke is polluting air and industrial waste is polluting the water.
- Multiplicity of transport vehicle has substantial increased noise and air pollution.
- Legal and institutional systems that promote unsuitable exploit.

Developing countries would take domestic measurers on saving environment and reducing emission but their first and foremost priority is social and economic development and the poverty alleviation of its people. In recognition of the fact that as countries develop, their emissions will increase; New Delhi has pledged that it will not allow its per capita GHG emissions to exceed the per capita emissions of the developed countries.

PART II

India has a rich developed environmental law which provides a frame work for regulatory use of the environment. The enactment of Water act, Air act, Forest conservation Act and many other provisions relating to environment are included in our constitution.

Environment (Protection) Act, 1986

In response to 1972 United Nations Conference on the human Environment at Stock Holm, India enacted, "The Environment

(Protection) Act 1986" and the corresponding 'Environment (Protection) Rules. This is a general enactment that empowers the central govt. to prevent, control and abate environment pollution. The Act has following important provisions:

- No person carrying on any industry, operation or process shall be permitted to discharge any environmental pollutant in excess of prescribed standard. (Section-7.1)
- The Central Government and its officers have the power to enter and inspect any place for the purpose of performing any function under the legislation (Section-10)
- The Central Government and its officers have the power to take samples of air, water, soil from factory or place for analysis. (Section-11)
- No person shall handle hazardous substances without complying the prescribed procedural safeguards (Section-8.2)
- The central government has the power to established environmental laboratories or recognizes any laboratory or institute as an environmental lab. (Section-12)
- The report signed by a government analyst may be used as an evidence of the facts stated therein any proceeding under the legislation (Section-14)

AIR (Prevention and Control of pollution) Act 1981

The salient features of this act are –

- The State Government after consultation with the State Pollution Control Board may prohibit the use of any fuel or appliances burning any material causing air pollution, in an Air Pollution Control area (Under Section-191).
- No person shall without the previous consent of State Pollution Control board establish or operate any industrial plant in Air Pollution Control Area (Sec 21.3).
- No person operating any industrial plant in any Air Pollution Control Area shall discharge the emission of any air pollutant in excess of the standard laid down by the State Pollution Control Boards. (Section-22.4)
- Any person empowered by the State Pollution Control Board shall have the right to enter any place for examining and testing any industrial plant, record, register, and document and to seize any

control equipment if he has reason to believe that it may create an offence punishable under the act (Section-24.5).

Wildlife Protection Act, 1972

This act adopted by all the states except J & K (which has its own act) governing the wildlife conservation and protection of endangered species both inside and outside forest area. This act was amended twice till 1991.

The main features of the Act–

- No person shall hunt any wild animal except as provided under section 11 & 12 of this Act (Section 11 and 12 states that if chief wildlife warden satisfied that any person can hunt that animal which has become dangerous to human life.
- No person shall be a manufacturers or dealer in animal articles, dealer in rugs, skins, horns hair feathers, tooth or captive animal or meat.
- The central govt. shall constitute the central Zoo Authority consisting of a) chairperson, b) Ten members c) a member secretary (Section-38A).

Water (Protection and Control of Pollution) Act 1974

The main concern of the Act is to maintain or restore the wholesomeness of water. The salient features are–

- The state government has the power to collect samples of water of any stream or well or any affluent being discharged into such a stream or well for analysis (Section-21[A]).
- The state pollution control board is empowered by the state government to enter any place for the purpose of performing any of the functions entrusted to it. (Section 23.5)
- No person shall establish any industry, operation or any treatment disposal system, which is likely to damage any sewage or effluent into stream or well or on land. (Section-25.7)
- No person shall use any new outlet for discharge of sewage (section-25.7)
- The Board can also levy and collect a cess on water consumed by industries and authorities for increasing its (Boards) resources under the water (prevention and Control of Pollution) Cess Act, 1977.

Forest (Conservation) Act, 1980

Although India had a forest Act in 1972, first related legislative after independence came about us as late as 1980 with the Forest (Conservation) Act, 1980. The salient features of the forest (conservation) Act, 1980 as amended in 1988 and the Forest (conservation) Rules, 1981 as amended in 1987 and 1989 are.

- The forest officers, with previous sanction of the State government, may stop any public or private way or water-course in a reserve forest, provided that an alternate way or water course already exists (section-5)
- The state government is empowered to protect and manage any portion of a forest, the cutting of grass and posturing of cattle in such forests. (Section 3.2)
- The state government is authorized to grant licenses to inhabitants living in the vicinity of protected forests to take trees, timber and frost produce (Section 35)
- The state government by notification in the official gazette may prohibit a) pasturing of cattle b) clearing of the vegetation c) breaking up for clearing of land for cultivation. (Section 25).
- An imprisonment of six months or more or fine of Rs. 500 or more or both is the penalty for any person who violates any of the provisions of this act or a) sets fire to such forest b) permits cattle to damage any reserved area c) violates any of the rules under section 32.

Besides New Principles and concepts like :–

a) ***Polluter pays Principle*** means that the person who is responsible for any hazardous and dangerous activity is liable to make good the loss caused to other person by his activity.

b) ***Precautionary Principle*** is a part of Indian Law. It is based on the principle that an ounce of prevention is worth a pound of cure. The application of this principle required either cautious progress until a development can be judged 'innocent.

c) ***Right to Information Act 2005*** also supports environmental planning and management because NGOs or even official bodies are how able to get information.

The National Environment policy 2006 of Ministry of Environment and Forests addresses the issue of global warming.

Following are the essential elements of India's fundamental strategy towards climate change mitigation.

- Identify key vulnerabilities of India to climate change in particular impacts on water resources, forests, coastal areas, agriculture and health.
- Assess the need for adaptation to future climate change and the scope for incorporating those in relevant programmes, including watershed management, coastal zone, planning and regulation, forestry management, agricultural technologies and practices and health programmes.
- Encourage Indian Industry to participate in the Clean Development Mechanism (C.D.M) through capacity building for identification and preparation of CDM projects.
- Participate in voluntary partnerships with other developed and developing countries to address the challenges of sustainable development and climate change.

Besides, India released its National Action plan on Climate Change (NAPCC) on 30th June 2008. It advocates a strategy that promotes firstly the adoption to climate change and secondly further enhancement of the ecological sustainability of India's development path. Eight National Missions are- a) National Solar Mission. b) National Mission on Enhanced Energy Efficiency. c) National Mission on Sustainable Habitat. d) National Water Mission. e) National Mission for sustaining the Himalayan Eco System. f) National Mission for green India. g) National Mission for Sustainable Agriculture. h) National Mission an Strategic Knowledge for climate change from the core of the National Action plan and represent multipronged long term and integrated strategies for achieving key goals in the context of climate change.

Strategies for Environment Protection in XIth five year plan (2007-12)

Planning Commission in its meeting held on Nov. 8, 2007 under the chairmanship of 'Hon'ble Prime Minister Dr. Manmohan Singh cleared the draft of XIth plan in which following targets fixed for environment protection.

- Increase in forest and tree cover by 5 percentage points.
- Attain WHO standards of air quality in all major cities by 2011-12.

- Treat all urban waste water by 2011-12 to clean river water.
- Increase energy efficiency by 20 percentage points by 2016-17

PART III

Climate change is no longer an environmental issue but it is related to global security and survival. Mere enumeration of the statutes and various constitutional provisions will not solve the problem. At the Copenhagen conference the Danish presidency of the CoP 15 invited some of the parties for a discussion on the relevant aspects of climate change and presented the results of the CoP in the form of "Copenhagen Accord" there was lack of political will amongst the countries to came to an agreement. So an outcome could not be reached at Copenhagen. The CoP didn't adopt the results of the discussion. They only took note of the "Accord". Even if we argue that the Indian diplomacy at Copenhagen was able to protect its right to develop, we can not ignore the predicted dangers that we face by maintaining the business usual path of development. There is a need to maintain a momentum post Copenhagen to reduce carbon footprint through emission reduction. ***Following steps can be taken to make our economy a law carbon economy.***

- ***Using renewable energy and low carbon technology*** by industrial houses will really help like Ambuja Cements Ltd. Is using biomass as the alternative source of energy in some of its plants Many companies and corporate houses such as Reliance, Videocon, Tata Shyam Group have installed wind energy, solar cells and bio fuels for ensuring constant supply to their plants.
- ***Adopting energy efficient technology*** such as using energy saving lamps (CFL) and other energy efficient electrical appliances like pumps and fans will help in energy conservation.
- ***Use of Environment-friendly sources of energy*** like LPG and CNG must be encouraged in place of petrol and diesel which emit huge amount of carbon-dioxide. Instead of using wood as a fuel, Gobar Gas plants may be developed as a source of domestic fuel in rural areas.
- ***Shift to organic farming*** will be better in place of excessive use of chemical fertilizers, insecticides and pesticides which causes loss of soil fertility. Organic farming promotes soil-health.
- ***Recycle of industrial as well as household wastes*** into compost and used as manure for organic farming rather than allowing them

to spread here and there for environmental pollution. Recycling paper also saves trees.

- ***Teleconferencing or videoconferencing instead of flying*** (traveling) by airplane is useful for saving time and money and also reducing carbon emissions. Big Indian corporate houses for conducting office meetings can go for them.
- ***Efforts at greening and green buildings*** help to perform social responsibility. 'Greening endeavors like afforestation or planting of trees, converting wastelands into plantation area help corporate houses to sequester CO_2 released in the atmosphere ACC cement has planted 3.3 million trees spread over an area of 1000 acres in mines, factories and residential townships. With growing environmental consciousness a number of Indian companies are trying to reduce their carbon footprint by setting up green office buildings. The confederation of Indian Industries (CII) has set up the Indian Green Building Council (I GBC).
- ***Clear cut fiscal Incentives*** should be introduced to encourage investment in green and environment friendly technologies.
- ***A carbon tax or pollution tax:*** should be levied by the government to reduce the consumption of fossil fuel.
- ***Fiscal Benefits like reduced excise duty*** should be provided to the industries using climate friendly 'greener' fuel like CNG.
- ***Introduction of carbon labels*** estimating the emissions created by producing, packaging, transporting and disposing of a product will give the companies an opportunity to show their commitment towards reducing CO_2 emissions embodied in their products to environmental conscious customers.
- ***Stringent laws and their strict enforcement*** is the need of the hour to save our land.

We, people of India and govt. of India should go hand in hand to save our earth and to maintain the quality of life for both present and future generations following our Vedic Mantra **'Vasudhaiv Kutumbkam'** i.e. the whole earth is my family.

References

- Deswal Surinder, "Environmental Studies" Dhanpat Rai Publishers New Delhi.
- Govt. of India, Ministry of Finance, "Economic Survey" 2006-07.
- Govt. of India, Ministry of Finance, "Economic Survey" 2009-10

- Govt. of India, Publication Division Ministry of Information and Broadcasting "India – 2010 reference Annual".
- Kumar D.N (2004) "The Scenario of Population Growth in India" Akanksha Publishing House, New Delhi.
- Pankaj Kumar "Vulnerability of India" Yojana June. 2008 Vol. 52
- Sen, Raj Kumar et al (eds) (2010); Environment and Sustainable Development in India, Deep and Deep publications, New Delhi.
- Shiju M.V "India Environmental Law and Climate Change" Yojana April 2010 Vol. 54.

9. Environment and Sustainable Development in the Era of Globalization

Shivani Kapoor
Praveen Sood

Globalization increases the pressure to pull resources from the environment and generate wastes. There are some very real threats to the environment associated with global economic growth. Yet without the increased income that will undoubtedly come with globalization, there is very little hope of making much progress improving global environmental quality. Especially for the poorer regions of the world, globalization is likely to make significant long-term improvements in environmental quality possible. Although there are real environmental risks associated with economic growth in countries that are currently poorly governed, there are even greater risks associated with scenarios of economic stagnation.

Key words: Globlization, Natural resourus, pollution, Environment Protection.

Introduction

It is tempting to take a simplistic view of globalization as a major risk to the environment. However, environmental protection requires extensive resources that our current global economy cannot afford. This paper acknowledges the risks of globalization but emphasizes that economic stagnation is risky as well. Globalization is a multifaceted phenomenon that will have diverse impacts on the world's environment that will change over time. Although development may entail an immediate loss of wild lands and increase in pollution in some countries, it will also provide future wealth to address environmental issues in the long run. In the long run, development is necessary to finance environmental protection. However, for environmental protection to be realized, it will also be necessary to develop institutional structures to protect the environment. In many countries, existing governments and institutions are inadequate. Globalization will put increasing pressure on weak governments and weak institutions. For global environmental protection to be realized requires a combination of economic growth and institutional reform.

Economic activity impacts the environment both as a source of inputs and as a destination for wastes. These two interactions lead to two distinct interactions: the effect of the economy on natural resources (sources) and the effect of the economy on pollution (wastes). Any changes in economic activity will inevitably affect both sources and wastes. As stated in many of the talks in this conference, globalization is intended to increase global economic activity through trade, investment, and multinational institutions. This growth of economic activity will be distributed across the planet reaching virtually everyone, even the most remote villagers. If successful, the "globalized" world economy is predicted to increase by six to eight times by 2100 according to the Intergovernmental Panel on Climate Change (IPCC, 1992). This increased wealth is not limited to wealthy countries; it is expected to be distributed across every country. In fact, most projections assume that developing countries will experience faster growth rates than currently industrialized countries. That growth inevitably will require inputs from the environment for food, energy, wood, and minerals. That means more mining, drilling, and possibly land conversion. A six to eight-fold increase in GDP also implies an increase in pollution at local, national, and global scales. Although the demand for inputs and potential pollution will increase with GDP, it is not expected that either inputs or wastes will increase in proportion to GDP. Most of the projected growth in the economy lies in sectors such as services that require people and human capital more than substantial natural resources. These sectors also involve less pollution. Manufacturing is expected to shrink as a fraction of world GDP and as it shrinks, pollution levels per unit GDP will decline. Second, technical change has led and will continue to lead to sharp declines in the amount of natural resources required per unit of output. For example, there have been steady and dramatic improvements in energy efficiency throughout the world. That translates into a steady decline of the energy needed per unit of output. The six to eight-fold increase in GDP expected by 2100,

Consequently, may require only a 2-4 times increase in energy and minerals as inputs. As these inputs decline per unit output, pollution will also decline. Even without abatement, potential pollution is expected to increase only 2-4 times over the next century (Houghton et al 1992). Nonetheless, even just a doubling of pollution and input requirements can have a large and deleterious impact on the environment. This paper discusses these environmental pressures on first, natural resources, and then, pollution. In this paper, we will look into the next century and predict how globalization will unfold and what it might mean for the environment. Although futurists may well tease us to speculate beyond 2100, our vision becomes ever hazier the further we look. We should be very cautious planning too far in advance. Before tackling future millenniums, will we get through even the next century? This we can indeed plan for. What will actually happen over the next century, however, depends upon both how the economy grows and how societies respond. As countries gain access to new technology and as they get wealthier, they can choose to abate the worst pollution and manage to protect their best environments. What actually happens to the environment depends upon not only market forces but also upon government response. The environmental impacts of globalization depend not only on the growth of economies but also upon how national governments and multinational institutions react to the increasing pressure from growth. Are the required governmental institutions in place in every country? Are international agencies empowered to protect international resources? Will governments behave efficiently? The environmental success of globalization hinges on institutions that are not yet completely in place and institutional behavior that is still not evident.

Natural Resources

Environmentalists should be concerned that the additional inputs required to fuel economic growth come at the expense to the environment. The doubling or quadrupling of inputs for food, energy, wood, and minerals will have environmental impacts. More mining, drilling, farming, grazing, and urban activities will put pressure on renewable resources such as land, water, and fish as well as nonrenewable resources such as fossil fuels. Even just a doubling of input flows will increase scarcity noticeably. This does not necessarily spell disaster as competitive markets are well designed to cope with scarcity. Nonrenewable resources such as fossil fuels will be efficiently distributed across time by market forces (Hotelling, 1931). The potential to sell to future generations at high prices because of ever more scarce resources will keep owners from squandering resources too quickly. Competitive markets will allocate private land efficiently across competing market uses. Many inputs will be managed efficiently by market forces alone. Prices will increase over time as resources become scarcer, but the available resources on the planet can support the growth expected well beyond the next century. Renewable resources such as land will become increasingly scarce. Different private land uses such as commercial forests, farms, and cities, will compete with each other for the available land remaining in the planet. These land uses will bid against each other and raise land prices across the world to reflect land scarcity. Increasing prices for scarce resources has led in the past to substantial clearing of wild lands for productive use. Europeans cleared substantial forests and grasslands, Americans planted their prairie, and low latitude countries have converted both their tropical forests and savannahs to cropland and pastures. However, increasing prices have helped guide society to protect the land uses of most value. Prices have served as an efficient mechanism to allocate land between different private uses. Land has been given to high valued uses such as cropland and taken away from low valued uses such as forests. Land intensive activities such as widen agriculture that were at one time efficient, will eventually be driven out because they do not return sufficient rent per hectare. Land intensive communal agricultural systems will also be driven out because they do not support intensive agriculture. The balance between forests and farms will continue to shift as some countries seek additional cropland at the expense of forest services. In Africa, Latin America, and Asia, there has been considerable forest clearing done to increase farmland. The balance between grazing and cropland may also shift as crops replace pasture as a source of animal feed. In the future, however, these land use changes will not be unidirectional. In most industrialized countries the rapid technological improvements in growing crops has reduced the land that is needed for cropland. If markets were allowed to operate efficiently, vast amounts of cropland would have already disappeared in the United States and Western Europe. Many countries in Asia have increased crop production dramatically without converting additional land to farms. The promise of new technology for growing crops is expected to make a vast difference in conserving land in the future. Although many environmentalists look upon this new technology with skepticism, advanced methods of crop production including new genetic varieties, could well save vast acreages of wild lands from turning into marginal farmland. Higher prices have had a dramatic effect on forestry as well. From the beginning of globalization, foresters have discovered a vast planet populated with over mature forest. In fact, there was such an abundance of standing trees, that the price of timber was too low to manage forests. Mature trees were consequently harvested without replanting. The mining of the world's forests has been going on for centuries. However, in the last fifty years, we have finally begun to run out of over mature forests. Prices for timber have risen reflecting this scarcity making it profitable to plant forests for timber. Timber management has consequently spread from its roots in Germany across the planet. This decade marks the beginning of a new era in forestry where the number of trees planted globally replaces the number of trees harvested globally. Forestry has finally become sustainable. Rising land prices, however, have created another battlefield perhaps of more concern to the environment. Rising land prices imply more rewards to developing land that was hitherto natural. Rising land prices put economic pressure on converting wildlands. Markets are effective at allocating land across developed uses, but markets are less suitable for protecting wildlands. Wildlands are a public good, enjoyed by many simultaneously. Markets will not efficiently protect wildlands from conversion because landowners cannot effectively collect payments for the jointly consumed services of wildlands. Wildlands will be undervalued by the market and converted to private uses unless governments intercede on their behalf. Globalization has the potential to destroy vast acreages of wildlands that may have very high social value but little market value. Currently, national governments jealously control these resources. However, there is deep concern that governments in developing countries may not reflect the will of their own people. There are many examples of government officials collecting vast rents from resources technically owned by the government. Partly this reflects a tendency to shift valuable market resources from public control to private citizens. This has happened in every country in the world.

The celebrated opening of the American West to settlers was just such a transition. There is ample evidence that market resources should be turned over to private hands for efficient management. However, long-term wildlands need to be kept in government hands because private owners do not have the incentive to protect them effectively. In some cases, private owners can create ecotourism sites and collect some revenues from wildlands. In general, however, these revenues are just a fraction of the social value of these resources. In the most favorable circumstances, owners are likely to collect only half of the social value of wildland services, the remainder being lost to uncollected consumer surplus and discouraged visitors. Market forces undervalue wildlands and cause excessive conversion to more destructive but more financially rewarding land uses. In order to protect wildlands efficiently, there must be effective governmental institutions in place. There is substantial evidence that not all countries have welldeveloped governmental institutions that will protect their own resources (Repetto and Gillis 1988). Globalization without effective national governments could wreak havoc on local environments through excessive depletion of wildlands. Without effective national governments everywhere in the world, globalization can destroy habitats that nations should preserve for their own sake. Wildlands also can have international value. Ecotourism is a rapidly growing dimension of travel. People are going all over the world to see rare and beautiful ecosystems and species. Over half of the visitors to many of the most valuable ecosystem destinations in the world are foreigners. Although it is reasonable to expect that national governments will protect resources of concern to their own citizens. It is not obvious why these same governments should protect resources of concern to other countries. This discrepancy is especially poignant for relatively poor countries.

Many of the most valued ecosystems in the world lie in the tropics. Although distant rich countries may treasure these resources, they are currently owned by relatively poor nations. Markets and on-site fees collect only a small fraction of the value that tourists place on these sites. For international sites to be protected, multinational mechanisms must be created to compensate domestic countries. The Global Environment Facility (GEF) is an example of an institution that could provide such payments. The GEF could be used as a funding mechanism by which rich countries pay annual compensation for the international value of specific wildlands that are preserved in poor countries. Rich countries could identify international sites that their citizens want to protect, quantify how much their citizens will pay, and provide annual grants to the host countries through the GEF that reflect these values. However, the current funding for GEF is woefully inadequate to actually pay for international values. Globalization increases the need to create effective multinational organizations and fully fund them.

There is an important role for government in the management of natural resources. However, governments sometimes interfere with the economy and actually make the environment worse. This statement is not referring just to developing countries. The massive crop subsidies in the United States and Europe have prevented cropland from shifting back to more natural uses. Not only have these programs cost developed countries huge welfare losses, but they have clearly reduced environmental quality in the process. Some might blame globalization for this as governments have created the subsidies allegedly to protect domestic farmers from imports. However, the technical change that has led to the vast increase in crop supplies actually emanates from these countries (especially the US). It is the domestic industry itself that has caused supply to increase so vastly, threatening to lower prices. Globalization is not to blame. In fact, the institutions intended to promote globalization, are working against these domestic policies. The World Trade Organization (WTO) is trying to reduce crop subsidies in the US and European Union (EU) to promote free trade. An indirect benefit of this effort is that it will improve the environment in the United States and Europe. The pressure from globalization institutions, in this case, is actually beneficial to the environment. The increasing pressure of scarcity puts increasing pressure on governments to be more efficient. Countries need to improve the way that they manage their market resources. Water is famous for being inefficiently allocated across users. The general rule is "first come first serve" rather than "delivery to the highest valued user". As water becomes more scarce because of increasing demand, this problem will get worse. It is critical that high valued future users such as urban and industrial users have a mechanism to buy water from low valued users. In most of the world, the lowest valued users are marginal farmers. In every country, new institutions need to be put in place to move water away from low valued uses to high valued uses. Economists have frequently called for water markets where users must buy the water from the government. However, many users believe they own the water they currently use and so avidly resist any attempt to make them pay. Alternatively, governments could recognize that current users do in fact have water rights. They could allocate permits to current users in proportion to existing uses. If they then made the permits tradable, high valued users could buy rights from low valued users. The efficiency of the market would come forth while protecting existing water rights. With globalization, there is ever increasing need for such institutional reforms in water.

Domestic countries must also improve forest management. In many parts of the world, the government is the official title-holder of forestland. Local people have overlapping rights to some forest uses, such as the collection of nontimber forest products, but the timber rights are generally federal. These overlapping rights create incentives to mismanage forests for either timber purposes alone on behalf of the government or nontimber resources alone on behalf of local people. Quite often, opportunities to maximize joint returns are lost as each owner seeks purely their own interest (Repetto and Gillis 1988; Houghton and Mendelsohn 1996). Property rights to forests whose primary value is market resources need to be unified in the hands of private owners. Who exactly will get these former public resources is an important question for every country to resolve. From an efficiency perspective, however, the critical issue is that the complete private rights for each piece of land be given to a single individual and not shared.

Another renewable resource in vast need of reform is fisheries. Commercial fisheries suffer because many fishermen share a common pool of resources. Because the future productivity of that pool is shared but the immediate gains of catching an extra fish are not, fishermen around the world were depleting fish stocks throughout the twentieth century (Hardin, 1968). This tragedy of the commons has been addressed by domestic regulations limiting aggregate fish catch out to the 200-mile limit from national seashores. However, there remain many examples of neighboring countries being unable to share the fish that swim across national borders with the US and Canada being a prominent example. Fisheries in international waters are also still at risk since the most effective regulations remain national. Despite these few remaining problems, most domestic fisheries are now protected from overfishing.

Although overfishing is largely under control because of regulations, fisheries around the world remain inefficient. In almost every fishery in the world, there is excess harvest capacity. There are more fishing boats than are needed. The common property fishery creates an incentive for fishermen to increase their harvesting capacity to try to get a larger share of the available fish. The net result is that they spend as much on harvesting capacity as the fish are worth. The excess capacity forces regulators to close fisheries prematurely. More importantly, the excess harvesting capacity wastes the entire potential value of fisheries worldwide. The biologists who regulate the world's fisheries have done a reasonably good job of protecting fish populations but not the potential wealth of fisheries. The potential value of the world's fisheries is being wasted because the common property regulated system creates too much harvest capacity. Globalization did not create this problem but it has done nothing to address the problem either.

Pollution

The doubling to quadrupling of physical inputs to the global economic system as the world economy grows over the next century will result in a corresponding increase in potential pollution. The pollution will come at local, national, and global scales. Given the anticipated increase in economic activity across the globe, there is every reason to expect that potential pollution will grow in every country. Not only the industrialized countries, but virtually all countries, are likely to increase their potential pollution over the next century. Harmful local and regional pollutants such as sulfur dioxide, small particulates, nitrogen dioxide, and carbon monoxide will all potentially increase in the atmosphere. If they increase, they will set in motion a new wave of premature mortality, serious illness, minor health effects, loss of visibility, material damage, crop loss, and ecological damage in the regions in which they are emitted. In addition, there is a host of fertilizers that potentially will be released into the world's aquatic ecosystems causing oxygen levels to fall enough to kill fish and damage water resources. Heavy metals and other lethal chemicals may also be released into water systems threatening the drinking supplies, irrigation water, and fish supplies of future generations. Dangerous pollutants may be dumped in nearby soils and slowly but surely emanate across the landscape making areas unfit for human habitation or farming. Finally, global pollutants will be released into the earth's atmosphere potentially destroying the upper ozone layer or blanketing the planet in greenhouse gases. Globalization and the resulting economic growth pose real threats to the earth's environment.Whether these threats are realized remain distinctly in the hands of government.If governments take no action, markets will encourage no abatement. Pollution affects many victims simultaneously. It is simply too costly for victims to organize themselves to effectively pay for pollution as a group. In places without government regulation, there is simply no effort made to abate. Given the increase in economic activity expected from globalization, the pollution problem we already face will get substantially worse. The damages from potential pollution at all spatial scales are serious and cannot be ignored. As globalization pushes economic development into the far corners of the earth, every society faces this problem. Examining the historical rolc that pollution has played in industrial countries provides one pattern that might unfold in the world as a whole.

Countries such as the US, Japan, Canada, and Western Europe rapidly increased pollution as their economies industrialized. Markets put very little pressure on firms to abate pollution; contrary to the prediction of Coase who argued victims would pay firms to abate (Coase 1968). Because pollution is jointly consumed by many victims, a large number of victims must coordinate in order to present a payment to a firm to abate. Victims fail to pay for abatement resulting in market failure. In the United States, individual states provided some meager regulation against pollution but states are too small to regulate effectively. A great deal of the damage from pollution falls outside of state boundaries so there is only a limited incentive for states to abate to protect their own citizens. Further, industries can easily move beyond state boundaries to avoid regulations. States consequently were always fearful that if they engaged in stringent controls, they would lose jobs and prosperity to neighboring states. Pollution in the industrialized countries continued largely unabated until the 1970's when federal regulations began. At first, even federal efforts were somewhat ineffective as many old plants were grandfathered (exempted) so that they did not have to install expensive controls. However, over time, federal regulations have gradually been tightened to the point that overall emission levels are actually declining despite continued economic growth. The overall historical pattern has been one of increasing and then decreasing emissions over time. This same pattern could well be repeated by developing countries. At first, countries will industrialize and create polluting factories across their landscape. They might abate but only modestly given their low incomes. Low incomes imply low values on environmental benefits. Minimizing abatement plus environmental costs implies only low investments in abatement at first. These investments need to be made where they will do the most good, which will be in the population centers of these countries. As countries gradually become wealthier, they can afford to spend increasing amounts on pollution control. They will gradually tighten regulations forcing ever more abatement. Eventually, a maximum emission is reached and thereafter emissions fall. Pollution will follow a hill-shaped path over time, increasing at first and then declining. This path implies a period where pollution worsens causing serious health effects. But life spans are likely to continue to increase in these countries because of the effects of higher incomes and growing health services. Ultimately, this path leads to more economic development and more pollution control. Although this is not the only feasible path that countries can take, this path balances wealth and pollution protection at every moment. The hill-shaped path way well be the path most countries will want to choose. The most effective institutions to regulate pollution have been federal. However, even federal regulations have the potential to be ineffective. For example, some pollution crosses national boundaries. Following self-interest, national regulatory agencies may take a blind eye towards impacts that fall outside their national boundaries. Although this is in principal a problem, there are many examples where bilateral agreements between countries have in fact resolved these problems. With respect to transboundary pollution, Coase's promised solutions have come to pass. There have been negotiations between polluting and neighboring victim countries. Victim countries do negotiate with polluting countries and encourage efficient outcomes. For example, both Canada and Mexico have negotiated with the US to reduce air and water pollution respectively. When the emitting country is wealthy, neighboring victims appear to be able to encourage the emitting country to pay for abatement. When the emitting country is relatively poor, the victim countries have sometimes volunteered to pay as in Sweden and Poland and Japan and China. A problem remains, however, for countries that do not have any effective environmental regulations or institutions (Esty and Mendelsohn 1998). Globalization could well cause highly polluting firms to congregate towards countries that have minimal environmental regulations. Such "pollution havens" could save firms vast sums of resources required for abatement. Whether firms will actually be attracted to pollution havens, however, remains unanswered. On the one hand, polluting firms could save on abatement costs. However, on the other hand, governments that cannot create environmental institutions also fail to create other required conditions for successful business operations. Such governments are likely to have problems with lawlessness, property rights, and dependable services.

Polluting industries tend to be capital intensive. The costs of operating in primitive conditions may far exceed any abatement savings. Empirical attempts to see if polluting firms are attracted to national pollution havens have consequently yielded mixed results (Jaffe et al 1995). It may well be the case thateconomic development is more closely tied to effective government rather than loose regulations. The pollution haven problem may never come to pass. Another important problem that globalization will exacerbate is global pollutants. Global pollutants are emissions that damage global not local resources. Increases in economic activity in any location will increase potential global emissions. Unfortunately, individual countries have little incentive to control global pollutants such as CFC's or carbon dioxide because they must pay for the abatement costs themselves but the benefits are shared around the globe. Global pollutants are one example where the scope of national environmental institutions is inadequate given the scale of the problem. Despite this general limitation of national governments, there has been substantial progress fighting CFC's, which were accused of creating the upper atmosphere ozone hole over the poles. The Montreal Protocol is widely recognized as a very successful agreement to limit CFC emissions. Prior to the Protocol, CFC's emissions were visibly reducing stratospheric ozone. This loss of stratospheric ozone was predicted to have serious health effects on high latitude residents and ecosystems from increased ultraviolet light. The damages of CFC's were expected to be substantial. In contrast, the cost of abating CFC's turned out to be relatively modest once substitute chemicals were discovered. Given the very favorable cost benefit ratio, countries quickly agreed to the CFC ban and emissions have been dramatically reduced. A similar success story has eluded the control of greenhouse gases. The cost of abating greenhouse gases has turned out to be quite substantial as countries are reluctant to give up burning the cheap fossil fuels that lead to greenhouse gas emissions. In turn, the benefits of controlling greenhouse gases have turned out to be murky. At least initially, warming the planet may not be unilaterally harmful. The emissions over the next few decades are expected to cause only small harm along the low latitudes and likely benefits in the higher latitudes (Mendelsohn et al 2000; Mendelsohn 2003). The incentive to engage in crash programs to control greenhouse gases in the near term is limited. However, as emissions continue to grow with higher world economic activity, concentrations are expected to accumulate. Eventually, further warming is going to become clearly harmful and efforts to control greenhouse gases must begin in earnest. Currently, the institutional mechanisms to control greenhouse gases are too weak to be effective. Strengthening international agreements to limit emissions or creating international organizations that help control greenhouse gases will eventually be necessary. In fact, the very international organizations that encourage globalization may eventually be called upon to address this serious environmental issue as well.

Conclusion

Globalization is a threat to the environment. The increased economic activity from globalization will create pressure for more inputs from the environment and create more waste to be emitted into the environment. A projected increase in world GDP of 6- 8 times by 2100 will likely increase input needs by 2-4 times. Potential pollution is expected to increase by 2-4 times as well. This increased economic activity will pose real threats to the environment.

Will the environment, however, necessarily deteriorate? There is a risk involved with economic progress that overall conditions will worsen and that whatever gains are being offered, they will be overshadowed by environmental catastrophes. However, that risk is very small. Although there will be challenges associated with increased growth as cited above, there will also be vast resources. As the countries of the world grow wealthier, there is every reason to believe that they will become increasingly environmentally sensitive. Governments in the future will be able to afford abatement of pollution. They will be able to afford to control activities that damage the landscape. They will be able to conserve valuable wildlands. It is unlikely that development will be perfectly efficient and that governments will always be effective. However, a future world with vast additional wealth is far more likely to protect the environment than the world is today. The greatest risk to the environment is a stagnant world economy. If globalization is stopped, international trade shrinks, and the world's economy comes to a standstill, there will be little resources left for environmental protection. A smaller global economy may use less input and may generate less pollution, but the environment would not improve. Poor countries would have few resources to address the problems they already face. There would be little chance to improve either human or natural conditions. Economic stagnation poses a far greater risk to the environment than economic growth. A stagnant world economy is very likely to continue to eat away at the world's resources without many technological improvements. A stagnant world is very likely to decide abatement is too expensive to take seriously. A stagnant world is very likely to decide that wild lands need to be developed. A stagnant world is exactly what Malthus examined when he predicted all future economic growth would be overcome by population growth. Global economic growth implies new challenges for protecting the environment. More would have to be done to reduce global inputs and increase abatement. But global economic growth also implies new opportunities. With increased wealth, countries could afford to protect their resources more carefully. Countries could afford to spend more on abatement. All countries would have to develop effective governing structures, but with the increased wealth they could afford them. The risks for the environment seem far greater if growth is stopped. People who are interested in protecting the environment simply cannot afford to stop economic growth. Long term environmental protection requires economic growth. Environmentalists should focus their energy on developing effective institutions to guide globalization, not creating chaos to stop growth.

References

- Coase, R. 1968. "The Problem of Social Cost" Journal of Law and Economics 3: 1-44.
- Esty, D. and R. Mendelsohn. 1998. "Moving From National to International Environmental Policy", Policy Sciences 31: 225-235.
- Houghton, J., B. Callander, and S. Varsey (eds). 1992. Climate Change 1992: The Supplementary Report to the IPCC Scientific Assessment. Cambridge University Press, Cambridge, UK.
- Hardin, G. 1968. "The Tragedy of the Commons" Science 162: 1343-48.
- Hotelling, H. 1931. "The Economics of Exhaustible Resources" Journal of Political Economy 39: 137-175.
- Houghton, K. and R. Mendelsohn. 1996. "An Economic Analysis of Multiple-Use Forestry in Nepal", Ambio 25: 156-159.

- Jaffe, A., S. Peterson, P. Portney, and R. Stavins. 1995. "Environmental Regulation and the Competitiveness of US Manufacturing: What Does the Evidence Tell Us?" Journal of Economic Literature 33: 132-163.
- Mendelsohn, R., W. Morrison, M. Schlesinger and N. Adronova. 2000. "Country- Specific Market Impacts from Climate Change", Climatic Change 45: 553-569.
- Mendelsohn, R. 2003. "Assessing The Market Damages From Climate Change" in J. Griffin (ed) Global Climate Change: The Science, Economics, and Politics Edward Elgar Publishing, UK, pp. 92-113.
- Reppetto, R. and M. Gillis (eds), 1988 Public Policies and the Misuse of Forest Resources. Cambridge University Press, Cambridge, UK.

10. Performance and Pattern of Economic Growth in Utter Pradesh: Challenge of Sustainability

S.K. Mishra

Growth performance in the state was better in pre-reform period and was worse during post reform period than that of the country. With induction of state level reforms, growth rate of the state tended to catch up with that of the country. However, the higher growth rate supported by service sector alone cannot take care of poverty reduction. The growth strategy has to incorporate broad based commodity sector in its fold for making the growth inclusive and sustainable. It needs to improve quality of life of the common people by giving them health, education and skill to ensure their participation in the process of growth.

Keywords: Skilled labour, Intensive production, Convergence Effect, Quality of Life, Demographic dividend, Commodity sector's share, Inclusive and Sustainable growth.

Introduction

India is now emerging as one of the fastest growing economies of the world and being considered, along with China, as engine of growth for the world economy, particularly in the post recession era. However, the pattern of India's economic development differs from other countries of the world. First, tertiary sector became larger than the secondary sector at an early stage of development in India. Secondly, agriculture with less than one fifth share in GDP continues to provide employment to nearly

60 percent of labour force. Lastly, India's international competitiveness is in high skilled rather than low skilled labour intensive production (Bhattaharya 2007).

Economic growth has shown signs of revival even in Uttar Pradesh. The State's growth rate in the Tenth Plan is estimated at 5.3 percent which is two and half times greater than that in the Ninth Plan. Consequently, the State government looks to be confident of attaining the growth rate of 10 percent targeted for the Eleventh Plan. However, the service sector has been the engine of growth for both India as well as for Uttar Pradesh.

The new trend and patterns of higher economic growth rate is not sustainable and has attracted number of serious queries. First and fore most of them is whether growth rates and prosperity in different regions would eventually converge or not (Bhattacharya and Sakthivel, 2004). The second one is whether this growth strategy would benefit majority of the people, if not to every one. Lastly, whether this growth strategy would ensure sustainability.

In this back drop, the present paper has following objectives:

1. To observe trend of economic growth of Uttar Pradesh and that of India during pre-reform and post reform period.
2. To enquire into link between economic growth and betterment in social sector.
3. To study the pattern of changes in sectoral structure and its impact on the state economy.
4. To find out role of other non-economic factors in development in U.P. and India.

The study is based on secondary data collected from different sources.

Table-1: Annual Economic Growth Rate in U.P. and India (in percent)

Five Year Plans	NSDP U.P.	GDP India	Difference in percentage point	Per capita Income		Difference in percentage point
				U.P.	India	
V Plan (1974-79)	5.7	5.3	+0.4	3.3	2.9	+0.4
VI „ (1980-85)	8.7	5.3	+3.4	6.3	3.1	+3.2
VII „ (1985-90)	5.7	5.8	-0.1	3.3	3.6	-0.3
VIII „ (1992-97)	3.3	6.8	-3.5	1.4	4.9	-3.5
IX „ (1997-2002)	2.0	5.6	-3.6	-0.4	3.6	-4.0
X „ (2002-07)	5.3	7.7	-2.4	3.3	6.0	-2.7

Source: Draft Eleventh Five Year Plan, U.P.

Table -1 clearly shows the difference in trend of growth in U.P. and India in pre and post reform period. Average annual growth rate in U.P. during one and a half decade prior to the reform was higher than that of India. The growth rate in U.P., during the last one and a half decade is far lesser than that of India. The growth rates in India in the post reform three five year plans were higher than that in pre reform three five year plans. Contrary to this, growth rates in U.P. in pre reform three five year plans were much higher than that in post reform three five year plans. Short fall in growth rate in U.P. was severe during 1997-2002. State level reforms were introduced a decade later, i.e. in new millennium. It seems to have benefited the state. The year wise growth rate at current prices of the state has picked up to the above 8 percent. It ended up at 14 percent in 2006-07 which is very close to that of the country, i.e., 15.8 percent.

The causes of disparate growth performance of U.P. vis-à-vis India in 90s may be observed in pattern of employment and availability of finance.

Table -2: Availability of Employment and Finance in U.P. and India

Year	Proportion of Main Workers to Total Worker (U.P./India)	Proportion of Agricultural Workers to total worker (U.P./India)	Per Capita Bank Credit (U.P./India)	Credit Deposit Ratio(U.P./India)
1991	0.87	0.73	0.30	0.52
2001	0.77	1.13	0.27	0.49
2006	-	-	0.25	0.56

Source: Different issues of Statistical Outline of India

The table-2 obviously shows that workers got concentrated for their employment in less income generating sector, i.e., agriculture in U.P. vis-à-vis that in India. Relatively growing dependency on agriculture, the sector well known for availability of seasonal employment, in the state led to fall in proportion of main workers and consequently to fall in proportion of state income to that of the country.

The last two columns in the table show the declining proportion of per capita bank credit and credit-deposit ratio of the state to the country. The declining financial facilities in the state hampered the growth of activities in non-agricultural sector accounting to slower rate of economic growth in the state.

Substantial increase in proportion of credit-deposit ratio between 2001 and 2006 seems to have boosted the pace of economic growth in the state during this period.

Table-3: Inter Sectoral Changes in Income in U.P. and India

Year	Uttar Pradesh		India		Structural Gap between U.P. and India (Percentage Point)	Ratio of number of poor in U.P. to India
	Share of Commodity Sector (Agr.+ Ind.) (in percent)	Share of Service sector (in percent)	Share of commodity sector (in percent)	Share of Service sector (in percent)		
1980-81	67.6	32.4	64.3	35.7	3.3	
1984-85	66.2	33.8	63.5	36.5	2.7	0.175(1983)
1989-90	62.4	37.6	60.7	39.3	1.7	
1994-95	62.2	37.8	58.1	41.9	4.1	0.168(1993-94)
1999-2000	59.1	40.9	50.1	49.9	9.0	0.153(1999-2000)
2005-2006	53.4	46.6	44.3	55.7	9.1	0.257(2004-05)

Source : Computed from Draft Eleventh Five Year Plan, U.P.; K.Sundaram et al., 2003 and P.K. Chaubey, 2008.

The table-3 shows that there was structural convergence in the economy of U.P. and India in pre reform decade. The post reform decade shows sign of rapid structural divergence. Reforms adopted at the state level after a decade led to stability in extent of structural divergence. In the last five years, share of services increased by 5.7 percentage point in NSDP of U.P. and by 5.8 percentage point in GDP of India. Thus, U.P. started catching up with the Indian trend of growth and of pattern of changes in sectoral structure after the year 2000.

The last column of the table gives very pertinent information regarding impact of structural changes on poverty. Proportion of poor in U.P. to that of India (15 major states) tends to decline even during the period of disparately slow growth rate of the state in 90s.

This may be owing to continuity of large share of commodity sector in total production. However, after the year 2000, higher service sector growth and gross avoidance of commodity sector in U.P. like that in India led to increase in proportion of number of poor people in U.P. to that of India. This may be termed as 'Convergence Effect '.

Table- 4: Changes in Proportion of Human Development Index of U.P. to India

Year	HDI(UP/India)	Literary (UP/India)	Life Expectancy (UP/India)	Per Capita Income (UP/India)
1981	.844	.76	.90	.78
1991	.824	.80	.94	.74
2001	.822	.88	.95	.57
2006	-	-	.94 (2002-06)	.50 (2008-09)

Source: Computed from Madhusudan Ghosh, 2006 and Economic Survey 2010-11.

The table-4 shows comparative changes in economic and non-economic indicators of development in U.P. and India during Pre reform and Post reform period. It is obvious that HDI ratio of U.P to India fell down more sharply in pre reform than post reform period. The slow decline in ratio was attributed to steep rise in literacy ratio and moderate rise in life expectancy ratio in post-reform period in Uttar Pradesh. However, decline in ratio of per capita income was very sharp during post reform period. This reveals that social development in U.P. was not much hindered by its slower economic growth. This may also be one of the reasons for structural divergence (less growth but more poverty reduction in U.P. vis-à-vis India) in U.P. during first decade of the economic reform.

The Prospect

Despite all these features, relatively poor quality of life in U.P. remains the great challenge before the current wave of economic growth. It can be observed by the table-5.

The table apparently demonstrates that proportion of number of districts in the state to that of the country with respect to care of women and children is measurably lower. Proportions number of districts with less than 28 percent of women receiving antenatal checkups and with less than 40 percent contraceptive prevalence rate in the state to that in the country are higher by 21.99 and 19.7 percentage points respectively. Like wise, proportion of numbers of districts with under five mortality rate128 and above and with less than 23 percent full immunization coverage in the state to that in the country is higher by 14.3 and 6.82 percentage points respectively.

Table-5: Quality of Life in U.P. and India

Quality of Life Indicators	U.P. (70 Districts)	India (593 Districts)	Ratio of U.P. to India (11.8%)
Number of districts with under five mortality rate128 and above	42 (60%)	161 (27.1%)	26.1
Number of districts with not more than 29 percent Households having any toilet facility.	43 (61.4%)	297 (50.01%)	14.48
Number of districts with less than 23 percent full Immunization coverage (12-23 months)	27 (38.6%)	145 (24.48%)	18.62
Number of districts with less than 28 percent of Women receiving three antenatal checkups	49 (70%)	145 (24.45%)	33.79
Number of districts with less than 40 percent Contraceptive prevalence rate	46 (65.7%)	146 (24.6%)	31.5
Number of least developed districts.	38 (54.29%)	150 (25.30%)	25.33

Sources : Computed from F. Ram and Chander Sekhar, 2006 and Economic Survey 2005-06.

The Demographic Dividend

With such a huge differentials in quality of life between U.P.and India, continued higher fertility rate in the state depict grim picture for future economic growth of the state. Difference in Wanted Fertility Rate between U.P. and India was 0.70 in $NFHS_2$ (1998-99) which fell down to 0.40 in $NFHS_3$ (2005-06) estimates. However, this difference in TFR in U.P. and India was 1.14 and 1.10 respectively in the two survey periods. To NFHS estimates, more than 83 p.c. currently married men and women desire to limit child bearing with 2 children in India. These figures for the state are 68.9 percent for men and 64.2 percent for women. Difference in demand for family planning between the country and the state during the two survey period has narrowed down sharply. But, the gap in demand satisfied is still glaring. This all show need of intensive endeavour at both motivational and operational level in family planning at the state level.

Uttar Pradesh, with more than 16 percent of India's population, has 9.4 percent of the Universities and Deemed Universities, 7.8 percent of the research institutes and only 46 percent of the professional/vocational institution to that of India. Illiteracy and gender disparity in the state abound. Therefore, rapidly growing number of under educated, under skilled and under nourished boys and girls, prospective labour force in the state can not accrue desired demographic dividend as it is expected in our economic survey of 2006-07.

Other Non-economic Factors of serious concern:

Inferences

Following inferences may be drawn from the above discussions:

- Growth performance of Uttar Pradesh in pre-reform decade was better than post- reform decade. However, it improved in new millennium.
- Growth performance in the state was better in pre-reform period and was worse during post reform period than that of the country. However, with induction of state level reforms growth rate of the state tended to catch up with that of the country.
- The growth performance may be attributed to availability of non-agricultural employment and credit facilities.
- Share of commodity sector in total out put in U.P. continued to remain at higher level vis-à-vis that in the country that led to reduction in proportion of poor in the state to that in the country in the first decade of reform. However, the situation reversed after the first decade. It shows that current pattern of growth is delinked to poverty reduction.
- Despite poor performance of U.P. in terms of income growth, the state is performing better with regards to social development measured in terms of literacy and life expectancy.
- Level of quality of life in the state measured in terms of level of care of women and children and sanitation, is far worse in the state vis-à-vis in the country.
- Population control needs special attention in view of probable failure in accruing demographic dividend to the state.

Conclusion

Economic growth rate of India and U.P. is not sustainable, because of disparity in growth rate in different sectors. The higher growth rate supported by service sector alone can not take care of poverty reduction. The growth strategy has to incorporate broad based commodity sector in its fold for making the growth inclusive and sustainable. It needs to improve quality of life of common people by giving them health, education and skill to ensure their participation in the process of growth.

References

- Bhattacharya B.B. and Sakthivel (2006), 'Regional Growth and Disparity in India, March 6, p. 1071.
- Bhattacharya, B.B.(2007), 'Education, Skill Formation and India's Economic Development,'Presidential address at 90th annual conference of IEA, October 25-27, University of Kashmir, Srinagar.
- Chaubey, P.K. (2008),'Poverty in U.P. : up or down', paper presented at National Seminar held on 25-27, March, GIDS, Lucknow.
- Draft Eleventh Five Year Plan (2007), Government of Uttar Pradesh, State Planning Commission, Lucknow, June
- Ghosh, Madhusudan (2006), 'Economic Growth and Human Development in Indian States,'EPW, July 29 – August 4, p. 3323
- GOI, Economic Survey, Ministry of Finance, 2005-06, 2006-07 and 2010-2011.
- Ram, F.and Chander Sekhar (2006), Ranking and Mapping of Districts, IIPS, Mumbai.
- Statistical Outline of India, 2004-05 and 2008-09, Tata Services Limited, Mumbai.
- Sundaram et al. (2003), 'Poverty in India in 1990s', EPW, April 5-11, p. 1387.

Section-II
Sustainable Development and Water Management

11. Water Balance Processes through Rainfall Interception in Dipterocarp Forest of Lake Chini, Pahang Malaysia

Mohd Ekhwan Toriman, Mushrifah Idris, Joy Jacqueline Pereira, Muhammad Barzani Gasim, Nor Azlina Abdul Aziz & Nor Rohaizah Jamil

In the hydrologic cycle, raindrops that fall on the forest canopies do not necessarily reach the soil surface. Part of the rainfall is captured by the crowns and other surfaces of trees, which then evaporates back to the atmosphere. The amount of water that is subtracted from the total rainfall in this manner is known as interception. Water drips down the forest canopy via two processes: stemflow and throughfall. Stemflow refers to the amount of rainwater that drips down tree stems and branches to the ground. Throughfall, on the other hand, refers to the penetration of rainfall through forest canopy via spaces between leaves or dripping from the leaves, twigs and stems. Both these components were measured in a 100 metre x 100 metre interception plot located in a secondary tropical forest at Lake Chini. Thirty tree samples were used and each tree was identified as regards species, family, diameter breast height (DBH), and canopy size and canopy density. In this study, interception data were collected during every rainfall event in November and December 2007. This interception study provided essential information on how the function of the forest could affect the crucial hydrological cycle within this forest ecosystem, and also the wetland water balance. Indirectly, the interception process that took place in this study area could reduce and control land erosion effectively by controlling the total amount of rainfall generating surface run-off. Apart from that, the interception process could also supply and replenish water to the soil and ensure that the water balance in the forest area was maintained.

Keywords: Interception, Secondary forest, Water balance, Ecosystem

Introduction

The study of interception refers to the study of loss of rainfall before it can reach the earth's surface. Interception is part of the hydrologic cycle that is so important to the water balance of this earth. The amount of rainfall which an area receives depends on whether there are objects covering the area, such as a forest, hill, and building. The trees in a forest act as cover and affect how much rainfall the area receives. Trees act as a filter too. In either role, trees will intercept part of the rainfall, reducing directly the flow of water to the earth's surface. Interception by trees is done through their leaves, canopies, branches, stems and twigs. Trees have various methods of intercepting rainfall. It was for the purpose of understanding these methods that this study was carried out in the secondary forest area of Lake Chini in Pekan, Pahang.

In general, trees affect the distribution of rainfall because they will intercept some of the rain that falls on their canopies before it can reach the ground. This process contributes towards maintaining the forest temperature, slowing down the surface run-off and balancing the water cycle in the atmosphere. The amount of water intercepted by a tree depends on the structure of its branches, forks and the density of its foliage. Therefore, different species of trees will intercept different amounts of rainfall. Some of the intercepted rainfall will not fall to the ground but will evaporate directly off the various tree surfaces back to the atmosphere. The amount of rainfall that is lost through this process is called net interception loss (Zulkifli, *et al*, 2003). Some of the intercepted rainfall will flow from the leaves to the branches, down the trunk and to the ground. This process is called stemflow. Some of the rain that falls onto the canopies of trees finds its way through the leaves and twigs of canopies and the spaces between canopies before falling to the ground. Rainfall that falls without being intercepted by plants is named throughfall (Mohd Ekhwan Hj Toriman & Shukor Md Nor, 2004). Both these processes determine the amount of rainfall that reaches the ground, which in turn contributes towards the moisture content of the soil.

Precipitation begins at high ground. The forests in the highland areas will capture the rainfall and this water will flow slowly down to the plains. Natural interception processes will occur in these forest areas. Some of the rain will be intercepted by the leaves and other parts of the tree and evaporate back to the atmosphere. The rest will either flow from the leaves of the canopies to the branches, down the branches to the trunk and then to the ground, or drip down the leaves and twigs of the canopy and then to the ground. Rain falls to the ground via the canopies, therefore, in either of two methods, by stemflow or by throughfall (Kunkle, 1974).

The amount of water that reaches the surface of the earth from the atmosphere can be measured either by net precipitation or net rainfall collected. The percentage of interception can be calculated from the amount of collected rainfall. With reference to Hoover (1953), the intercepted amount of rainfall is equal to the total gross rainfall less total stemflow and less total throughfall. The components of this equation are influenced by the gross amount of rain that falls at a certain time.

An increase of green areas will increase the processes of evapotranspiration and interception by the canopies of trees (De Roo, *et al*, 2001). Forest studies made in tropical and temperate areas showed the relevance of interception and evapotranspiration processes in the control of amount of surface water, by reducing and slowing the time taken by the water to be absorbed into the soil (Lai & Osman, 1989; Manokaran, 1979; Bruijnzeel, 1990).

According to Wan Ruslan (1994), the amount of rainfall measured in a certain area depends on the existence of surface cover such as a forest, hill, building, road or other man-made covers. These covers, be they natural or man-made, will intercept part of the rainfall in a certain area. In a forest area, the canopies of the trees will intercept the rain before it can reach the ground. Therefore, interception refers to loss of rainfall before it reaches the ground.

According to Herwitz (1987), stemflow depends greatly on the species of the tree. A tree with smooth branches and trunk will allow rainwater to flow down them in greater amounts and in a shorter time than a tree with rough bark. The angle the twigs make with the branches and the angle the branches make with the trunk too will determine the amount of stemflow. An angle of 60 degrees will result in a stemflow of 80 percent of rain that falls on the tree. It depends also on whether the branches are wet or dry. This means that insufficient rainfall will cause insufficient wetness to allow rainwater to penetrate the canopy for a flow to happen (Lull, 1964).

According to Sharifah Meryam (1998), throughfall is rain that falls straight to the ground through spaces within a canopy, or rain that drips from leaves or branches straight to the ground. The rain falls on the tree or canopy, collects on the leaves before falling to the ground as throughfall. Water collected on the leaves falls to the ground in a number of ways, for example when the tree is shaken by the wind.

Forests must be protected for they are an important factor in determining how much rainfall an area receives. They are necessary to ensure the processes of interception; throughfall and stemflow take place in the hydrologic cycle of a forest ecosystem and are kept in equilibrium. The measure of intercepted, throughfall and stemflow amounts are therefore important for the understanding of a forest's functions in the hydrologic cycle. Studies have shown the adverse effects on the equilibrium of water as a result of failure in the process, for example an increase in surface water run-off, soil erosion and the frequency of floods. This interception study was carried out in the secondary forest of Lake Chini to show the importance of this process in controlling the problems mentioned.

Concept of Study Framework

Interception is an important factor in the calculation of the amount of water stored in the forest area because it affects the total amount of precipitation that reaches the earth's surface (Kittredge, 1948). From this aspect, forests will be able to control the amount of river water by keeping part of the precipitation from the atmosphere before it can reach the earth's surface. In an indirect way, interception will be able to assist in preventing damage and destruction from, for example, soil erósion, landslide, landslip, fault and the battering effect of rain by reducing its kinetic energy (Wan Ruslan, 1994).

According to Wan Ruslan (1994), the amount of rain that falls in a certain area depends on the existence of covers such as a forest, hill, building, road and other man-made covers in the area. Covers, whether natural or man-made, will intercept part of the total rainfall in the area. Rain that falls in a forest area will first be intercepted by the canopies of the trees before it can reach the ground. In this context, interception refers to the loss of rainfall before it can reach the ground.

An increase of green areas will increase the processes of evapotranspiration and interception by canopies of plants (De Roo, *et al*, 2001). Forest studies in tropical and temperate areas have shown the relevance of interception and evapotranspiration processes in the control of the amount of surface water, by reducing and slowing the time taken by the water to be absorbed into the soil (Lai & Osman, 1989; Manokaran, 1979; Bruijnzeel, 1990; Mohd Ekhwan & Syukor, 2006). This study conducted in the secondary forest of Lake Chini is important because the area of study is a reserve area under the protection of the Pahang State Government. Any change in the form of land-use in the area will affect the rate of entry of surface water run-off into the lake, with the danger of affecting the water equilibrium of the lake. Interception is also important in controlling the effect of sedimentation in the lake, where sedimentation is a problem.

Study Area

The Lake Chini secondary forest, located in the Lake Chini basin, is a dynamic natural forest ecosystem encompassing an area of 5,105.23 hectares. The basin is located at latitude 3° 15' 40" N and longitude 102° 45' 40" E. There are two dominant secondary forests in this valley, one where logging had been carried out and the other which had been cleared by inhabitants of the area who wanted to improve their lot by growing crops on the cleared area. This farmland was subsequently abandoned and a secondary forest grew in its place. FELDA is active in the basin with its rubber and oil palm plantations. However, in the south-western part of this lake, a primary forest exists. This forest is a protected area and has been gazetted by the state government as a national park, the Lake Chini State Park.

Having a tropical climate, the area is hot and humid year-long. Rain falls regularly. Humidity is high, being more than 80 percent. Average yearly temperature is also high at 28°C. This results in a high rate of water evaporation back to the atmosphere. This in turn results in frequent rain. Wind velocity in the area is low at 0.5 m/s (Nor Aseken Binti Rosley, 2007).

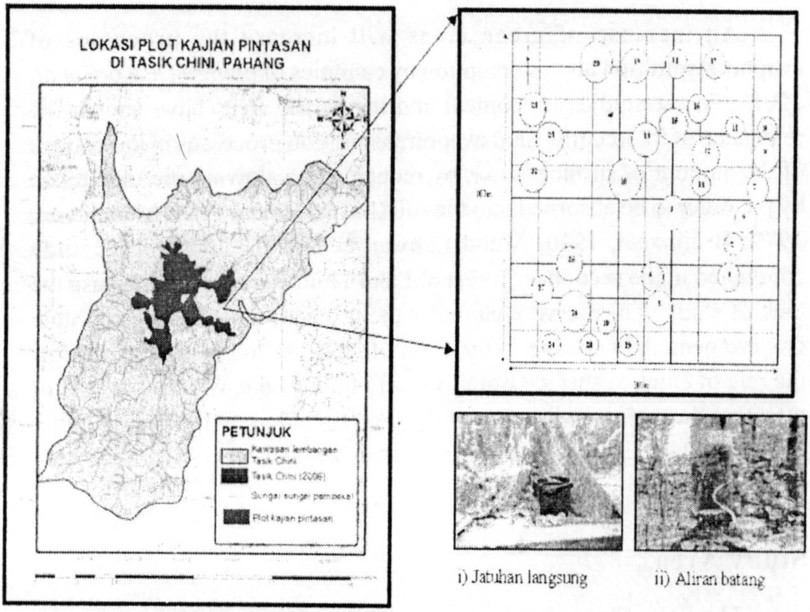

Fig. 1 Location of plot of Interception Study at Lake Chini, Pahang

Key
Lake Chini basin area
Lake Chini (2006)
Tributaries
Plot of interception study

(i) Throughfall
(ii) Stemflow

Figure 1: Map of Lake Chini and area plot of interception study in the Lake Chini secondary forest, Pekan, Pahang

Study Methodology

In this interception study, a total of 30 trees identified as samples were used. Each tree of a minimum 20 cm diameter was attached a stemflow measurement set (Diagram 1(ii)). Each sample tree in this plot of 100 m x 100 m (Diagram 1), located in the Lake Chini secondary forest, is a native species. Black bin throughfall measurement sets (Diagram 1 (i)) were placed under the thinnest canopies in the plot. The physiological characteristics of each tree sample (Schedule 1) in the plot were determined through observations made with botanical faculty members of UKM.

Schedule 1: Tree sample physiological characteristics

S.N.	Species name	Height (m)	DBH (cm) DBH (cm)	Canopy size
1	*Teijsmanniodendron Simplicifolium*	22	43.4	Moderately dense
2	*Terminalia calamansanai*	23	35.7	Moderately dense
3	*Shorea sp*	47	72.0	Dense
4	*Cryptocarya rugulosa*	25	47.7	Moderately dense
5	*Horsefieldia superba*	21	51.1	Moderately dense
6	*Pentaspadon motleyi*	22	39.8	Moderately dense
7	*Pentaspadon motleyi*	26	37.7	Moderately dense
8	*Gymnacranthera forbesii*	24	48.9	Sparse
9	*Xanthophyllum amoenum*	27	30.7	Moderately dense
10	*Koompassia malacceasis*	43	82.9	Dense
11	*Canarium megalanthum*	26	33.5	Moderately dense
12	*Dialium laurinum*	40	67.2	Moderately dense
13	*Ochanostachys amentera*	24	33.7	Moderately dense
14	*Shorea* sp	35	65.6	Moderately dense
15	*Dialium wallichii*	35	56.0	Moderately dense
16	*Daryodes rugosa*	27	37.5	Sparse
17	*Syzgium cumingiana*	30	38.5	Moderately dense
18	*Cyathocalyx pruniferus*	26	31.7	Sparse
19	*Shorea balanocarpoides*	40	69.0	Moderately dense
20	*Shorea lepidota*	41	73.0	Moderately dense
21	*Pentaspadon motleyi*	44	61.5	Moderately dense

22	*Scorodocarpus borneensis*	31	62.7	Sparse
23	*Ochanostachys amentera*	34	56.0	Sparse
24	*Shorea macroptera*	22	27.5	Sparse
25	*Canarium megalanthum*	25	56.2	Moderately dense
26	*Pentaspadon motleyi*	46	80.1	Moderately dense
27	*Shorea maxwelliana*	32	43.7	Moderately dense
28	*Shorea maxwelliana*	21	36.1	Moderately dense
29	*Syzgium clariflora*	24	35.8	Moderately dense
30	*Shorea maxwelliana*	28	34.8	Moderately dense

**** Field work source 2007***

Result and Discussion

The total stemflow is greatly influenced by such factors as rainfall period, total rainfall, how long it lasted, its intensity, and presence of wind. In general, the total throughfall obtained can be seen in the absolute amount through increase in wind speed (Lawson, 1967). The total throughfall is low if the wind velocity is low and *vice versa*. If wind velocity is increased, the total throughfall will also increase (Low, 1972). During the period the study was made, the wind velocity was low. As a result, the total throughfall obtained too was low as wind velocity did not play a part during this period. Figure 2 compares total gross rainfall, throughfall rate and stemflow rate on the days observed.

The study made in the Lake Chini secondary forest showed conclusively that the stemflow readings were greatly influenced by gross rainfall because they were high. Diagram 3 (a) correlates the relationship existing between these two. The correlated coefficient value obtained is 0.9494 (R). The significance level for this variable is 95% with a relationship value of 0.9013 (r^2). The coefficient value obtained is 0.9832 (R) where the points occur in an orderly straight line showing a continuous relationship between these two variables (Diagram 3 (b)). This indicates that the interception rate is influenced by rainfall in the study area. Should the rainfall amount increase, the interception amount too will increase, in line with the increase in total rainfall. The relationship value is 0.9667 (r^2) and the significance level for this variable is 98%.

Comparison of gross rainfall, through fall and stemflow (Nov 2007)

Depth
Date/day

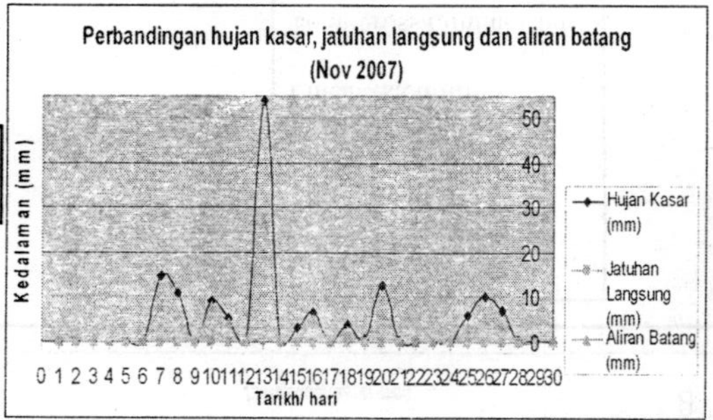

Comparison of gross rain fall, and stemflow (Dec 2007)

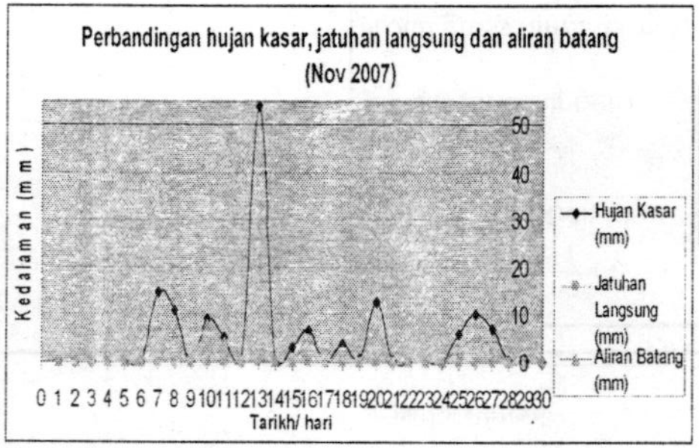

Gross rain (mm)
Throughfall (mm)
Stemflow (mm)

Interception rate (mm)
Date/day

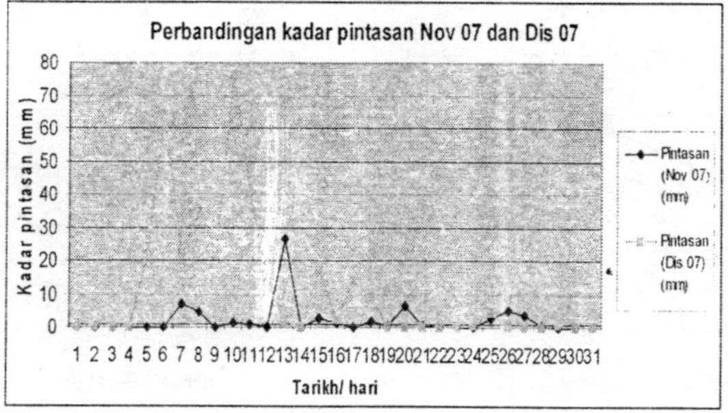

Interception (Nov 07) (mm)

Interception (Dec 07) (mm)

Comparison of Interception rates in Nov and Dec 07

Figure 2. Comparison of Interception rates in study plot

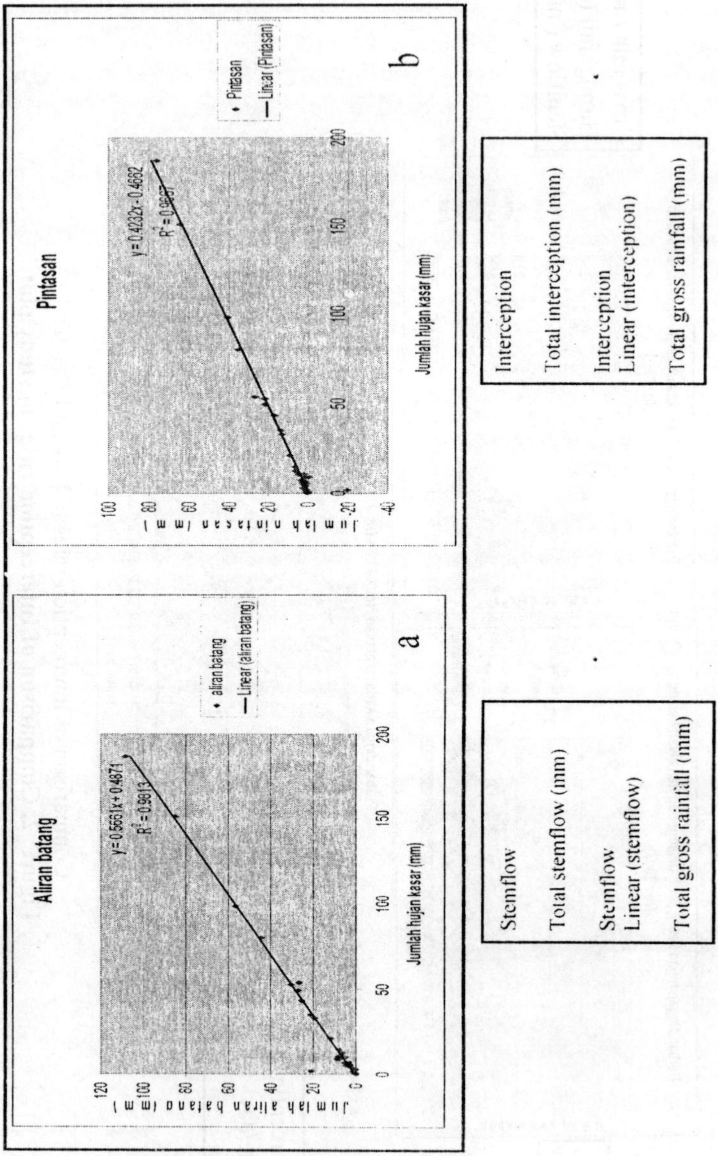

Figure 3: Relationship between total rainfall and stemflow

The nature of the bark of a tree, whether it is smooth or rough, affected the amount of stemflow collected. The amount collected from a smooth-barked tree was more than that from a rough-barked tree. This is because the rough bark contains holes that have a tendency to absorb more water and thereby reduce the stemflow (Helvey, 1967). According to Kittredge *et al* (1948), a tree with smooth bark has a low absorption rate and has the ability to allow water to flow faster because the water is not hindered by holes in the trunk. In the study conducted in the Lake Chini secondary forest, it was found that sample trees 27, 28 and 30 (*Shorea maxwelliana*) had the most collection of stemflow because their trunks were smooth and their twigs thin. These twigs were capable of catching rainfall with the result that more rainwater was collected. The fourth tree (*Lauraceae*) had a large trunk diameter (72 cm) but the stemflow was small, compared to another sample tree (*Gymnacranthera forbesii*) with a trunk diameter of 48.9 cm which collected more stemflow. Sample tree 20 was a tree of great height with a smooth trunk. A large amount of stemflow was collected from this tree.

Twigs and branches are important factors in determining the amount of stemflow collected. Branches and twigs attached at an angle make it easier for more rainwater to flow to the trunk faster when compared with the case of branches or twigs that are attached horizontally (Wan Ruslan, 1994). The density of a canopy too plays a role in determining the amount of stemflow. A dense canopy will capture more rainwater than a sparse one. In the study, sample tree 6, a *Pentaspadon motleyi* species, had a large stemflow amount even though the tree stood at an angle. This condition in fact made it easier for rainwater to flow faster down the tree.

Total interception is measured using the following equation (Mohd Ekhwan & Shukor, 2006):

$I = R - (Sf + Tf)$ where ....(1)

I = interception
R = direct rainfall
Sf = stemflow
Tf = throughfall

From the study carried out in the secondary forest of Lake Chini, it was found that interception was determined more by stemflow than throughfall. This was proved by the fact that stemflow collections were much more than throughfall amounts. The study showed the highest total interception of 76.21 mm was obtained from total gross rainfall of 187 mm on 10.12.07. On 29.11.07, 21.12.07 and 22.12.07, the lowest total interception of 0.02 mm was shown from gross rainfall of 0.2 mm. The total throughfall was 9.41 mm while stemflow contributed 17,112.78 mm. This showed that stemflow was the major contributor in the interception process in the study area. The thickness of the canopies that covered almost the entire study plot was the reason. The results of study are summarised in Schedule 2.

Schedule 2: Percentages of throughfall and stemflow in the interception process in the secondary forest of Lake Chini

	Gross Rainfall (mm)	Throughfall (mm)	(%)	Stemflow (mm)	(%)
Nov-07	144.6	0.81	44.509	80.24	55.491
Dec-07	844.4	8.6	39.654	509.56	60.346

Studies showed tha. the total interception obtained in tropical forest areas was greater than those obtained in temperate forest areas. That the highest total interception was obtained in tropical areas is justified because the trees there tend to grow closer to one another, are of differing heights and sizes resulting in many layers of dense canopies and undergrowth (Smith, 1972). This statement was confirmed by the study conducted in the secondary forest of Lake Chini which possessed characteristics similar to those described by Smith. These forest characteristics resulted in the interception components, throughfall and stemflow, registering high amounts. The forest under study was a tropical forest where the canopies of the trees were dense, which greatly influenced interception.

Other factors such as season, rainfall intensity, period of rainfall and wind velocity have an effect on interception. According to Kittredge *et al* (1948) and Pavari (1959), certain factors have immediate effect on evaporation, for example atmospheric pressure, humidity, wind velocity, a moderate climate and the amount of sunshine. In the study, it was found that the seasons greatly influenced interception. November and December are rainy months. The intensity of the sun during the rainy months is naturally less than that during the hotter season, resulting in less evaporation.

Conclusion

The results of the study showed that the roles played by throughfall and stemflow in the hydrology system influenced the interception rate in the study area. This demonstrated that the role played by stemflow influenced interception rate to a greater extent than throughfall. This study also demonstrated the importance of the forest area of the lake basin and that any minor change in land-use could influence the function of the forest as a water catchment area, apart from causing a more damaging impact should the process of interception by the forest fail. The incidence of surface water run-off and the transport of sediment into the lake are influenced by this forest function.

Acknowledgement

This study was wholly financed by the Science Fund 06-01-02-SF0220 research grant with the co-operation of the Lake Chini Research Centre (PPTC) of UKM.

References

- Aldridge, R & Jackson, RJ. 1973. Interception of Rainfall by hard Beech (Nothatagus Andy D Ward & Stanley W. Trimble. 2004. Environmental Hydrology. London. Lewis Publishers *Applied Hydrology*, McGraw Hill, Mew York.
- Azmi Nordin. 1983. *Interception, Throughfall and Stemflow by Araucaria cunninghamii (AIT) and Pinus Caribaea (MOOR)*. Project paper. Faculty of Forestry, Universiti Putra Malaysia.
- Baharuddin Kasran. 1989. *Rainfall Interception in Dipterocarp Forest of Peninsular* Journal of Hydrology. 3:723-729.
- Bo Myeong Woo, Tae Ho Kwon & Kyong Ha Kim. 1989. *Rainfall Interception Loss from Canopy in Forest Catchment*. IHP-UNESCO-FRIM Regional Seminar on Tropical Forest Hydrology. 4-9 September 1989, Kuala Lumpur.
- Gash, JHC, Wright, IR & Lloyd, CR. 1980. *Comparative Estimates of Interception Loss from Three Coniferous Forests in Great Britain. J Hydrol* 48:89-105 .
- Gran, RF & Wilson, CC. 1994. Some Components of Rainfall Interception. *Forestry* 42:90-899.
- Hamilton, EL & Rowe, PB. 1949. *Rainfall Interception by Chaparal in California*.
- Helvey, JD & Patric, JH. 1965. Canopy and Litter Interception of Rainfall by Hardwoods of Eastern United States. *Water Resource* 1: 193-206.
- Helvey, JD & Patric, JH. 1967. Interception by Easter White Pine. *Water Resource*.
- Hoover Mid. 1953. *Interception of Rainfall in Ayoung Loblolly Pine Plantation South*.

- Hydrology, 4-9 September 1989, Kuala Lumpur.
- Jackson, WM. 1942. *The Interception of Rain and Snow by a Forest of Young*
- Lai FS & Osman S. 1989. *Rainfall Interception, Throughfall, Stemflow in Two Acacia*
- Low KS. 1972. Interception Loss in the Humid Forested Areas (with Special References to Sg Lui Catchment). *Malay National* 25: 104-111.
- Lull, HW. 1964 Ecological and Silvicultural Aspects. In: Chow, VT. *Handbook of Malaysia*. IHP-UNESCO-FRIM Regional Seminar on Tropical Forest.
- Malchanov, AA. 1963. The Hydrological Role of Forest. Russians by the Israel *Mangium Stands in Kemasul Pahang*. IHP-UNESCO-FRIM Regional Seminar .
- Mohd Ekhwan T & Shukor Md Nor. 2006. An analysis of Rainfall Interception on the Selected Experimental Plot of Pangkor Hill Reserved Forest. *Journal Wildlife and National Park.* December 2006. 169-178.
- Raudkivi, AJ. 1979. *Hydrology, An Advanced Introduction to Hydrological Processes and Modelling*. New Zealand. University of Auckland. California National Forestry. *East.*

12. Effective Measures for Water Harvesting and Ground Water Recharging for Sustainable Development of Ground Water Storage in Kerala

P.Suseela and Helen

Kerala is frequently facing severe droughts followed by acute drinking water scarcity for the last two decades. The rivers hardly contain any water during six months in a year. Only a few reservoirs get filled up even in the monsoon. During the summer, water level goes down to the silted up bottom in many cases. More and more wells are getting dry in summer. Due to steep topography, high population density, non-availability of waste lands and financial constraints; conventional method of storing water in the open reservoirs by building a dam across the river is not practical. The study revealed that subsurface dam, percolation pond and ground water recharging through wells/pits are simple, cheap and ecofriendly and are effective measures for raising the ground water table in Kerala. In coastal areas, recharging ground water through wells and rain pits using harvested rain water, helped to provide fresh water in the wells.

KeyWords: Ground water recharge, Subsurface dyke, Subsurface dam, Percolation pond.

Introduction

With about 3000 mm average annual rainfall, 44 rivers, chains of back water bodies, reservoirs, tanks, ponds, springs and wells, Kerala is considered by many as land of water. But in Kerala, due to steep and undulating topography, rainwater is not much retained on the land in comparison with other states, there by obviating the advantages of having high rainfall (Padre, 2002). The rainwater reaches the sea within 24 to 48 hours due to steep gradient. Only 5% of the runoff is believed to be converted into the ground water resources. Hence, in spite of heavy rainfall, the state experiences severe draught during summer months due to inefficient utilization of rainwater (Basak, P. (1998). During the last decade, ground water level in different parts of Kerala has decreased by 30 cm to 3 m. The decline of water table has resulted decrease in well yield, failure of wells/tube wells, increase of pumping cost and a big threat to the environment. The problem of ground water depletion has arisen due to the development of ground water at a faster rate than the annual replenishment of recharge. Therefore steps must be taken to maintain hydrological equilibrium between annual replenishable recharge and ground water draft with an emphasis on conservation of rainwater and its harvest at the site of occurrence as much amount as possible (Dhruva narayana et al., 1997).

Due to steep and undulating topography, lesser depth of top soil, less humus content in the soil and forceful beating of raindrops; rain water is not much retained on the land. There is no scope for storing the monsoon water in dams and bunds as being done in other states due to Kerala's topographic features, high population density and lack of availability of wastelands. Considering all these facts, suitability of ground water recharging through water harvesting structures like subsurface dams/ percolation pond and also the effect of recharging the harvested roof top water through wells/ponds were studied.

Subsurface Dyke

Subsurface dyke is an impermeable layer that is built below the ground surface with an intention to obstruct the natural subsurface flow of water, thereby augmenting the recharge of ground water and raising the water table. In contrast to conventional method of storing water in the open reservoirs by building a dam across the river to collect the surface runoff, the subsurface dam arrests the subsurface flow of water and store it below the ground surface (Thomas, et. Al,1998). Subsurface dam structures are simple, cheap and ecofriendly.

(i) Selection of Site

Drinking water has become a scarce commodity along the Bharatapuzha and many wells have dried up in the villages along the course of river. Water scarcity begins in most places by March and in some places even as early as January. This scarcity lasts till the arrival of southwest monsoons in June. Hence, in order to reduce the water scarcity problem in Thrissur, Malappuram and Palakkad districts, check dams/ subsurface dams were proposed to construct across Bharatapuzha at 14 locations based on the hydrological and topographical conditions.

In the present study, subsurface dyke was constructed at Thrangali Kadavu in Palakkad district across Bharatapuzha, considering the local topography, lesser width of the river (only 300 m), soil properties, economy and benefits. Construction of dyke would permit augmenting water for irrigation (a lift irrigation scheme is situated approximately 105m upstream of the proposed weir line) and drinking (a drinking water scheme is also situated 1 km upstream of subsurface dyke).

In order to study the suitability of the site for constructing the dam, four drill holes were taken upto the clay level. The average depth of clay level was found to be at 3m. Rocky formation was found at a depth of 6m at the right bank and 7m at the left bank. Hence an impervious cut-off could be provided at relatively shallow depth.

Construction Details of Subsurface Dam

Subsurface dam was constructed by using sand-cement-bentonite grout to form impermeable barrier for arresting the subsurface flow. This was much cheaper compared to conventional concrete structures. Slurry tank was constructed near the trench for easy circulation of bentonite slurry into the trench (Fig.1). Slurry pit was lined with LDPE sheet (150 micron) and the sides were protected by sand bags. Initially bentonite was mixed in small quantities in 200 litre barrels and then filled the slurry in the slurry pit. Water was added to get the bentonite slurry having a specific gravity of 1.11(roughly 175 g bentonite is necessary to prepare 1cumec slurry). Slurry was mixed thoroughly by using a mud circulation pump keeping both the suction and délivery pipe inside the slurry pit. Bentonite slurry was prepared 24 to 48 hours before the construction of the slurry wall for proper mixing and expansion of bentonite

A trench having a width of 0.6m was constructed using poclain across the river. The sides of the trench were protected by pumping the prepared slurry into the trench through the canvas pipe during the progress of excavation. The excess slurry will flow back to the slurry tank through the return channel (Fig.1)

Preparation of grout was done as follows :

1. The excavated sand from the trench (trench made for constructing the subsurface dyke) was spread over an area of 7m x 7m
2. Clay lumps were broken into small pieces and spread over the sand layer
3. Cement was then spread over the clay soil and the roto tiller was then run to dry mix the material
4. Bentonite slurry of adequate bentonite content spread over the mixed material and roto tiller was run
5. The grout was prepared for the flowing consistency and was then fed by gravity into the tremmie.
6. The grout was mixed in the trench by mixing device

One m^3 of sand- cement- bentonite grout was prepared by using sand (1500 kg), cement (60 kg), bentonite (68 kg) and water (450 lit).

A sand - cement - bentonite wall having a length of 320 m [width of the river (300m) + anchorage of 10 m to the right and left bank of the river], average depth of 3m (stiff clay was present at an average depth of 3m) and a thickness of 0.6m was constructed. In order to protect the diaphragm wall from scour due to degradation of river bed downstream, an apron structure consisting of soil-cement has been provided and tied to the diaphragm wall. Soil cement concrete weir body was constructed by using 1600 kg of sand, 120 kg of cement, 120 kg of clay for 1 m^3 soil cement concrete. At the bottom of the river, slurry wall was protected by providing gabbion work. Gabbion work was provided for a length of 3 m on the upstream side and 12 m on the downstream side (Fig.2). Gabbion work is the arrangement of layers of packed rocks in well-protected GI net work. Both left and right bank of the river was also protected by providing gabbion work for a length of 25m on upstream and downstream side upto a height of high flood level.

Performance Evaluation

The performance of the subsurface dyke was studied by monitoring the water level in the adjoining wells within a distance of 1km upstream and downstream side of the dyke, water source used for lift irrigation scheme and well utilized for drinking water supply, before and after the

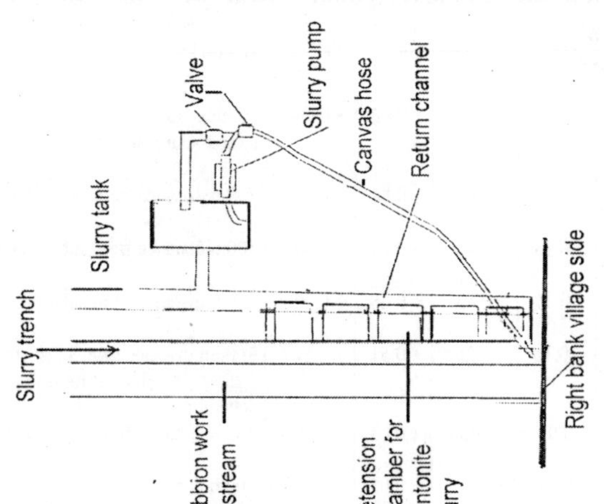

Fig.1. Process of circulation of slurry into the trench

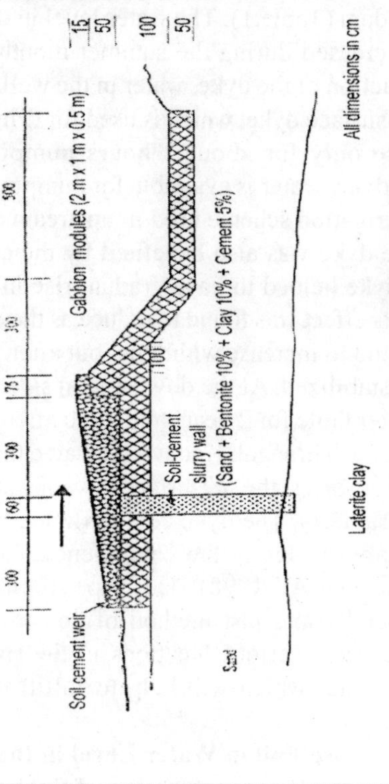

Fig.2. Cross section of sub-surface dyke across Bharatapuzha

construction of the dam (Table.1). The water level in the adjoining wells was found to be increased during the summer months by about 2 to 3 feet. Before construction of the dyke, water in the well, constructed 1 km upstream of the subsurface dyke, which is used for drinking water supply (W_6) was available only for about 2 hours pumping. But after the construction of the dyke, water is available for pumping continuously for 24 hours. The lift irrigation scheme (105 m upstream of the weir line) in the upstream of the dyke was also benefited by the construction of the dyke. Subsurface dyke helped to have gradual rise in water table at the upstream side and its effect was found to reduce as the distance increases. Water level was found to increase with time, but after four years, ground water table almost stabilized. At the downstream side, there was a slight reduction in the water table for 2 years and there after this effect was not seen. The experience at Thrangali Kadavu indicated that subsurface dyke is an efficient barrier to arrest the subsurface flow of water and to conserve the ground water (Table.1). The dyke could efficiently arrest the rate of depletion of drainable water in the catchment area due to the steep topography (Thomas, et. Al, 1998). This type of check dam is cheap, economical, ecofriendly and fast method of construction. Hence, this type dam can be used at various locations in the rivers of Kerala for proper water harvesting; which will help to fulfill the water needs in summer.

Table.1 Average Rise/Fall in Water Level in the Wells during Summer months after construction of Subsurface dyke

Well No.	1999	2000	2001	2002	2003	Remarks
W1	+ 0.19	+ 0.45	+ 0. 75	+ 0.92	+ 0.96	105m distance from subsurface dyke at U/S (water source using for lift irrigation)
W2	+ 0.15	+ 0.35	+ 0.68	+ 0.88	+ 0.91	300m distance from subsurface dyke at U/S
W3	+ 0.10	+ 0.18	+ 0.46	+ 0.73	+ 0.80	500m distance from subsurface dyke at U/S
W4	+ 0.06	+ 0.15	+ 0.43	+ 0.68	+ 0.70	700m distance from subsurface dyke at U/S
W5	+ 0.05	+ 0.10	+ 0.25	+ 0.51	+ 0.51	1000m distance from subsurface dyke at U/S (well used for drinking water scheme)
W6	- 0.25	- 0.05	+0.025	+ 0.04	+ 0.10	100m distance from subsurface dyke at D/S
W7	- 0.20	- 0.03	+0.06	+ 0.05	+ 0.13	300m distance from subsurface dyke at D/S
W8	- 0.15	- 0.05	+0.38	+ 0.06	+ 0.11	500m distance from subsurface dyke at D/S
W9	- 0.05	+0.03	+0.05	+ 0.0 8	+ 0.075	700m distance from subsurface dyke at D/S

Percolation Pond

Percolation ponds are small water storage structures constructed to harvest runoff from the catchment area and impound for longer time to facilitate percolation of impounded water into the soil substrata vertically and laterally, thereby recharging ground water storage in the zone of influence of the ponds. These structures will help to increase the yield of wells, especially on the downstream side of the pond. Water stored in the Percolation ponds can be used for irrigation and drinking purpose of the animals. They will also check soil erosion and wastages of runoff during flash floods and keep the drought at bay. A reconnaissance survey was carried out for the selection of suitable site for the construction of percolation pond. Based on the reconnaissance survey, an existing small pond in the Kerala Agricultural University main campus, which is located in the rubber estate was selected for the study.

During the summer season, the water level of three wells lying at the down stream side of this small existing pond was found to increase when it was deepened and widened to get a volume of 379 m^3. On utility point of view, these wells are very important. One well is using for drinking water supply to the staff quarters. Water of the second well, in the estate factory, is using for making rubber sheet. The third well in the central nursery is using for irrigating the nursery plants. With this in mind, this pond was selected and it was deepened and widened using JCB during 2004. Total expenditure involved for making the percolation pond was Rs. 60,000.

The present water-spread area of the pond is 1608 m^2 and average depth of the pond is 4.47m. The capacity of the pond at full level is 7195 m^3. The contour map of the percolation pond is shown in Fig.3. The catchment area of the pond is having a moderate slope towards the pond. The pond mainly receives water from a small channel which collects surface water from a catchment area of 10 ha on the upstream side of the pond. The topography of the terrain is moderately sloping towards the pond. The area consists of red gravelly soil for a depth of 1 to 3m and is followed by weathered rock upto a depth of 3 to 5m. Drainage channels are constructed to divert the runoff flowing towards the pond. Since the pond is situated in a rubber estate, the ground is fully covered by the dried leaves and grass. Hence silting of the pond is comparatively very small.

The storage in the percolation pond is increased by surface runoff from catchment area, by direct addition from rainfall and by ground water appearing as effluent seepage. The loss in storage is caused entirely due to percolation and evaporation. The percolation of water will mainly occur to the downstream side of the pond (the topography of the land is moderately sloping towards the downstream side). Hence, in order to observe the ground water fluctuationand also to study the effectiveness of the percolation pond in increasing the yield of wells, two observation wells were constructed at the upstream side and two at the downstream side of the pond. All the observation wells were installed touching the hard strata, and projecting about 1m above ground level for easy observation. The performance of the percolation pond was studied by monitoring the water level in the observation wells and selected wells both at the downstream and upstream side. During the rainy season, there wasn't much variation in the water level of both observation wells and selected wells lying at the upstream and downstream side of the pond. But during summer season, there was marked variation in the water level of both observation wells and selected wells lying at the upstream and downstream side of the pond. This may be due to the fact that, during the rainy season, the water table rises to nearly 1 to 2m below the ground surface. Higher water table during the rainy season nullifies the effect of percolation pond. But, water table gradually recedes after the cessation of rainfall. So water level in the wells of upstream side of the percolation pond decreases. Whereas, the water stored in percolation pond keeps on feeding the wells lying on the downstream side of the pond. Hence, a marked variation between the water level in the wells at the upstream side and downstream side of the percolation pond is noticeable (Table.2).

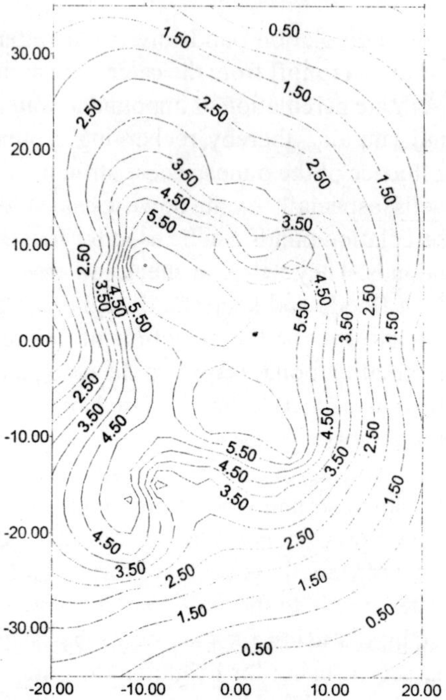

Fig.3.Contour map of the Percolation Pond

Table.2 Details of Water Level in Summer and Rainy Season

Well no.	Particulars	Distance from the bund of pond	Water level from the datum (m)	
			Rainy season	Summer season
Down Stream Side				
B4	Observation well at 6m D/s of the pond	7	0.25	1.235
B3	Observation well at 30 m D/s of the pond	17m	0.05	3.625
W1	Estate Office Well	208m	0.3	5.5
W2	Near Central Nursery	278.1	0.3	3
W3	Well Near to Pond University Boundary	500m	0.8	2.11
W4	Well at Nursery Office	600	0.2	2.12
W5	Well at Quarters	550	0.3	3.9
Upstream side				
B1	Observation well at 6m U/s of the pond	10.2	0.32	2.685
B2	Observation well at 30 m U/s of the pond	30	0.45	4.155
W6	At Chirakkakode Gate	129	0.45	6.35
W7	Well at School Side	186	0.9	4.5
W8	Well Near Boundary of KAU	243	1.5	7.4
W9	Near Chirakkakode road	325	0.45	5.7
W10	Well Near to Olericulture Farm office	250	0.2	10.25

shallow ground water aquifers. Because of the increased availability of water, the hours of pumping and hence the area that can be irrigated also increased.

Ground Water Recharging through Open Wells/ Rain pits

The total length of the state is about 700 km with an average width of 65 km. Within this narrow strip of land, there are areas lying below the sea level and peaks ranging from 2500 to 3000m above MSL. Kerala is identified as the state with the highest well density in India having around 48 lakh wells which amount to Rs. 9600 crores @ Rs. 20,000 per well. However, Kerala is frequently facing severe draughts followed by acute drinking water scarcity for the last two decades. In Kerala, 60% of the total rain is received in about 120 days which has a maximum intensity of more than 8 cm/hr. The downpour during monsoon sometimes reaches more than 20 cm/day, which is even beyond the control. The state has a surplus of 8506 million cubic metres of water in monsoon and a deficit of about 7142 million cubic metre in summer. The variation in spatial and temporal distribution of water causes frequent floods and droughts in Kerala. About 80% of its water gets wasted and drains into the Arabian Sea, in the absence of any conscious effort on in-situ rain water harvesting and storage. Discharge of untreated effluents into surface water streams and lakes by industries resulted not only in contaminating these surface waters but also the ground water bodies. In coastal areas, over exploitation of ground water resulted in seawater intrusion thereby rendering ground water bodies saline in these areas. Identification and promotion of simple, reliable and environmental friendly technologies are necessary to overcome the above problems. Reviving the traditional practices of water harvesting using scientific method is potential option.

Ground Water Recharging Helped to Reduce the Salt Problem

Alapuzha district is a beautiful place of kerala state which is blessed with lot of back waters. Even though, the district is blessed with plenty of water, fresh water availability is very poor. In most of the places, ground water is salty and soil is acidic. Five years back, Mr. Samuel Mathew, who was working in Dubai, purchased a plot in Aroor panchayath (Pattanakkadu block, Cherthala District) and constructed a new house. He dug a well for his domestic and irrigation purpose. But the water was salty and hence they have to go more than 300 m for collecting fresh water from a public water tap. A canal is flowing with salty water along the side of his house. He contacted the research station and requested advice to solve the problem. Based on that, we visited the site and advised to harvest the roof water and to recharge in the well after filtering the water and also to make some more provisions to collect the rain water directly for recharging the ground water. Hence action was taken to harvest rain water from the roof top of 1200 ft^2 and the complete water was used for recharging in the well and an existing tank (since the bottom of the tank is having full of cracks, the rain water is penetrated through the pores and recharged the ground water). The ground water level in this area is just 2 m below the ground surface even in the summer months. After two years, he got 1.5m depth of fresh water in the top of the well and the water in the bottom layers remained salty (for this, he had to spend around one thousand rupees only). Since the density of fresh water is less than the salty water, fresh water remained at the top level. Due to this action, his brother staying near to his house also got benefited. This inspired the neighbors and hence they also gave some provisions for recharging the ground water with fresh water. Ultimately after five years, 6 to 8 families staying in that belt got benefited. Now they are getting 4 m height of fresh water in their wells.

Ground Water Recharging Helped to Raise the Water Table

The well in the PRC, vellanikkara used to dry up in the month of February - march. In 2004-05, as a part of the plan project on performance evaluation of water harvesting and ground water recharging structures, roof top water harvesting system was constructed for the office building of PRC, Vellanikkara (Concrete flat roof – 100 m^2) and the sloped roof (92 m^2) of WMRU, Vellanikkara, which were provided with ferrocement tanks of 10,000 lit capacity. The overflow water from each ferrocement tank was utilized to recharge into the existing well at PRC, Vellanikkara. Three rain pits of size 2 m x 1.5 m x 1 m was also constructed around the well at a distance of 1.5 m distance from the well. These rain pits were filled with pebbles and sand upto a height of 65 cm. Earthen bund (30 cm height) was made around each pit which was stabilized by growing grass. This helped to prevent the entry of muddy water into the rain pits. These types of ground water recharging measures helped to increase the water level in the recharged and adjoining wells (Table.3). The water level in the well was found to increase from 2005 to 2008 and thereafter water level was almost stabilized. Ground water recharging through W_1 helped to increase the water level in other wells also.

Table.3 Average Depth of Water Level in the Wells

Well	Depth of Water in the Well (cm) during the month of March						Remarks
	2005	2006	2007	2008	2009	2010	
W1	Empty	32	52	85	79	89	Well in which recharging was done
W2	" "	26	48	80	75	81	50 m distance from W1
W3	25	38	50	63	58	65	72 m distance from W1
W4	40	48	59	72	68	80	150 m distance from W1
W5	30	38	45	79	67	72	125 m distance from W1

Conclusion

Construction of subsurface dyke, percolation pond and ground water recharging through wells are effective measures for raising the ground water table in limited area. In coastal areas, recharging ground water through wells and rain pits using harvested rain water, helped to provide fresh water in the wells. All these measures are cheap and ecofriendly and are giving positive results within 3 to 4 years.

References

- Basak, P. (1998). Water resources of Kerala myths and realities. Water scenario of Kerala. State committee on science and technology and Environment, Govt. of Kerala publication. pp: 1-6.
- Druvanarayana V.V., Sastry, G and Patnaik, U.S. 1997. Watershed Management. ICAR publication, New Delhi
- Padre, S. 2002. Rain water harvesting. Altermedia, Brahmaswom Madam Bldg, Thrissur, India. p 115.
- Thomas, J., Joy, P.P. and Samuel, M. (1998). Subsurface dike for ground water Conservation. Proceedings of tenth Kerala.

13. Knowledge Dissemination for Ground Water Recharge in Urban Area Using Spatial Decision Support System

Saikiran Tharimena
Katpatal Y. B.

Water is a vital component for the survival of Humans. Most of the water that is directly accessible is either saline or is polluted in some way. The rapid industrialization and drastic increase in population has led to pollution of the surface water to such extents that purification and use of these waters has become too expensive. People have hence started to exploit the ground water which is regarded as one of the purest forms of water. Out of the total available water on the earth only 1.6% is available as ground water which is 20% of the world's fresh water. The exploitation of the ground water has in time led to its depletion. The urban infrastructure has reduced the space available for the recharge of the ground water and hence alternative artificial techniques have to be employed to recharge the ground water.

In the present study in Nagpur city, it is evident that there is an increase in demand for water and people are exploiting ground water to its limits with almost 'no recharge' to it. The aim of the present study is to develop a methology to assess the potential zones of ground water recharge and the method to be used for ground water recharge using the roof top rain water. GIS is one of the most emerging and advanced tools for mapping and to develop decision support systems for urban planning. In the present study overlay analysis of various coverages was carried out using weight factor for coverage.

Different maps in the form of coverages viz. land use map, geology, pre monsoon and post monsoon ground water fluctuations for past five years, digital elevation model and rainfall intensities. Various primary, secondary and tertiary parameters have been derived to precisely map the final theme of the study.

By carrying out overlay analysis, a final governing theme was generated which can further be used in decision support systems for generating decisions based on the rainfall intensities and roof top area to predict and design an appropriate method of artificial ground water recharge. This paper also discusses the methodology involved in predicting the optimal depth for ground water recharge for both dug wells and bore wells in an urban setup and also the potential zones in the urban setup for recharge of both ground water table and the deep aquifers.

This paper is restricted to show the possibility and applicability of GIS for generating decision support systems for ground water recharge in urban areas and the generation of coverages to be used in overlay analysis, and the methodology used in such studies. Spatial Decision Support Systems (SDSS) are one of the most advance and user friendly ways for the dissemination information to the users.

Key Words: Ground water recharge, SDSS, Rain Water, Urban area planning, ThematicMap.

Introduction

It is quite evident that the natural ground water resources in the world are depleting at a much greater rate owing to the over exploitation of the source and lesser ground water recharge. Ground water recharge occurs primarily through natural infiltration of the surface water. But the depleting ground water has alarmed the government agencies to look for techniques of recharging the ground water. In most of the countries Artificial Ground Water Recharge is carried out through various well established techniques viz. tube wells, injection wells, dug wells, recharge pits etc.

In a rapidly growing urban infrastructure, the space available for natural recharge of ground water is decreasing at a greater rate forcing the use of artificial techniques for recharge. It is apparent that a large quantity of the rain water is discharged into streams and is let out into the ocean. Rain water is considered to be the primary source of fresh water for most of the world, but not many measures are being taken to harness this, rather pure source of water. But the water crisis in most parts of the world has forced the government agencies to find means to harness this source and hence rain water harvesting techniques has been developed to collect and store this rain water for use.

Not much attention is being paid and action being taken to disseminate this crucial piece of information on ground water recharge to a common man. Since every nation today is taking initiative for rain water harvesting, it is essential that sufficient information has to be provided to the common man about the utility of this rain water for the purpose of ground water recharge and how each individual can contribute for the replenishment of the fast depleting ground water source.

Materials and Methods

Ground water recharge can be best studied on a regional scale rather than a micro level scale owing to the fact that the ground water storage or its flow is not confined to a specific location but varies with varying geology, slope, and other parameters. The primary source for the spatial study is the high resolution LISS IV data of IRS-P6 sensor pertaining to the year 2010 having a spatial resolution of 5m. The Nagpur urban area (falling under the Nagpur Municipal Corporation (NMC) limits) has been demarcated as the study area (Figure 1). Other data that was used for the study are the geology maps of the region which is used to generate the profile of the land, ground water level fluctuations in observation wells for the past five years and the topology maps of NMC.

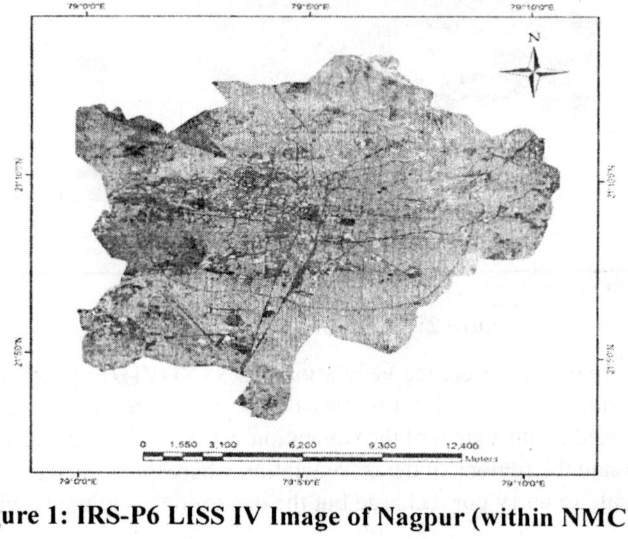

Figure 1: IRS-P6 LISS IV Image of Nagpur (within NMC limits)

The primary task was to project the satellite data to a reference coordinate system, UTM WGS84 44N in this case. Extensive survey of the area was carried out and the precise location of institutions (includes colleges, government offices, public utility and research organizations) was used to digitize them on the image. Land use classification to a large extent governs the economics of the extent and the system / methods to be adopted for carrying out recharge. The study focuses more on this consideration and hence advanced image processing softwares; Erdas Imagine and eCognition Developer were used to precisely classify the image in to the 10 land use classes (Figure 2).

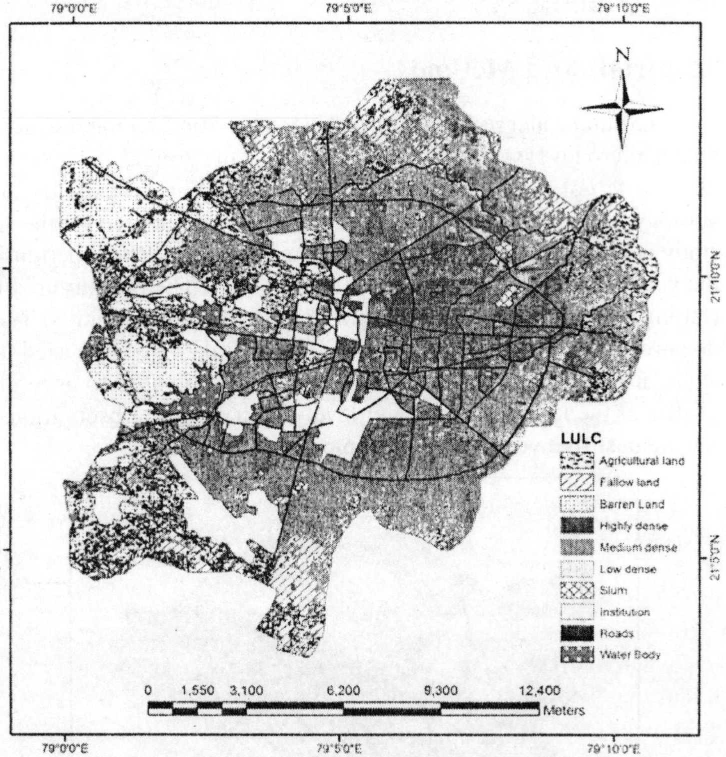

Figure 2: Land-Use Classification

Normalized Difference Vegetation Index (NDVI) (Figure 3) of the LISS IV data was carried out to classify the study area into four basic classes based on the extent of the vegetation in the area. This is necessary to demarcate the regions of dense vegetation where the conditions might be adequate to carry put recharge but the economics may not be viable.

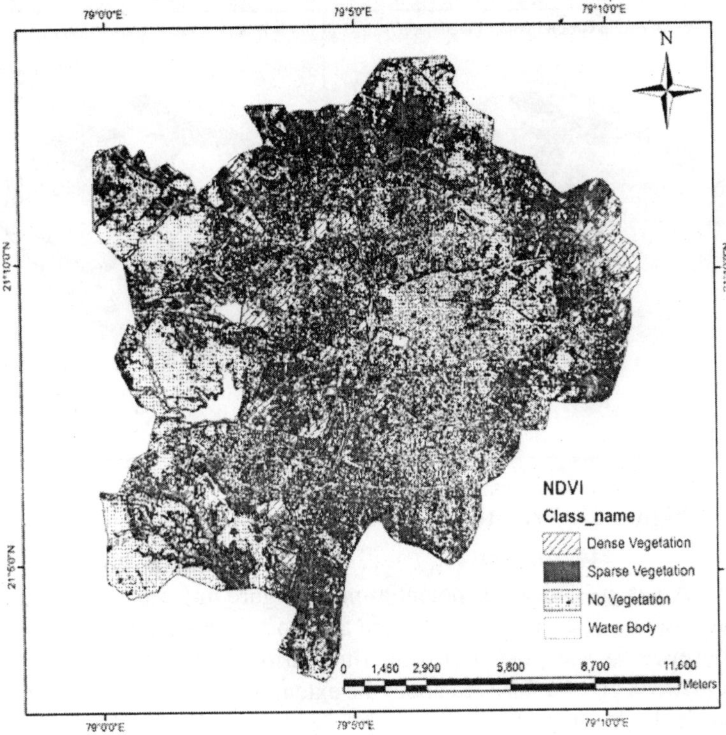

Figure 3: NDVI of Study Area

Digital Elevation Model of the area was generated using the digitized contours of the study area. The geology map of the study area was used to generate the profile of the land. A major part of the study area falls in the Basaltic region. Geologically, vesicular basalt is the zone of ground water or which has the potential to hold ground water. The DEM when overlaid on the profile of the land can be used to obtain the optimal depth to carry out recharge. Based on the information on the depths to the aquifer generated from the overlay of the DEM on the profile of land, slope map can be generated to be used to determine the probable flow of ground water.

The pre-monsoon and the post monsoon ground water levels for the past five years was used to generate ground water fluctuation slope maps (Figure 4a-4b). This map gives a holistic idea about the ground water fluctuations and the potential zones where the recharge can be carried out to yield better recharge results.

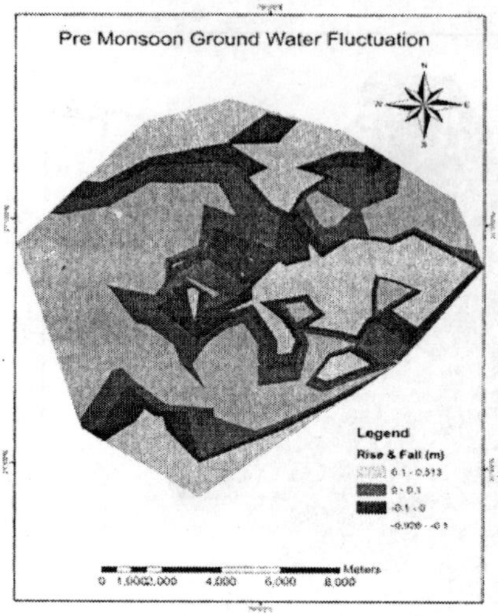

Figure 4a: Pre-Monsoon Fluctuation

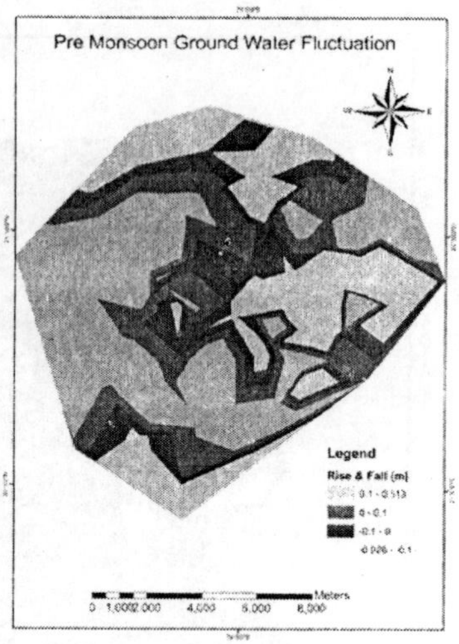

Figure 4b: Post-Monsoon Fluctuation

A net ground water potential map (Figure 4c) was generated by using the net rise or falls in the pre and post monsoon data. A simple difference in the post monsoon fluctuation and the pre monsoon fluctuations gives a fair idea about the extent of recharge taking place at a particular location. Since in a pre-monsoon data the ground water level is usually low and after the rains, i.e. in the post monsoon data, there is an increase in the water level in the observation wells. This increase in water is due to the recharge to the ground water that has taken place during the rains. A difference of these values gives a fair idea about the potential to recharge in that area. A positive value is indicative of very good amount of recharge that has taken place in that area and lesser values show relatively lesser amounts or no recharge.

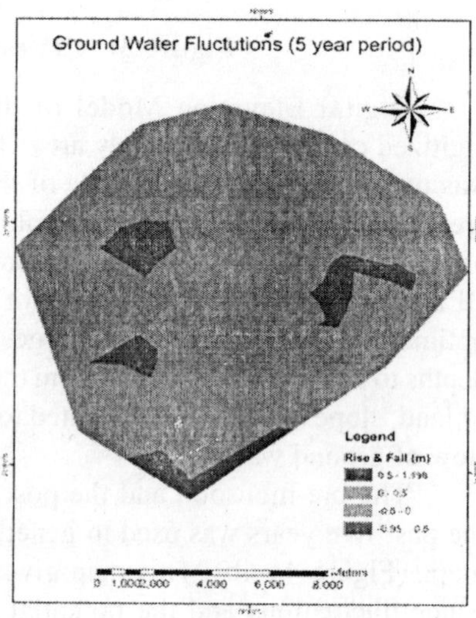

Figure 4c: Net Ground Water Fluctuations

These thematic maps form the basis for further analysis where based on the extent to which each of the themes contributes to the process of recharge, weights are assigned to each of these themes to generate the final thematic map which will be used as a base map for the development of the SDSS. To attain a more realistic result for the study, the weight factor generation can be taken a step ahead by assigning the layer a weight factor and further the classes can be assigned a secondary weight factor depending on the contribution of the feature of the layer for ground water recharge. The generation of weights for various classes was achieved using the Importance Matrix method (Analytical Hierarchical Process developed by T.L. Saaty; 1980). The importance matrix was first generated for assessing the superiority of any individual parameter over other. Here, the theme coefficient was generated which was later used for calculation of the weights of classes within each parameter. Then, for each parameter, a matrix was generated to assess the importance of each class within the thematic parameter over the other classes of the theme. Here, the class relative coefficients were generated. A nine point scale is the basis for the pair wise comparison of the criteria used in the Analytical Hierarchy Process. For each pair of criteria C_{ij}, a value U_{ij} can be found to express the difference in importance between pairs of criteria when $U_{ij} = 1$ for all criteria, and $U_{ij} = 1/U_{ij}$ for each pair under comparison. The U_{ij} values together form a square pair wise comparison matrix. In this method a pairwise comparison matrix was prepared for each map using importance scale method suggested, and this matrix was solved using eigen vector method. In this method, the basic input is the pairwise comparison matrix "A" of order n x n constructed based on Saatys scaling ratios.

$A = [a_{ij}]$ where i, j = 1, 2, 3....... N

The matrix "A" possesses the reciprocal relationship. This is explained as

$A_{ij} = 1/a_{ji}$

Mathematically the equation of matrix can be defined as

$(A - I\,i)\,X = 0$

I = Identity matrix of order n x n

i = Eigen Value

X = Eigen Vector

This resulted in relative coefficients of different classes. The theme coefficients were also calculated by the same method. These weights were multiplied with theme coefficients and applied to linear summation equation

$$_{1}^{m}W = \sum (x_i y_i)$$

W = Final weight for particular class within a theme for final overlay at a location

Xi = relative coefficient for class (Rcf)

Yi = theme coefficient (Tcf)

m = total no. of themes

The weights for various classes and themes were derived using importance matrix as shown below (Table 3). Weights for various individual classes in a theme were also generated. The weights were then converted into a percentage scale to be used in the overlay analysis.

Table 1

Weights for Governing Themes

	gwtpot	geo	slope	lulc	ndvi	Weight
Gwtpot	1	5	7	9	9	0.521885
Geo	0.2	1	4	8	9	0.23961
slope	0.142857	0.25	1	7	6	0.134297
Lulc	0.111111	0.125	0.142857	1	8	0.076819
NDVI	0.111111	0.111111	0.166667	0.125	1	0.027388

Gwtpot – Ground Water Potential

Geo – geology

Lulc – Land use – Land cover

NDVI – Normalized Difference Vegetation Index

The methodology involved in generation of the base thematic map for the use in an SDSS is generated using weighted overlay analysis. One of the most widely used GIS software; ArcGIS was used to carry out this analysis. A common spatial analysis query is to identify the suitability of each cell location relative to specific criteria. The criteria can be economics, preferences, or risks. Is the specific location suitable to carry out recharge? Or which method is economically viable for recharge at the specific location?; Questions such as these are answered by the Suitability models. The Weighted Overlay generates a suitability map based on the inputs.

Overlay analysis combines multiple attribute fields in a table using a common measurement scale and weights each field according to its importance. Generally the weighted overlay analysis uses raster datasets which have been reclassified into integers from float points (if any). In the present study weighted overlay of the attribute data was performed since this method yields realistic results as it uses the classes for the analysis which forms the basis for deciding the method of recharge. Every class in the attribute table and the ranges of slopes in case of the slope maps has been prioritized based on their contributions to recharge. For example in the land-use map, the classes have been assigned a value from 1- 9 with the institutions having the highest factor 9 and slums the least. Similar prioritization of the classes has been done for other thematic maps.

In the suitability map thus generated, higher values indicate that a location is more suitable. The scale has to be understood and the same scale should be used for all the maps in the input to avoid ambiguous outputs.

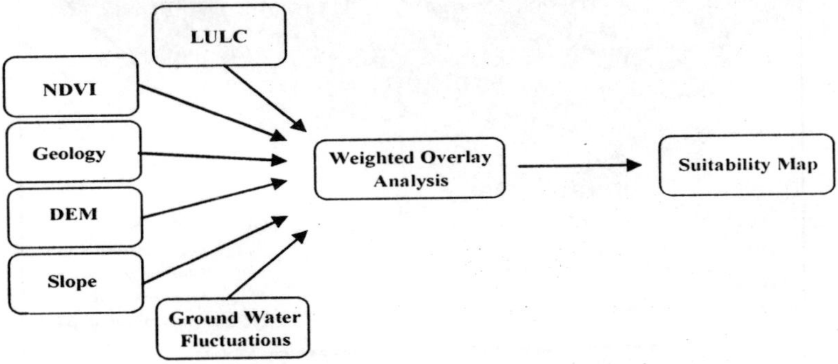

Figure 5: Weighted Overlay Analysis Procedure Flow Chart

Results and Discussion

The suitability map generated from the weighted overlay analysis (Figure 6) shows the feasibility of ground water recharge in the urban setup of Nagpur city. The analysis has been carried out using the parameters that best affects recharge in an urban setup. The landuse classes taken into consideration plays a vital role in the process due to the lack of open space required for carrying out recharge.

The map has been classified into four categories based on the generated output. These four classes are classified on a regional scale into highly potential zones, medium high, medium and low potential zones. The classification is based on the scheme that the difference of the post monsoon data and the pre monsoon gives a general idea about the extent of recharge taking place at a particular location due to natural recharge after monsoon season. The area where there is a net fall of more than 50 cm can be best augmented with water using artificial ground water recharge techniques. in these areas recharging the aquifers at lower levels by means of dug wells is more preferable. The areas showing a net rise in

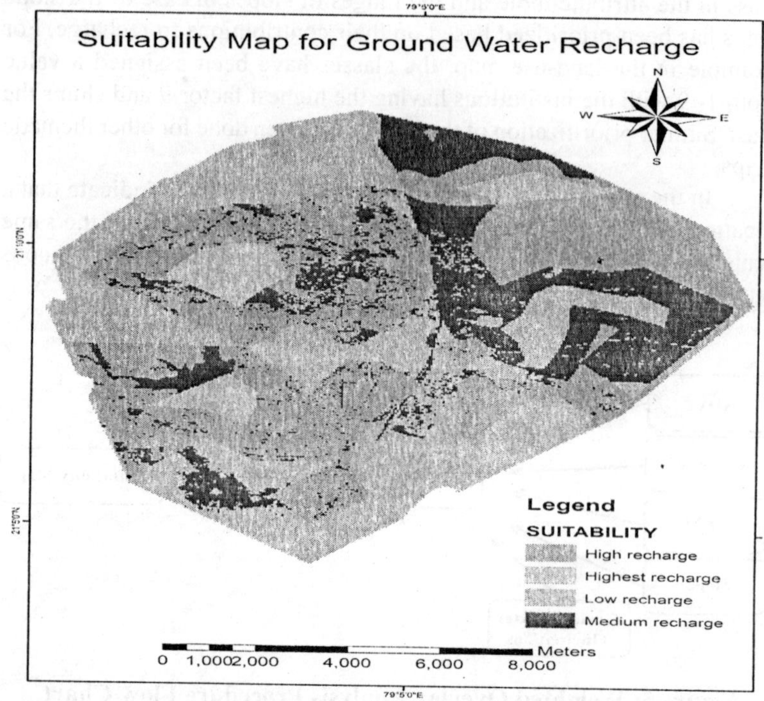

Figure 6: Suitability Map Generated from Weighted Overlay Analysis

the water level indicates recharge occurring through natural system. In such areas it is quite feasible and preferable to recharge the deeper aquifers.

The suitability map has been generated taking into consideration the factors governing recharge in an urban setup. It is very essential to note that the ground water recharge scenario varies a lot on a regional scale and the factors that affect recharge are numerous. The recharge depends largely on the lineaments, type of geology, weathering depth, recharge scenarios in specific areas etc. In this study only a few parameters are taken into consideration because the purpose of the study was to give a holistic approach to the study on a regional scale. It is the intent of the study that this vital piece of information can be used on by people to take initiatives to carry out recharge at an individual level. Such a regional study gives quite satisfying results as far as recharge scenario is concerned because the weights were derived using importance matrix and the highest weight was assigned to the net ground water rise – fall data (this tells the information about the extent of recharge). This gives a fair idea about the extent of recharge that can be done in areas where there is net fall in ground water level so as to augment the resource and the areas showing net rise in ground water level are the areas where natural recharge is occurring after the monsoon and hence in these localities the deeper aquifers can be augmented. This map becomes the basis for the SDSS.

Spatial Decision Support System (SDSS) refers to those systems which are based on the use of GIS technology. GIS provides an important source of tools and techniques which can usefully be incorporated in a Decision Support System (DSS) which makes use of geographic or spatial data. SDSS is an interactive, computer-based system designed to support a user or group of users in achieving a higher effectiveness of decision making while solving a semi-structured spatial decision problems.

As mentioned earlier in the paper, the ground water resources are depleting at a faster rate. The need for recharge of this valuable source of water is essential. Rain water is one of the purest forms of water which is available and a large quantity of this rain water is left to flow into drains. Such water can be augmented and put to use to recharge aquifers on a larger scale by government agencies. Every person should feel responsible and take initiative to be a part of this system to replenish the reserves. The development of SDSS which will basically be a web-based GIS system is an effective way of disseminating information to the general public. Artificial recharge techniques have been developed to address these needs. A SDSS could provide scientific information to the common man about ground water recharge. In an urban setup, the availability of open space is scarce and hence the techniques of artificial recharge viz. dug wells and tube wells are best suited techniques which could be effectively used. Based on the information of the average precipitation in an area and the total precipitation water that could be collected from a roof top could be used as additional inputs in a SDSS which, based on the suitability map could analyze and suggest recharge technique that can be most effectively put into use at a specific location.

The design and development of such a SDSS and the methods to effectively incorporate the suitability map in the SDSS is beyond the scope of this paper. Extensive research and development has been carried out by the authors to develop SDSS to address the specific needs of ground water recharge in an urban setup.

Conclusion

From the generation of the thematic maps that have been used in the study, it is quite evident that Remote Sensing and GIS offer a promising spatial approach to find solutions to ground water recharge related issues in an urban area. Many techniques have been developed and put into practice by the government agencies but enough is not being done to check the over exploitation of the ground water resource. The exploitation, since it difficult to check in an ever increasing urban area, can be countered by the recharge of the aquifers. Municipal bodies can enforce directives to incorporate rainwater harvesting methods and recharge techniques before approving any new constructions. A SDSS can prove to be a very effective tool for such agencies. Well established colonies can also be directed to setup common rain water harvesting sumps and carry out recharge in those localities. The ground water resource has to be replenished and preserved. The partial or complete depletion of it could lead to dire consequences. Every person or group of persons collectively should take initiatives to preserve the most precious source of pure water, the Ground water.

References

- Ground Water Assessment, Development and Management; K R Karanth (Tata McGraw Hill Education Privte Ltd.)
- Ground Water Hydrology; David Keith Todd (Wiley)
- Decision Support Systems in the 21st Century; George M. Marakas (Pearson Education)
- Concepts and Techniques of Geographic Information System; C.P.Lo & Albert K.W. Yeung (Prentice Hall India)
- Geographic Information Systems and Science; Paul A. Langley et al. (Wiley)
- Fundamentals of Remote Sensing; George Joseph (Universities Press)
- Remote Perceiving: A Step Toward a Unified Science of Remote Sensing; Robert R. Hoffman
- Decision by Objectives; Ernest H. Forman, Virginia Tech
- Analytical Heirarchy Process As A Tool for Selecting and Evaluating Projects; Palcic I. (University of Maribor) & Lalic B. (University of Novi Sad)
- Comparative Overlay Analysis Through Analytical Hierarchical Process to Delineate Groundwater Potential Zones Using Satellite Data; Yashwant B. Katpatal and Y.A. Dube, International Journal of Earth Sciences and Engineering

- Dissertation report on Designing Spatial Decision Support System for generating Ground Water Recharge Scenario in an Urban Area; Guided by Dr. Y. B. Katpatal, VNIT
- Manual on Artificial Recharge of Ground Water; Central Ground Water Board (CGWB) (Govt. of India)
- Rain Water Harvesting and Artificial Recharge to Ground Water; CGWB (Govt. of India) & International Hydrological Programme (UNESCO)
- Ground Water Assessment, Development and Management; K R Karanth, Tata McGraw Hill Education Private Ltd.
- Ground Water Hydrology; David Keith Todd, John Wiley & Sons.

14. Morphometric Analysis of Four Sub-Watersheds in Bah Area of Agra District Using Remote Sensing and GIS Techniques

A. K. Rao,
Zia-ur-Rahman,
Mohammad Sadique and Alia Yusuf

Morphometric analysis of four sub-watersheds namely Pidhaura, Batesar, Balapur and Pariar forming parts of Yamuna River around Bah Tahsil of Agra district have been carried out using Geocoded FCC of bands 2, 3 and 4 of IRS-1D, LISS-III, SOI toposheets number 54 J/5 and 54 J/9 and GIS softwares - ArcGIS, Arcview and 3Dem software for preparation of DEM. The present study shows that the terrain exhibits dendritic to sub-dendritic drainage pattern, stream order ranges from third to fourth order; drainage density varies slightly and has very coarse to coarse texture in Pidhaura and Balarpur sub-watersheds and fine drainage texture in Batesar and Pariar sub-watersheds. The other parameters determined and discussed include bifurcation ratio, elongation ratio, stream frequency, infiltration number, length of overland flow and relief aspects of the watersheds. The mean bifurcation ratio depicts that the basin falls under normal basin category. The elongation ratio shows that Balarpur sub-watershed mark elongation pattern whereas Pidhaura, Batesar and Pariar sub-watersheds possess circular shape. The present study concludes that remote sensing techniques proved to be a competent tool in morphometric analysis of drainage basin and channel network.

Keywords: Morphometry, Drainage Characteristics, Sub-watershed, Agra.

Introduction

The morphometric analysis of the drainage basin and channel network play a vital role in order to understand the hydro-geological behavior of drainage basin and expresses the prevailing climate, geology, geomorphology and structure etc. The relationship between various drainage parameters and the aforesaid factors are well recognized by Horton, 1945, Strahler, 1957, Melton, 1958, Pakhmode, *et al.*, 2003 and Gangalakunta, *et al.*, 2004. Recently many workers have used remote sensing data and GIS generated more precise data on morphometric parameters (Srivastava, 1997, Agarwal, 1998, Nag, 1998, Das and Mukherjee, 2005) and concluded that remote sensing has emerged as a powerful tool and useful in analyzing the drainage morphometry. The objective of the present study is to analyze the morphometric attributes of Yamuna river basin for the first time and no systematic work has been carried out so far around Bah *Tahsil* of Agra district.

Bah *Tahsil* is a part of Indo-Gangetic plains lies between the Yamuna and Chambal Rivers in the south-eastern part of Agra, located at a distance of 70 km from Agra city and about 55 km from Etawah district (Fig.1). The major part of the area is covered by Indo- Gangetic alluvial deposit of the Quaternary period, overlying the rocks, belonging to the Vindhyan Super group. The study area is included in the Survey of India toposheet number 54 J/5 and 54 J/9 between the parallels of 27^0 00′ and 26^0 50′N latitudes and 78^0 25′ and 78^0 45′E meridians of longitudes.

Physiographically, Agra district is divided into five distinct regions viz. Khadar lowland, Trans-Yamuna plain, Yamuna upland, South-West Upland and Yamuna Chambal Ravines. The study area is having a long irregular land, narrow towards the extremities and fairly wide in the centre. Its physical character differs greatly from the others. The level land between Yamuna and Chambal consists of mere ridges and narrow strips flanked on either side by ravines leading to the river.

The climate of the area is semi-arid and characterized by hot summer and temperature reaches as high as 47 °C during summers (April to June). The monsoon months (July to September) receive about 69 cm of rainfall annually. During winters (November to February) Agra district goes through comfortably warm, but temperatures below freezing point is also noted at many places during the night. Agra is also prone to dense fog during the winter months of December and January.

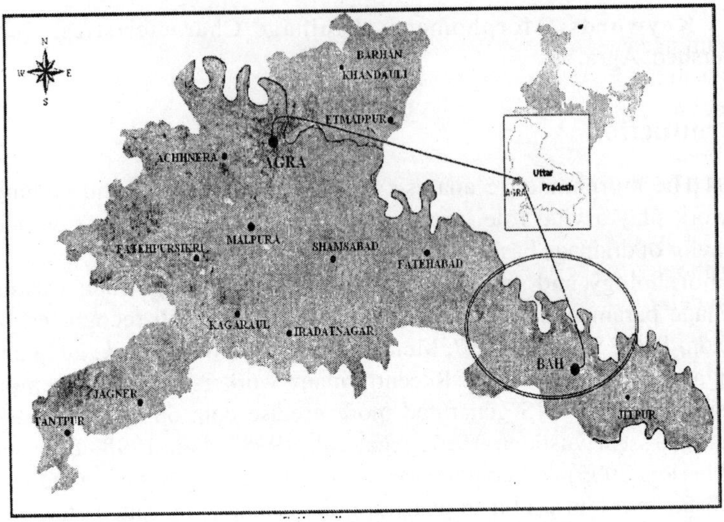

Fig-1 Location Map

Geologically the area is characterized by alluvium which is an admixture of gravel, sand, silt and clay in various proportions, deposited during the Quaternary period. The rock units belonging to the Vindhyan Super group represented by the upper Bhander and lower Rewa sandstones exposed in the west and southwest of Agra. The top soil is coarse and angular sand with small fraction of clay. However, in the study area the soil is fine loam in central part, sandy in the northern part and clayey in the southern part.

Methodology

The drainage map (Fig.3) of the study area is prepared using Geocoded FCC of bands 2, 3 and 4 of IRS-1D, LISS-III, March, 2003 and SOI toposheets on 1: 50,000 scale with sufficient ground truth. ArcView 3.2 software has been used for digitizing, morphometric analysis and to generate out put. Perimeter and area of the sub-watersheds are also calculated and cross checked from the online measurement tool and satellite imagery available on ISRO Bhuvan webpage (http://bhuvan2.nrsc.gov.in/) by overlaying the shape file (.shp) of the study area prepared in ArcGIS 9.2. The methodology for the computation of morphometric parameters is given in Table-1 and the results of the analysis summarized in details are given in Table-2. SRTM (Shuttle Radar Topographic Mission) 90m resolution data (freely available on *http://srtm.csir.cgiar.org/*) is processed in 3Dem and SAGA 2.0 and is used for the preparation of Digital Elevation Model to supplement the analysis (Fig. 3).

Morphometric Analysis

The present study integrated the use of remote sensing in morphometric analysis and the results of morphometric parameters have been determined are discussed and briefly described. The study of basin morphometry attempts to relate basin and stream network geometries to the transmission of water and sediment through the basin. The size of a drainage basin acts upon the amount of water yield, the length, shape and relief, affect the rate at which water is discharged from the basin and total yield of sediments, the length and character for stream transport and rate at which water and sediments are discharged.

In present study the morphometric analysis is carried out using mathematical formulae given in Table-1 and the results are summarized in Table-2.

Linear Aspects

The linear aspects determined include stream order, stream length, mean stream length stream length ratio and bifurcation ratio and the results of the analysis are given in Table -1and discussed below.

Stream Order

The designation of stream orders is the first step in drainage basin analysis and expresses the hierarchical relationship between segments. In the present study the number of stream ordering is carried out using the method given by Strahler (1964).

The order wise stream numbers, area and length of the four sub-watersheds are counted and presented in the Table - 2; indicate that Pidhaura and Batesar sub-watersheds are of second order while Balarpur and Pariar sub-watersheds are of fourth order.

The observation of stream order verifies the Horton's Law (1945) of stream number i.e. the number of stream segments of each order forms an inverse geometric sequence with order number geometric relationship is shown graphically in the form of straight line when the log value of these variables (stream order and stream number) and (stream order and stream length) on semi log plot (Fig.2a and Fig.2b).

Table- 1

Methodology adopted for the computation of morphometric parameters

S.No.	Morphometric Parameters	Formula/Definition
1	Stream order	Hierarchical Rank
2	Bifurcation Ratio (Rb)	**Rb** = **Nu / Nu+1** Where, **Nu**=Number of stream segments present in the given order **Nu+1**= Number of segments of the next higher order
3	Mean Bifurcation Ratio (Rbm)	**Rbm** = Average of bifurcation ratios of all orders
4	Stream Length (Lu)	Length of the Stream (km)
5	Mean Stream Length (Lsm)	**Lsm = Lu / Nu,** km Where, **Lu**=Mean stream length of a given order (km) **Nu**= Number of stream segments
7	Drainage Density (D)	**D= ∑Lu /Au)** km/km^2 Where,**Lu**=Total Stream length of all orders (km) **Au**=Area of the Basin (km^2)
8	Drainage Texture(Rt)	**Rt = ∑ Nu/P** Where, **Nu=** Stream Number **P** = Perimeter (km)
9	Stream Frequency (Fs)	**Fs** = **∑ Nu /Au** Where, **Nu**=Total number of streams in the basin **Au**= Basin Area (km^2)
10	Length of Over Land Flow (Lg)	**Lg = 1/ D×2** Km Where, **D** = Drainage density (km/km^2)
11	Form Factor (Rf)	**Rf = Au / Lb^2** Where, **Au**=Area of the Basin (km^2) **Lb**=Maximum Basin length (km)
12	Circularity Ratio (Rc)	**Rc = 4πAu/ P^2** Where, **Au**= Basin Area (km^2) **P**= Perimeter of the basin (km) **Π** = 3.14
13	Elongation Ratio (Re)	**Re= √Au/π / Lb** Where, **Au**= Area of the Basin (km^2) **Lb**=Maximum Basin length (km) **Π** = 3.14
14	Relief Ratio (Rh)	**Rh = H / Lbmax** Where, **H** = Maximum basin relief (km) **Lbmax**= Maximum basin length (km)
15	Ruggedness Number (HD)	**HD= H×Dd** Where, **H**= Maximum basin relief **Dd**= Drainage density
16	Relative Relief (Rhp)	**Rhp = H× (100) / P** Where, **H** = Maximum basin relief **P** = Perimeter of the basin (km)

Stream Length (Lu)

The stream length is computed based on the law proposed by Horton (1945) for all the counted and their lengths are measured. In the study area, total length of stream segments is maximum for first order and decreases as stream order increases. However in all four sub-watersheds the stream segments of various order, varies considerably (Table- 2).

Bifurcation Ratio (Rb)

The bifurcation ratio is the ratio of the number of stream segments of given order to the number of segments of next higher order. Horton (1945) and Schumm (1956) considered the bifurcation ratio as an index of relief and dissection. It has been found that the bifurcation ratios characteristically range between 3.0 and 5.0 for watershed in which geology is reasonably homogeneous and not suffered structural disturbances.

Bifurcation ratio for different sub-basins of the study area have been determined and tabulated in Table – 2. Semi log plots of stream order vs. stream number have been drawn and a straight line was fitted through these points (Fig.2a). The irregularities in the bifurcation ratios of the four sub-watersheds are possibly dependent upon the drainage basin (Strahler, 1952). The lower values of Rb show that the sub-watershed has suffered less structural disturbances (Strahler, 1964) and the drainage pattern is not distorted. Furthermore the low Rb values signify high drainage density and indicate areas with uniform surficial materials where geology is reasonably homogeneous. The mean bifurcation ratio, which is the average of bifurcation ratios of all orders is, calculated which varies from 4.48 to 5.80 (Table - 2).

Mean Stream Length (Lsm)

The mean stream length is calculated by dividing the total stream length of order 'u' and number of stream of segment of order 'u' (Table – 2 and Fig.2b). In the study area it is noted that Lsm varies from 0.37 to 4.55 km and Lsm of any given order is greater than that of the lower order and less than that of its next higher order in all the sub-watersheds except Balarpur sub-watershed which is possibly due to variation in the slope and topography.

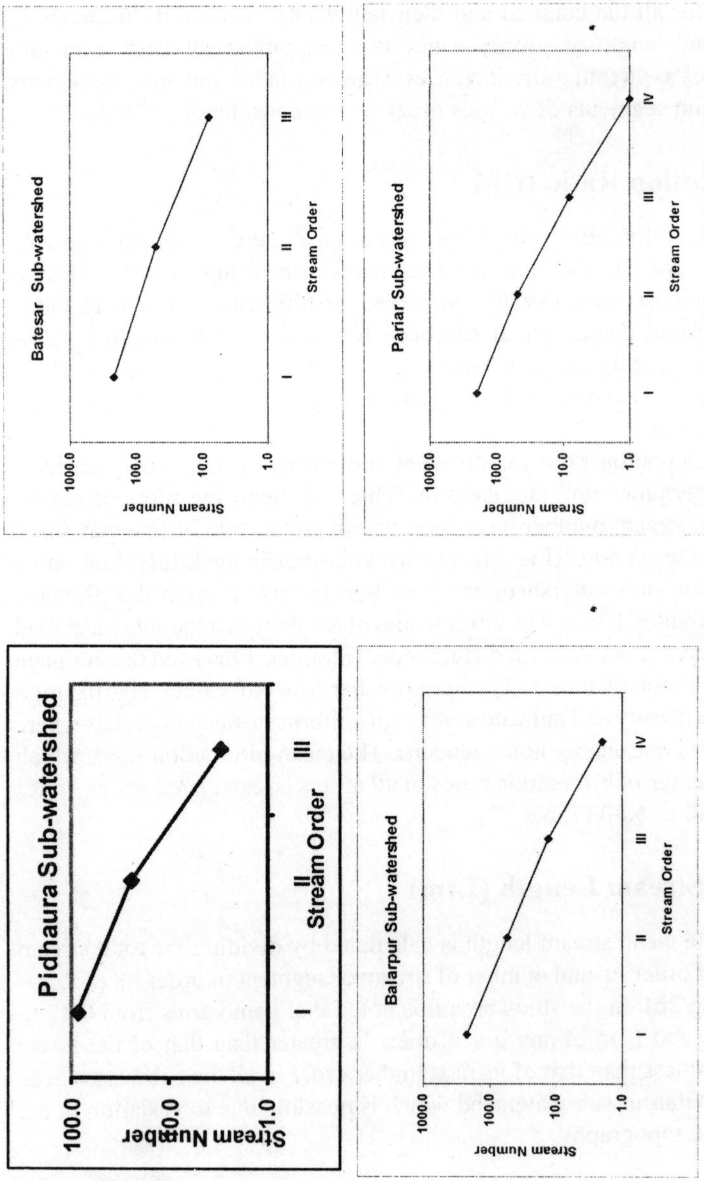

Fig.2a Semi Log Plots of Stream Order Vs Stream Number

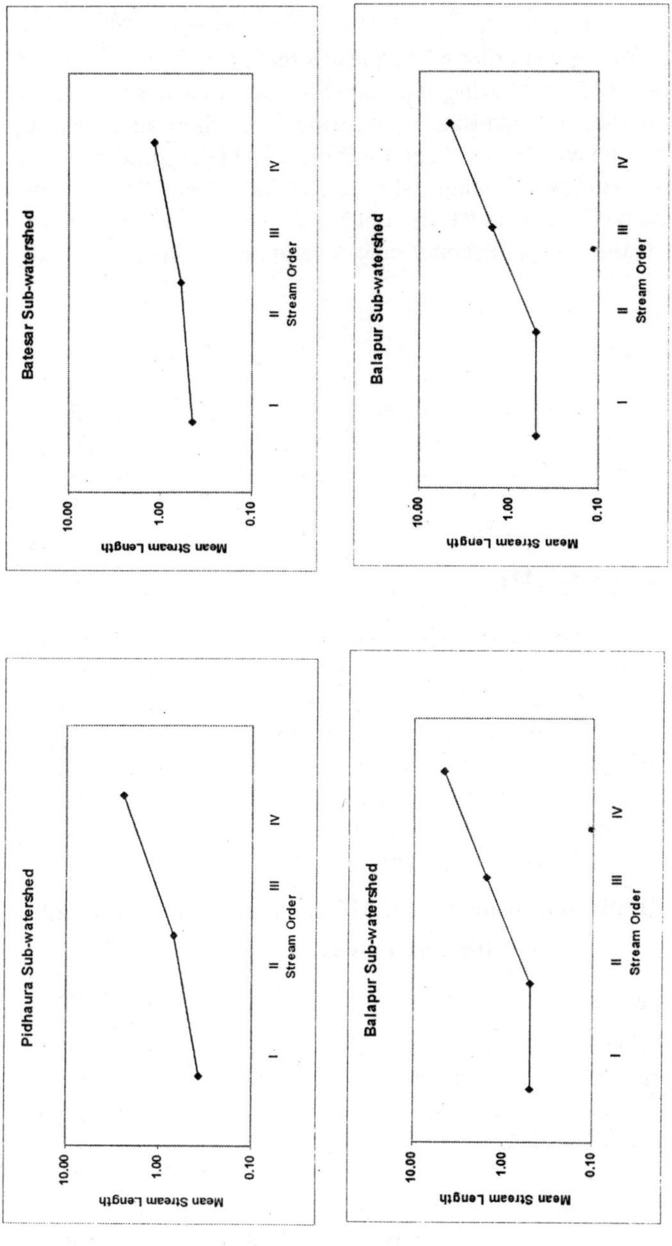

Fig.2b Semi – Log Plots of Stream Order Vs Mean Stream Length

Stream length Ratio (RL)

The stream length ratio can be defined as the ratio of the mean stream length of a given order to the mean stream length of next lower order (Horton, 1945) and having important relationship with surface flow and discharge. Pidhaura and Batesar sub-watersheds show an increasing trend in the stream length ratio from lower order to higher order indicate their mature geomorphic stage whereas, Balarpur and Pariar sub-watersheds show changes in RL from one order to another, indicate the late youth to mature stage of geomorphic development (Singh and Singh, 1997).

Aerial Aspects

The aerial aspects which were included, drainage density, drainage texture, stream frequency, form factor circularity ratio, elongation ratio and length of overland flow. The results are given in Table-2 and discussed in details.

Drainage Density (D)

Horton (1932) defines drainage density as the total length of stream of all order/drainage area and may be an expression of the closeness of spacing of channels. Langbein (1947) suggested that drainage density varies between 0.55 and 2.09 km/km^2 in humid region with an average density of 1.03 km/km^2, where high drainage densities usually reduce the discharge in any single stream more evenly distributing runoff and speeding runoff into secondary and tertiary streams. The low drainage density is favoured in regions of highly

Table - 2 Results of Parameters of Morphometric analysis of all the watersheds

Stream Order	Pidhaura		Batesar		Balarpur		Pariar	
	StreamNo	Bifurcation Ratio (Rb)	Stream No.	Bifurcation Ratio (Rb)	Stream No.	Bifurcation Ratio (Rb)	Stream No.	Bifurcation Ratio (Rb)
First	88.0	3.2	224.0	4.3	224.0	4.2	202.0	4.1
Second	25.0	8.3	50.0	6.0	53.0	3.7	48.0	5.3
Third	3.0	-	8.0	-	13.0	6.5	8.0	4.0
Fourth	-	-	-	-	2.0	-	1.0	-
Mean Rb (Rbm)		5.8		5.16		4.79		4.48

Stream Order	Pidhaura			Batesar			Balarpur			Pariar		
	Stream Length (km)	Mean Stream Length (km) (Lsm)	Stream Length Ratio (RL)	Stream Length (km)	Mean Stream Length (km) (Lsm)	Stream Length Ratio (RL	Stream Length (km)	Mean Stream Length (km) (Lsm)	Stream Length Ratio (RL)	Stream Length (km)	Mean Stream Length (km) (Lsm)	Stream Length Ratio (RL)
First	32.50	0.37	1.86	99.75	0.44	1.34	112.62	0.50	0.98	93.00	0.46	1.41
Second	17.25	0.69	3.49	29.60	0.59	1.95	26.15	0.49	3.10	31.00	0.65	1.72
Third	7.25	2.41	-	9.25	1.15	-	20.00	1.53	2.97	8.95	1.12	1.69
Fourth	-	-	-	-	-	-	9.10	4.55	-	1.90	1.90	-

S. No	Name of Watershed	Area (Au) (km^2)	Maximum Basin Length (Lb) (km)	Basin Perimeter (P) (km)	Elongation Ratio (Re)	Circularity Ratio (Rc)	Form Factor (Rf)
1	Pidhaura	44.00	10.00	33.25	0.74	0.49	0.44
2	Bateswar	84.62	12.40	34.50	0.84	0.89	0.55
3	Balarpur	136.38	19.00	60.00	0.69	0.48	0.38
4	Pariar	81.93	12.25	38.25	0.83	0.70	0.55

S. No.	Name of the Watershed	Drainage Density (Dd)	Stream Frequency (Fs)	Infiltration Number (If)	Length of Overland Flow (Lg)
1	Pidhaura	1.29	2.63	3.39	0.39
2	Bateswar	1.63	3.33	5.43	0.31
3	Balarpur	1.23	2.14	2.63	0.40
4	Pariar	1.65	3.16	5.22	0.30

Name of Watersheds	Max. Basin Relief (km)	Relief Ratio (m)	Relative Relief (Rhp)	Ruggedness Number (HD)
Pidhaura	0.017	0.0017	0.051	0.022
Bateswar	0.022	0.0018	0.064	0.036
Balarpur	0.02	0.0011	0.033	0.025
Pariar	0.038	0.0031	0.081	0.063

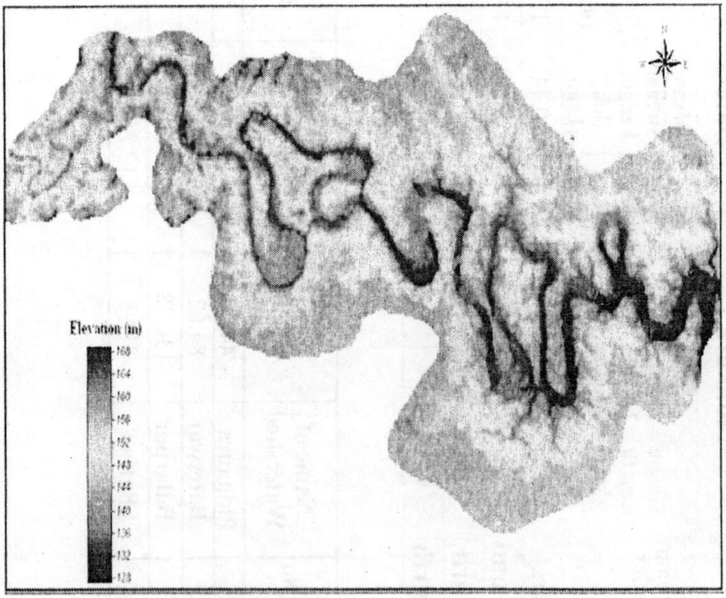

Fig.3 Digital Elevation Model of the study area

permeable subsoil material, under dense vegetation cover and where relief is low while high drainage density is favoured in regions of weak or impermeable subsurface materials, sparse vegetation and mountain relief (Chow, 1964). The low drainage density is also indicative of relatively long overland flow of surface water. The drainage density varies from 1.29 to 1.65 km/km^2, indicates clearly that the region has highly permeable subsoil, dense vegetation cover and low relief.

Drainage Texture (Rt)

Horton (1945) defines drainage texture as the total number of segments of all order per perimeter of that area. Smith (1950) classified drainage density into five different textures. The drainage density less than 2 indicates very coarse, between 2 and 4 is related to coarse, between 4 and 6 is moderate, between 6 and 8 is fine and greater than 8 is very fine drainage. According to this classification drainage texture of Pidhura and Balarpur sub-watersheds having values 3.48 and 5.54 respectively falls under very coarse to coarse drainage texture category while Baetsar and Pariar sub-watersheds having values 6.77 and 8.17 respectively (Table- 2). In the present study it is found that the drainage density is variable and suggests very coarse to coarse and fine drainage texture.

Stream Frequency (Fs)

Horton (1932) defined stream frequency as the total number of stream segment of all order per unit area. Table-2 shows that stream frequency for all four sub-watersheds of the study area show positive correlation with the drainage density indicate that stream population increases with the increase of drainage density in all four sub-watersheds.

Form Factor (Rf)

Horton (1932) define form factor as the ratio of basin area to square of the basin length and is a quantitative expression of drainage basin out line. The Rf values (Table- 2) in the area of interest varies from 0.38 to 0.55 in all the sub-watersheds indicate that the sub-watersheds are circular in shape.

Circularity Ratio (Rc)

Miller (1953) define circularity ratio as the ratio of the basin area to the area of a circle having the same circumference perimeter as the basin. Circularity ratio is dimensionless and expresses the degree of circularity of the basin. The Rc values in the Pidhura and Balarpur vary between 0.48 and 0.49, indicate that these watersheds are more or less elongated in shape, whereas Baetsar and Pariar, which show Rc values 0.71 and 0.89 respectively indicate that these are more or less circular Fig.3 (b) and are characterized by moderate to low relief and drainage system seems to be partially controlled by the structural disturbances (Table-2). This also indicates the tendency of small drainage basin in homogeneous geologic materials to preserve geometrical similarity.

Elongation Ratio (Re)

Schumm (1956) defined elongation ratio as the ratio of diameter of a circle of the same area as the drainage basin and the maximum length of the basin. A circular basin appears more efficient in the discharge of run-off than that of an elongated basin (Singh and Singh 1997). The Re values generally ranges between 0.6 and 1.0 over a wide variety of climatic and geologic types. Values near to 1.0 are the characteristics of the region of very low relief while values in the range of 0.6-0.8 are usually occur in the areas of high relief and steep ground slope (Strahler, 1964). These values are further categorized as (a) circular (>0.9), (b) oval (0.9-0.8), (c) less elongated (<0.7). The Re values in the study area vary between 0.69 and 0.84 (Table- 2), which indicates moderate to slightly steep ground slope. The lowest value in case of Balrapur sub-watershed indicate high relief and steep slope whereas sub-watersheds Batesar and Pariar having values 0.84 and 0.83 respectively indicate moderate to low relief and low slope.

Length of Overland Flow (Lg)

Horton (1945) defined length of overland flow as the length of flow path, projected to the horizontal of non channel flow from point on the drainage divide to a point on the adjacent stream channel. He further noted that Lg is one of the most important independent variable affecting both hydrologic and physiographic development of drainage basins. The length of overland flow is approximately equal to the half of the reciprocal of drainage density. This factor basically relates inversely to the average slope of the channel and is quiet synonymous with the length of sheet flow to a large degree. Table - 2 reveals that Lg is Baetsar and Balapur watersheds is less with high drainage density when compared with Pidhura

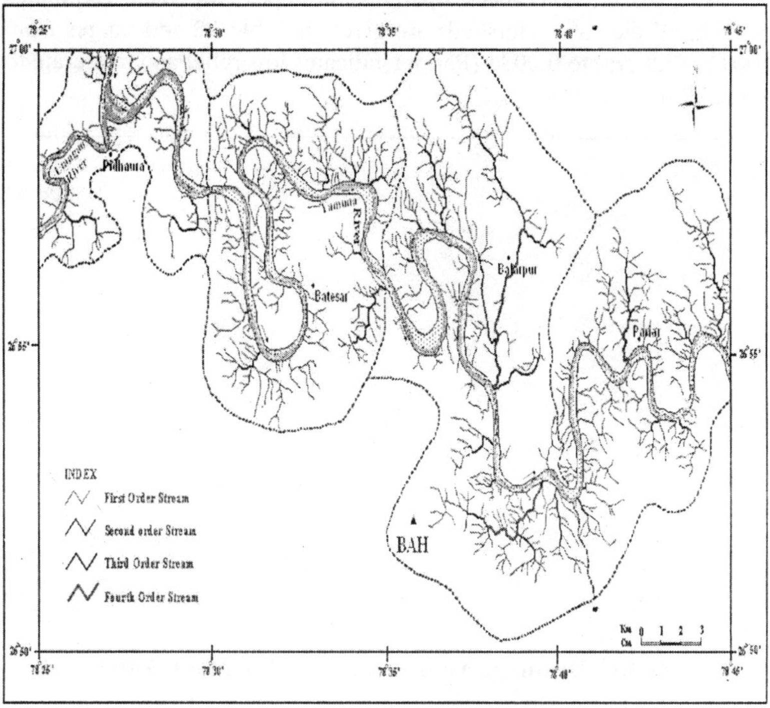

Figure- 4(a) Drainage Map of the Study Area

and Pariar watersheds. The computed value of all 4 watersheds varies from 0.30 to 0.40.

Relief Aspects

The relief aspects determined include, relief ratio, relative relief and ruggedness number and the results of the analysis are given in Table – 2.

Relief Ratio (Rh)

According to Schumm (1954) the relief ratio is obtained when basin relief 'H' is divided by the maximum basin length (Lb), results in a dimensionless ratio which is equal to the tangent of the angle formed by two planes intersecting at the mouth of the basin called relief ratio measures the overall steepness of a drainage basin and is an indicator of the intensity of erosion process operating on slope of the basin. The values of Rh in all the sub-watersheds are given in Table - 2 and ranges from 0.0017 (Pidhura) to 0.0031 (Pariar) indicates low relief and moderate to gentle slope.

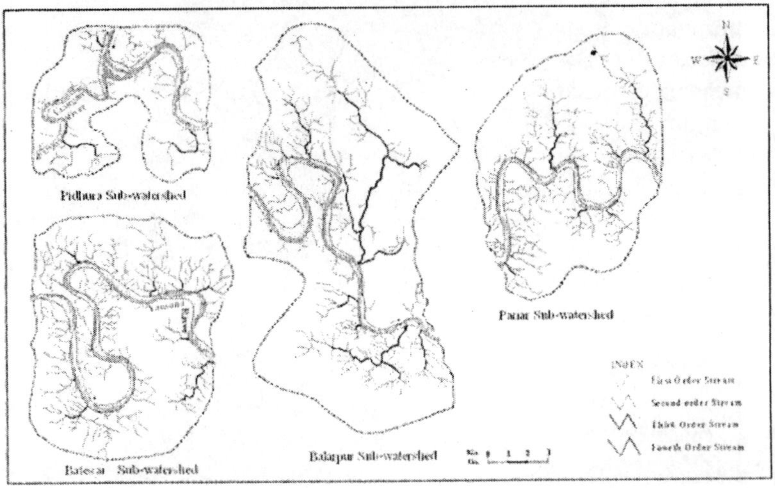

Figure- 4(b) Drainage Map Showing Splited Sun Watershed

Relative Relief (Rhp)

This term was given by Melton (1957) and in the present study it is noted that Balarpur sun-watershed has maximum relative relief while Pariar has the minimum value. Furthermore, visual analysis of the digital elevation model prepared from the SRTM data (Fig.2) indicate that the elevation varies from 128m to 167m which represent the land surface has gentle to moderate slope.

Ruggedness Number (HD)

It is the product of maximum basin relief (H) and drainage density (D), where both parameters are in the same unit. Extreme high values of ruggedness number occur when both variables are large, when slope is not only steep but long as well (Strahler, 1958). In the present study the value of ruggedness number (Table - 2) is low, indicates gentle slope of all four sub-watersheds.

Conclusion

The present study demonstrates that remote sensing techniques and GIS play a vital role for the preparation of updated drainage map in a timely and cost-effective manner and morphometric analysis. Detailed morphometric study of all sub-watersheds represents dendritic to sub-dendritic drainage pattern, indicating homogenous lithology and variations in the values of *Rb* among the sub-watersheds attributed to the difference in topography and geometric development. The *Fs* values for all sub-watersheds demonstrate a positive correlation with the *Dd* values suggesting an increase in the stream population with respect to increase in drainage density whereas the values of *Dd* show that the region has highly permeable subsoil, dense vegetation cover and low relief. Drainage texture varies from very coarse to coarse to fine and shape parameters of the sub-watersheds viz; *Re*, *Rc* and *Rf* values indicate that the Pidhura and Balarpur sub-watersheds are more or less elongated whereas, Baetsar and Pariar, are more or less circular. Relief aspects and visual interpretation of DEM of the study area indicate gentle to moderate slope, low run off and high infiltrations. Moreover, overlaying the shape file of the study area from the satellite imagery available on recently developed ISRO Bhuvan webpage is proved to be efficient technique for calculating some of the parameters of areal aspects using the measurement tool available on the same webpage. Furthermore, this study suggests that systematic analysis of morphometric parameters within drainage networks using a GIS can provide substantial value in understanding sub-watersheds' drainage characteristics and hydro-geological behavior of the area.

References

- Agarwal CS (1998) Study of drainage pattern through aerial data in Nahagarh area of varansi district, U.P. *J. Indian Soc. Remote Sensing* 26(4):169 – 175
- Das AK and Mukhrjee S (2005) Drainage morphometry using satellite data and GIS in Raigad district, Maharashtra. *Jour. Geol. Soc. India* 65:577 – 586
- Horton RE (1932) Drainage basin characteristics. *Trans. Am. Geophys. Union* 13: 350-360
- Horton RE (1945) Erosional development of stream and their drainage basin. Hydrogeological approach to quantitative morphology. *Bull.Geol. Societ. Am* 56: 275-370

- Melton MA (1958) A derivation of Strahler's Channel-ordering system. *Jour. Geol.*67:345-346
- Melton MA (1958) Correlation strucrture of morphometric properties of drainage system and their controlling agents. *Jour. Geol.* 66: 442-460
- Morisawa M (1985) Geomorphology text books: rivers, forms and process, Chapter-5, " Structural and lithological control"
- Nag SK (1998) Morphometric Analysis using remote sensing techniques in the Chaka Sub-basin, Purulia district, West Bengal. *J. Indian Soc. Remote Sensing* 26(1 and 2): 69-76
- Pankaj A and Kumar P (2009) GIS-based Morphometric analysis of five major sub-watersheds of Song River, Dehradun district, Uttarakhand with special reference to landslide incidences. *J. Indian Soc. Remote Sensing* 37(1):157-166
- Pathak BD (1968) Geology and groundwater conditions in Bah Tahsil, Agra, district with special reference to soil erosion problems. *Indian Minerals* 22 (2): 71-79
- Schumm SA (1956) Evolution of drainage system and slope in badlands of Perth Amboy, New Jersey. *Bull.Geol.Soc.Am.* 67: 597-46
- Singh S and Singh MC (1997) Morphomertric Analysis of Kanhar River Basin. *National Geographical Jour. of India* 43 (1):31-43
- Sinha BPC (1969) Geohydrological studies in parts of Agra district, U.P. *Indian Mineral* 23 (4):
- Strahler AN (1945) Quantitative geomorphology of drainage basin and channel network. In: VT Chow (ed), Handbook of Applied Hydrology Magraw Hill, New York section-4-II
- Strahler AN (1952) Hypsometric (area-altitude) analysis of erosional topography. *Bull. Geol. Am.* 63: 1117- 1142
- Strahler AN (1956) Quantitative slope, analysis. *Bull.Geol. Societ.Am.* 67: 571-596
- Strahler AN (1957) Quantitative analysis of watershed geomorphology. *Trans. Am. Geophys. Union* 38: 913-920
- Vittala S Srinivasa (2004) Morphometric analysis of sub-watersheds of in the Pavagada area of Tumkur, district, South India using remote sensing and GIS techniques. *J. Indian Soc. Remote Sensing* 32(4): 351 -362.

15. Issues and Challenges on Water and Sanitation Management:

A Case Study of Tuensang Town, Nagaland, North-East India

Soyhunlo Sebu
Sangyu Yaden

The expansion of town area and rapid growth of population is dramatically increasing the per capita consumption of water. Population pressure has become a challenge to the authority to provide potable water and sanitation systems. PHED, the competent authority to supply safe drinking water and proper sanitation system to citizen has failing due to management failure, lack of public support, unsuitable topography for laying pipeline, pilferage and sabotage etc. The district received 1268 mm of rainfall annually. If this rainfall is harvest in state- of -art infrastructure it can minimize the water scarcity in dry season. Sanitation system is another issue which needs to be ponder. Absence of good sanitation system may be attributed to lack of awareness and economic problem of the citizen. In this situation water borne related diseases cannot be neglected as it is leading to health problem. Conflict and quarrels between Government and villagers over water is a rising phenomenon. In most cases government have to take permission from the villages authority to tap water from river as Article 371 (A) of the Indian Constitution empowered Naga's authority over land and its resource. This paper while highlighting the reason for water scarcity and sanitation problems gives some suggestions for the management.

Key Words: Water, Sanitation, Population, Scarcity, Management

Introduction

Historically, civilization around the world have largely evolved and developed where water is available as most human activities, depend on water. The present century is marked by over exploitation of natural resources in the name of development at the cost of future generation. Tuensang town is one of the oldest in Nagaland. It is bestowed with enormous water resources. However, demand and need for fresh water and proper sanitation in all the sectors is increasing tangentially which is placing a challenge before the government. Public Health Engineering Department (PHED) is making all afford to provide sufficient water and hygienic sanitation to the citizen but it faces hindrances due to the attitude of people and the topography of the area. Scarcity of water is the emerging environmental problems in the town. Even those who live in areas of high rainfall in India often face drought because landscapes have been denuded. Soil is compacted and most rainfall runs off before it can sink into the ground Inderjeet (2005). Usage of groundwater is limited in the state. However, it is estimated that groundwater meet most of the household water requirement in district headquarters.

The lack of adequate sanitation and safe water has significant negative health impacts. According to World Health Organization around 700,000 Indians die each year from diarrhea. UN Commission for Sustainable Development, second session, New York, 1994 noted as follows; as the crisis approaches and as water resources becomes scarce, the risk of conflict over them will become greater. After 2025 AD climate change could also make conditions worse if precipitation amounts decrease in the major food producing regions and evaporations rates increase. The bulk of increase in food production has to come from irrigated lands and this, in turn, will require more money to be spent on long distance water transfers, dams and the like, should the resources be available. The increasing size and number of cities will create a much bigger pollution load unless sanitation systems are provided. Urgent and decisive action must begin now if impending water crisis of a national proportions later in the 12th century – are to be avoided during the next 30 years. Water forms an important part of the environment and plays a vital role on the quality of life. The maintenance of quality of water has become a subject of great concern in view of the water crisis. The state government has adopted 'Water Prevention and Control of Pollution Act' during 1997. So far the Act has little impact on the matter.

Methodology

The study is empirical by nature, as such extensive observation and survey was conducted. Methods like informal interviews and focus group discussion were done with the local people to generate information's on the topic. The secondary sources were also widely referred and collected from various research reports, article, census publications, daily newspaper and records relating to the topic done by various authors.

Objectives

1. To assess the water resources shortage problem in Tuensang town.
2. To identify problems relating to sanitation system.
3. To prepare integrated water resources and sanitation management strategy.

Result and Discussion

Tuensang district is one of the oldest districts in Nagaland. Over the decades, the district has gradually diminished in size with the carving out of Mon, Longleng, and Kiphire districts. The district is situated between 25°6′ to 27°4′ North latitude and 93°2′ to 95°15′ East longitude with an altitude ranging from 800 to 3500msl. Tuensang remained totally unadministered in pre-independence days. The British had declared it an excluded area. In 1948, for the first time, an administrative centre was established at Tuensang and in 1954 the area including Mon, Longleng and Kiphire district was brought under the North East Frontier Agency (NEFA) as the Tuensang Frontier Division. In 1957, the Tuensang Frontier Division was merged with the Naga Hills District of Assam to form an administrative unit called the Naga Hills Tuensang Area (NHTA). As a result of the 16 point agreement signed between the Naga People Convention and the Government of India, NHTA become the full-fledged state of Nagaland on 1st December 1963. At present Tuensang town is headquarter of Chang, Sangtam, Khiamniungan and Yimchunger tribes.

Tuensang district headquarter is situated at an altitude of 1372 msl. It has a population of 36810 and 6920 households which is divided into 15 sectors. The town exhibits a remarkable topographic diversity and is dissected by a number of seasonal and perennial rivers and streams. Despite of plentiful water resources citizen face water problem and depends mostly on ground water and private water supply. The sudden shortage of water can largely be attributed to the increasing population and management failure. The demand for water has gone up disproportionately especially during dry season – February to May. According to survey an estimate of 70 per cent of urban households reported being connected with PHED (Public Health Engineering Department) pipeline but it has being observed that water supply from PHED is insufficient for daily household use. The survey observed that most of households depend on Ground water and Private Water Supply for their daily household requirement as PHED water supply is insufficient and irregular. The survey indicates that 43 per cent of urban household depends on Ground water, 31.7 per cent on Private Water Supply and 25.1 per cent on PHED. There are two Private Water Supplier in the town which is supplying water to 31.7 per cent of households.

Government water work personnel encounter difficulties while implementing and lying the pipe with the local people and topographic. It is reported in most cases the surrounding villages do not allow PHED to tap water from the rivers, streams and lake in their territories. The problem can be attributed to Article 371(A) of the Indian Constitution which empowered the landowners with rights over land and its resources and as such, attempts to convince the communities to share the water resources met limited success. It is observed that villagers could rather choose to let the water flow down the rivers and streams than share with their neighbors. In some cases department is reluctant to allow even a water pipe line to pass through village land. The latest controversy is between government (PHED) and Tuensang village. It is reported that pipeline used by the villagers was connected in 1981 which frequently got damage due to expiry of pipe. The villagers cautioned that if the authorities failed to take any positive steps within the stipulated time, it would disconnect water pipelines whose sources were extracted from Tuensang Village River. PHED has two water sources Viz., Latak and Photung which is 31 Km and 49 Km respectively from the reservoir. Distance and topography of the region is also an obstacle in maintains and management of pipelines. It is reported that most of the pipes are damaged and derailed by the landslide during rainy season. The misunderstanding between consumer and government cannot be neglected. The consumer except that it is the water works personnel who should do everything to deliver the water, as an obligation. Whereas the government policy is that water works personnel to implement water supply scheme. According to PHED, it is not the job of the water works personnel to prevent pilferage or sabotage or wasteful use, except keeping vigil on the installation, proper distribution and sensitizing the consumer on water.

Water supply and sanitation is a state responsibility under the Indian Constitution as such PHED government of Nagaland is entrusted with the responsibility to provide proper sanitation. The lack of adequate sanitation and safe water has significant negative health impacts. According to hospital sources around 133 people got admitted with water borne diseases such as diarrhea, dysentery, typhoid etc from January to June 2010. The survey indicates that only 42.3 per cent of the unban households have proper sanitation and 57.6 per cent of urban household still lack proper sanitation. It is a great concern for health because adequate sanitation system is considered to be the most important factors in a civilized society. Despite of been the oldest town in the state sewerage system is still in pathetic condition. It is observed that sewerage system in Tuensang town is going from bad to worse as most of the drainage was reported to be constructed a decade back and not well maintain.

Realizing the fact that water and sanitation systems are the basic amenities for human environment, some suggestions to improve and manage are mentioned below, First, proper distribution of water resources can bring solution to water shortage. Secondly, PHED water supply pipe should be properly maintain, Thirdly, implement Action Policy for land owners and government conflict, fourthly, State -of- art reservoir and more water augmentation programmes, fifthly, greater emphasis need to be given to rainwater harvesting, sixthly, more awareness programmes for proper sanitation system and the consequence of water borne diseases should be initiate.

Conclusion

Tuensang town is experiencing fast rate of growth in population. The huge population and congested houses is a challenge to the authority to provide adequate potable drinking water and proper sanitation. PHED is making a huge affords to provide sufficient water to the people but is face with problems like ownership and topography. In many occasion PHED failed to convince the landowners to share water. It has been observed that most of the pipe line are damaged and derailed by the landslide during rainy season. Urban citizen mostly depend on Ground Water and Private Water Supply for their daily household water requirement. The lack of adequate sanitation has significant negative impact on health. The town is facing with increasing water borne diseases. It is a great concerned for health because water borne diseases are directly link with inadequate sanitation system. Sewerage system is in pathetic condition because it is going from bad to worse as most of the drainage was constructed one decade back. The water resources and sanitation system are facing silent death. It is high time that authority implement effective action plan policy for both potable water supply and sanitation system.

References

- Directorate of Soil and Water Conservation (2003). *State Land Use Board (SLBU)*. Government of Nagaland.
- Groundwater Information Booklet. (2007). *Wokha district. Nagaland Central Ground Water*. Public Health Engineering Department. Government of Nagaland.
- Jeet Inder(2005). *Groundwater Resources of India. Occurrence, Utilization and Management.* Mittal Publication New Delhi-110059 (India).
- Directorate of Information and Public Relation (2007) .*Nagaland Basic Facts*. Government of Nagaland. Kohima.
- Department of Planning and Coordination (2004). *Nagaland State Human Development Report*. Department of Planning and Coordination. Government of Nagaland.
- Raghunath, H.M (1987). *Groundwater, Hydrogeology, Groundwater Survey and pumping tests*. Rural water supply and irrigation system. John Wiley and Sons. New York, Chichester, Brisbane, Toronto, Singapore.
- State of Environment Nagaland (2005). *Nagaland Pollution Control Board.* The energy and Resource Institute. New Delhi. Ministry of Environment and Forests. Government of India.

16. A Need of Drinking Water Supply in India:

Implementation of Various Policies and Programmes of Government and International Organizations

Amar Prakash
Deepali Singh

Now – a – days the world has become too small, a place to live because of various technological advancements. Problems like pollution are no longer the concern of individual nation. The entire global community is exercised over it. The air we breathe and the oceans that sustain as in various ways belongs to the world and therefore the responsibility for keeping them pollutions free must necessarily he shared by all of us. Water pollution is running India's famous lakes in to dirty ponds and its major river into threatening the health of Milhous of rural folk at the receiving end of the city wastes.

Key Words: Water Supply, Drinking water, Global warming, Climate change, Globalization.

Introduction

Fresh water, which we have always thought to be available in abundance, is very scarce in comparison to total water present on earth. India's 94% of rural population and 91% of urban areas in the country have access to safe drinking water, the fundamental right of any individual. India has 4 %of world's water availability and 25 %of the world's population to sustain of the limited water resources available for human consumption 70% is not potable. As far as availability of safe drinking water is concerned it ranks 120th in the list of 122 countries. Annually 37.7 million people are affected by water borne diseases 1.5 million children are estimated to die of diarrhea and 62 million people are the suffering from florosis due to excessive fluoride in ground water. In India approximately 6,700 crore is annually spent on treatment of water-borne diseases. It is heartening to note that while only 38% of households had access to drinking water in 1981. That figure has gone up to 78% by 2001 much of the progress has been in rural areas. With the number of households having access to drinking water going up from a mere 26% in 1981 to 73% in 2001 . Rural India has more that 700 million people residing in about 1.42 million habitations spread over 1.42 million habitation spread over diverse ecological regions. Meeting drinking water needs of such a large population can be a daunting task especially with non-uniformly in level of awareness, socio-economic development, education, poverty, practices and water availability.

Table-1
Water Availability in the World

Water Source	Water Availability(in Percent)
World Water Resource	97.2%
Ocean	2.15%
Polar and Glacier under Ground water 8000 mt. depth	0.31%
Below 800 mt. depth	0.31%
Soil Moisture	0.005%
Surface Water Lake	0.017%
River and Stream	0.001%
Atmosphere	0.001%

Generally water contains iron, calcium, magnesium, manganese, silica, fluoride, nitrate, phosphates, sulphates and chlorides. When the quality of these parts increase then they affect the body systems and cause destruction of health. Arsenic salts can create cancer, cadmium affects kindly while barium carbonate has effects on veins never and heart. The presence of manganese in water causes loss of memory, impotence and eye diseases. Vanadium spoils the fertility of the soil and creates cancer in the body while beryllium gives dives inflammation in pulmonary tissue. Mercury poisoning systems include loss of vision, Hearing and of intellectual abilities. There is no treatment for it and damage is permanent.

The subject of rural water supply vests with the Panchyati Raj Institution (PRIS). The panchyats are to play major role in providing safe drinking water and managing the systems and sources in their respective areas. They can be involved in the implementation of schemes, particularly in selecting the location of hand pumps, stand posts and spot sources in operation and maintenance (O&M) etc. Moreover, government of India emphasizes on empowering and capacity building of the PRIs to enable them for discharging their responsibilities in drinking water supply. It has been realized that to strengthen the socio-economic conditions rural India, mere administrative decentralization or increases investment is not enough. The power of people's participation has been recognized and brought to the fore. Despite good investments and improvement in the rural water supply and increased outlay by the government, particularly in the last one decade. General satisfaction is rather limited at the community level. The change in technology – earlier in an impressive increase in the total rural water supply coverage. However, the availability of potable drinking still not satisfactory. Though about 1 lakh habitations are covered every year. The number of problem habitations has not decline proportionately. Hence, government of India realized that sustainability of sources and system is key for people's satisfaction. System are falling idle and into disrepair. This is due to the perception of the rural people that water is a social right to be provided by the government, free of cost. The government tried to drive home the principle that water is an economic and social good and should be treated as such. It should be managed at the lowest appropriate with users involved in the planning and implementation of projects. With this aim in view government of India has brought about policy changes by introducing reforms in the rural drinking water supply sector.

During the 60 years, India has invented a huge amount in harnessing water resources for irrigation, domestic use and non-agricultural activates. The public sector has invested some Rs. 1000 billion in water resources development and it is reported that a whooping Rs. 32000 crore has been spent on rural water supply since 1954. Government of India and various state governments have implemented plenty of water schemes in various matrices under bilateral and multilateral augment as well. Despite a large investment in water resources conservation, irrigation infrastructure development and investment in bringing about sustainability of drinking water supply sources and systems. The net result is inadequate water supply for various needs. The imminent water crises in India is being ascribes-

- The growing gap between the scarcity values of water and realizes value under current pattern of water use and management.
- Weak government and misgovernment of water resources by the state and
- The absence of clear- cut property regime or ownership of water.

Traditionally water was perceived to be a free good available from nature –existing institutional, legal and governance framework relating to acquisition, distribution and utilization of water have not helped in changing this perception. A stage has come where the sovereign right of states on water or any natural resource for that matter is disputed that role of a government to assume this right has become questionable and that it needs to be dispassionately defined through consensus among stakeholders. A review of research – based literature on water resource in India brings to surface several new concepts that were not in use a decades and a half ago. There concepts are viewed as evidence s of change after deregulation and liberalization of the economy which have facilities privatization of water sources/ services and co modification of water. Whether this change implies development or retrogression needs to be analyzed and empirically verified. The new concepts that have emerged in the past decades is -

- Competing demand for water.
- Right-need dichotomy in water.
- Privatization of water resources.
- Privatization of water services.
- Public private partnership(ppp)
- Water markets.
- Commercial supplier of water.
- Environmental and ecological concerns with reference to water.

Government's Programmes for Rural Drinking Water Supply

Clean drinking water is a necessity of life. Supply of clean drinking water in the rural areas has always been one of the highest priorities of the government. A technology mission on drinking water named " National Drinking Water Mission" (NDWM) was launched in 1986, which subsequently was rechristened as " Rajiv Gandhi National Drinking Water Mission "(RGNDWM) in 1991 with there key objectives:-

- Providing safe drinking water to all villages.
- Assisting local communities to maintain sources of safe drinking water in good condition.
- Giving special attention for water supply to Scheduled Castes and Scheduled Tribes

Integrated developments of rural areas are one the abiding tasks before the government of India. The National Common Minimum Programme (NCMP) of the central government reiterates the cardinal importance of villages to the overall development of the country and commits to work towards developments of rural areas. This for various reasons could not keep pace with urban areas in the past in conformity with this commitment of the government. The Ministry Rural development accords foremost priority to development in rural areas and eradication of poverty and hunger from the face rural India. A number of initiatives have been taken in the recent years for creation of social and economic infrastructure in rural areas to bridge the rural-urban divide as well as to provide food security and fulfill other basic needs of the rural populace. For the Tenth five plan. The allocation of funds for rural development programmes has been enhanced to Rs. 76.774 crore as against Rs. 82.874 crore in Ninth plan. The minister of rural development has been mandated to provide safe drinking water in all rural habitations by 2004. To achieve this objective many programmes like Accelerated Rural Water Supply Programme (ARWSP) and Prime Minister's Gramodaya Yogana- Rural Drinking water (PMGY-RDW) have been implemented to resolve drinking water crises inrural habitations which effective from 1 April 2002. The prime minister's Gramodya Yojana- Rural drinking, water (PMGY-RDW) has been transferred to the rain water harvesting sustainability of sources and community participation. The Central government has achieved considerable sources in meeting drinking water needs of the population over the year, with an investment of more than 40,000 crore on rural drinking water supply. 91.03% of habitations have been fully covered (FC) with drinking water facilities: 7.93% are partially covered (PC)

To build up a human resources base of trained personnel to serve the needs of the rural water supply and sanitation sector. The National Human Resource Development Prgoramme (NHRDP) was launched by the government in 1994. The progemme aims at training beneficiaries especially women at the grass-roots level. It also aims at empowerment of Panchyati Raj Institutions to enable them to take up activities related to water supply. The objective of the Research and Development (R&D) programme is to provide cost-effective technologies in rural water supply and sanitation programme. It has been realized that innovative and appropriate R&D input is an essential component for sustained water supply programme. R&D programmes focus on sustainability issues, control of water quality problem, reviving traditional technologies.

In India, sanitation coverage in rural area is only 20% as results a large number of people practice open defecation, this is the root- cause of various water – borne diseases and incidence of diarrhea cases. In addition to the health hazard, lack of sanitation facilities causes a great inconvenience to the people particularly women in the rural area. Rural water supply is a state subject. State has been taking up projects and schemes for the provision of safe drinking water from their own resources. However, recognizing the importance of providing safe drinking water in rural habitations government of India has been providing financial assistance to state governments. Rajiv Gandhi National Drinking water Mission has been launched by the government of India in 2005 to be implemented in a period of four years from 2005-06 to 2008-09. Rural drinking water is one of the six components of Bharat Nirman. During the Bharat Nirman period, 55.067 un-covered and about 3.31 lakh slipped – back habitations are to be covered with provisions of drinking water facilities and 2.17 lakh quality- affected habitations are to be addressed for water quality problem. The Central Government announcement a scheme in 2008 in which 4.455 million wells be taken up for recharging groundwater in 110 blocks of the identified states for the benefits of farmers. The scheme has been implemented before monsoon of 2008. The Artificial Recharger of Ground Water through wells scheme was launched keeping in view the concerns of over exploitation of groundwater resource in the country as well as to ensure sustainable water resource management and irrigation facilities in the affected rural areas . About 80% of these groundwater stresses areas- over- exploted, critical and semi critical areas are located in hard rock areas in the states of Andhra Pradesh, Gujarat, Madhya Pradesh, Maharashtra, Rajasthan and Tamil Nadu where rapid decline of ground water levels have been observed on long term basis. It is a state sector scheme and will be implements by the respective state government in association with the Panchyati raj Institution (PRIs), the Central Ground Water Board, The NABARD and NGOs.

Table-2 Data of present utilization and future requirement of water: 1985 and 2025(mhm)

Water use by purpose 2025	Utilization 1985			Future Estimated Requirement		
	Surface Water	Ground water	Total	Surface water	Ground water	Total
Irrigation	33.14	17.34	50.48	61.17	34.27	65.54
Net consumption	0.79	0.06	0.85	3.53	0.48	3.96
I - Domestic & Municipal supply	0.22	0.08	0.28	1.21	0.43	1.64
II - Industrial Use	0.14	-	0.14	0.82	-	0.82
III - Thermal Power Generation	0.43	-	0.43	1.50	-	1.50
Pisiculture	-	-	-	2.79	-	2.79
Forestry	-	-	-	2.21	-	2.21
Livestock need	0.49	-	0.49	1.18	-	1.18
Total Demand	34.42	17.40	51.82	70.88	24.80	95.68
Total utilizable flow available	68.41	35.58	103.99	68.41	35.58	103.99
Demand utilizable flow(in Percent)	50.3	48.9	49.8	103.6	69.7	92.0

Source: Ministry of Water Resources, New Delhi, 2006

Water, the precious drop is the lifeline of human lives and the world needs to wake up to the ground reality that day this nature's gift dry up from this planet unless we take good care to nature and replenish its sources. Global warming and climate change have immediate and dire consequences for the availability of water. By 2025, nearly 2 billion people will be living in countries or regions with absolute water shortage, where water resources per person falls below the recommended of 500 cubic meters per year.

Water scarcity already affects every continent and four out of every ten people in the world. The situation is getting worse due to population growth, urbanization and the increase in domestic and industrial water use. Water scarcity forces people to rely on unsafe sources of drinking water. Poor water quality can increase the risk of diseases including cholera, typical fever, salmonellas, other gastrointestinal viruses and dysentery, not only are there 1.1 billion without adequate drinking water, but the United Nations acknowledges 2.6 billion people are without adequate water for sanitation. Many countries do not have sufficient water demand, with the result that aquifer depletion due to over extraction is common. Moreover, the scarcity of water is accompanied by deterioration in the quality of available water due to pollution and environmental degradation.

Global Water Shortage

The amount of water in the world is finite. One third of the world's population lives in water – stressed countries now. By 2025, this is expected to rise to two- third s. There is more than enough water available in total, for overtone's basic needs. The UN recommended that people need a minimum of 50 liters of water a day for drinking, washing, cooking and sanitation. In 1990, over a billion people did not have even that providing universal access to that basic minimum worldwide by 2015 would take less than 1% of the amount of water we use today. But we are a long way from achieving that. The total demand of domestic water is in 1995 and 2025 under the three scenarios. Globally on an annual basis for human use 12.5 to 14 billion cubic meter water is considered available:-

Table-3

The Total Demand of Domestic Water in 1995 and 2025

Year	Percapita Availability of Water
1989	9000cu.mt/ year
2000	7800 cu.mt./ year
2025	5100 cu.mt./ year

Even the amount of fresh water/ capita would be enough to meet human needs, if it was properly distributed. In India the pumping of underground water is now estimated to be double the rate of aquifer recharge from rainfall. The International Water Management Institution, The world's premier water research group estimates that India's grain harvest could be reduced by up to one fourth as a result of aquifer depletion in the country adding 18 million people per year. This is not good news. In India, overall water demand will increase from 552 BCM to BCM by 2025, which will require the use of all the available water resources in the country (World Dank 1999) of the present water usage 92% is devoted to agriculture with roughly 3% used in industry and only 5% for domestic purpose like drinking water sanitation (WRI 2000). Demand from the industrial and domestic sectors is expected to increase with the growing population, urbanization and industrialization. For world that is 70% water things are drying up fast. Only 2.5% of water is fresh and only a fraction of that is accessible. At present, 1.1 billion people lack access to clean drinking water and more that 2.4 billion lack adequate sanitation. By 2025, two thirds of the world's population may be living in countries that face serious water shortages unless we take stringent measures today.

Process of Globalization and Water

Globalization process are said to have made way for global corporations as well as the domestic private traders to draw ground water for a variety of commercial purposes over water resources . It also has a substantial impact on the livelihoods of the poorest people for whom land and water are the prime livelihood resources. Therefore, it has become a serious concern as matter of citizen's rights and it terms of deprivation and environmental damage.

Domestic Water Consumption under Business- as – Usual, Water Crisis and sustainable water use scenarios, 1995 and 2025 –

Table-4
Water Use Scenarios of World 1995 and 2025

Region/ country	Domestic Water Demand (KM3)			
	1995 Baseline-estimates	2025 Projections		
		BAU	CRI	SUS
Asia	79.1	156.7	113.0	143.9
China	30.0	59.4	42.3	54.3
India	21.0	40.9	27.7	42.0
Southeast Asia	13.9	30.4	23.6	23.8
South Asia excluding India	7.0	16.2	11.1	15.3
Latin America (LA)	18.2	30.7	24.7	22.8
Sub- Saharan Africa	9.5	23.9	15.2	23.8
West Asia / North Africa	7.1	13.1	9.9	11.2
Developed Countries	58.7	68.6	62.8	65.8
Developing Countries	110.6	221.0	159.7	198.7
World	169.3	289.6	222.5	264.5

Sources: Impact – Water assessments and projections 2002.

Notes:– BAU, indicates Business –as-usual scenario; **CRI,** Water crises Scenario: **SUS** , Sustainable water Use Scenario.

The international observance of World Water Day is an initiative that grew out of the 1992 United Nations Conference on Environment and Development (UNCED) in Rio de Janerio. The Unite Nations General Assembly designated 22 March of each year as the World Day for water as a resolution. A resolution is adopted by a This world day for water to be observed starting in 1993 in conformity with the recommendations of the United Nations Conference on environment and development . The goal is to raise awareness of the world water crises and heir goal of reducing by half the percentage of people who lack access to safe (clean) drinking water and basic sanitation. The World Health Organization (WHO) has certified India as free from guinea worm disease. This announcement was made by WHO, Geneva, in February 2000. Guinea worm is water borne disease caused by dracunculiasis mesinensis. Government of india took efforts to control this disease since early eighties. However, by the 2000, the disease was completely national institute of communicable Diseases (NICD).. The main control measures taken were closing down of step wells and providing tube wells and piped water supply. The United Nation predicts that by 2025, two thirds of the population will experience water shortage, affecting lives and livelihood of 1.8 billion. According to the UN world water scarcity in World Economic Forum in 2008, UN Secretary General Ban KI- Moon recommended that water scarcity should be at the top of the international agenda. One programme winning quick supports is dubbed WASH – for Water, Sanitation and Hygiene for all – global effort that provide water services and Hygiene training to everyone who lacks them by 2015.

Today, water is the top most agenda on development dialogue all over the world. The problem human race face with water is that it either it is unmanageably ABUNDANT OR CRITICALLY SCARE. No water, the condition of water being plentiful is for a few months in a year "India, with 16% of the world's population and 2.45% of the land mass has only 4% of water resources ". There is a completing demand for water amongst purposes such as drinking irrigation and industrial use. Water is in deed a RENEWABLE RESOURCES. The per-capita water availability of water for India in 2001 was estimated to be half its. The crises f water affects the lives of millions of people today. In 15 states, the ground water levels have been falling at raging between five and seven % and some are expected to sun dry as early as 2015 because of overdraft and misuse. It is reported that every year we draw twice as much as we replenish it. By the year 2025 the demand for fresh water is expected to rise to 56% above what currently available water can deliver, if current trend persist.

The earth is a beautiful and fascinating planet of universe. It has provided optimum conditions for development of life and evolution of myriad species of planets and animals including human being. The resources and environment of the planet have provided us great heights in arts, science and technology and other pursuits but unlimited material comforts have led to callous exploitation of environmental resources making impossible to provide wholesome living conditions. In fact the wholesome living conditions mean environment which is hygienically safe and is also aesthetically acceptable and provides us sufficient clean water the soil and other natural resources as necessary for our modern living.

References

- Ramasamy R. Lyar, "Water changing a course for future – I" Economic , PoLITICAL Weekly , March 31,2001.
- Agrawal and Narayan S. , " Making water every body's", Business No- center for Science and envirnmet, 2002.
- Maude Barlow , " Blue Gold – The Fight to stop the corporate theft of the worl's water", New Delhi- New Press- APRIL 2003 .
- Lennert Nilsson , " Making Rural water supply schemes work Government of Tamil Nadu", Damida 2002.
- NAAS – Emerging issues in water management- The question of ownership polcy paper 32 national academies of agricultural science, New Delhi June-2005.
- Dr. V.KP. Kudesia, 'Water Pollution" Pragati park ashan, Merut, 1998.
- Arihent, completion in focus , June -2009
- Book India -2009.
- K.R. Gupta, "Rural Develop in India" Atlentic publishers, New Delhi -2004.
- Satish Tiwari, " Rural Developent (E.d) " Anmol Publication , New Delhi-2000
- Kurukshetra, An Development Journal of Rural , 2009
- P.C. Bancl, "Water Management in india ", 2004.

17. Meteorological Drought Risk Mapping in Vaippar Basin

K.Shimola
M.Krishnaveni

Drought is a normal, recurrent feature of climate, although many erroneously consider it a rare and random event. In this study, Vaippar basin is selected as study area for studying meteorological drought pattern in the basin and it is located in the southern part of TamilNadu. The daily rainfall values of 12 raingauge stations were collected from State Statistical department. The daily rainfall values was converted into monthly and then converted into annual rainfall data. Based on the IMD method of classification, the percentage of deviation of annual rainfall for all stations is calculated. The India Meteorological Department considers the rainfall to be deficient if the deficit is more than 20% of the seasonal normal. Further, the value of 'Di' was reclassified, based on the ranges suggested by Indian Meteorological Department, into four such as no drought, mild drought, moderate drought and severe drought. Drought severity maps are generated using Arc View/GIS by summarizing the percentage of occurrence of droughts in areas within the study basin and helps in mitigation and preparedness. The major advantage of a GIS is that it allows identifying the spatial pattern of drought severity. Drought severity index of each rain gauge stations was generated within the basin. The temporal variation of drought over the basin was also studied. The results of this study show that the maximum severe drought was occurred during 1985-86. Similarly, during 1990-91 many part of the basin were experienced Mild drought in the Vaippar basin. The most areas values of moderate drought were occurred in the year 1995-96. Drought severity index for each category is calculated by the number of occurrence divided by total year taken for the analysis. The spatial distribution of weighted drought risk index is used to produce drought risk map of Vaippar basin. Major part of basin is covered under the moderate drought. Also it is found that the usage of GIS is very help in mapping the spatial pattern of drought severity and also to identify the drought risk areas of the basin. The results of the study can be used by the local inhabitants to alleviate the continuation of the drought phenomenon. The study on drought spread and intensification is for planning, managing and combating the drought.

Key Words: Drought, IMD method, GIS

Introduction

Total water available on this earth remains constant but its distribution with respect to time at a place is highly variable leading to hydrological extremes of drought and flood. Drought is an inevitable part of Earth's climate that occurs regularly with no clear warning and without recognizable borders, causing billions of dollars in loss annually for the farming community (*Kagon*, 2000). It occurs in virtually all-climatic zones, but its characteristics vary significantly from one region to another. Drought should be considered relative to some long-term average condition of balance between precipitation and evapotranspiration (i.e., evaporation + transpiration) in a particular area, a condition often perceived as "normal". Drought results from a long continuous dry weather and lack of rainfall, which causes exhaustion of soil moisture, depletion of surface and ground water storages and reduction and eventual cessation of stream flow. Consequently, drought analysis is generally carried out based on lack of rainfall [meteorological drought] or stream flow/surface and ground water storages [hydrological drought] or with respect to the supply and crop water requirements [agricultural drought]. Government of India is identifying drought prone areas and spending huge amounts for drought management measures. But the detailed drought severity assessments are not made to inform the farmers about the quantum of deficiency of water for their planning purposes. Drought risk is a product of a region's exposure to the natural hazard and its vulnerability to extended periods of water shortages (D.A Wilhite 2000). Drought severity can be assessed in meteorological, hydrological and agricultural contexts. Drought assessment involves with spatial and temporal water related data. Meteorological drought is a situation when the actual rainfall is significantly lower than the climatologically expected rainfall over a wide area. The meteorological drought results in hydrological drought with marked depletion of surface water and consequent drying up of inland water bodies. Meteorological analysis comprises of estimation of degree of dryness in term of percentage reduction from the long-term average annual or seasonal rainfall. Meteorological drought is a situation when the actual rainfall is significant lower than the climatologically expected rainfall over a wide area. Monitoring, assessment and mitigation of drought impacts are the most important elements of drought preparedness plans (*Wilhite and Svoboda*, 2000). One of the aspects of global climate change scenarios in Vaippar basin is the increase frequency and severity of droughts. This project aims to study the meteorological drought occurring in the Vaippar basin based on Indian Meteorological Department (IMD) method. The objectives of this study are: to analyze the rainfall data of Vaippar basin; to calculate the drought index using IMD method; and to develop the drought severity map for Vaippar basin using GIS.

Meteorological Drought Study in India

Based on recent experiences, the National Drought Mitigation Center NDMC has learned that one way to identify appropriate drought mitigation actions is to conduct an overall risk analysis as part of drought planning (Wilhite et al.2000). Government of India is taking several measures in identifying the drought affected areas and action plans in mitigating the drought risk. According to the Irrigation Commission, the areas which received less than 750 mm of rainfall per annum are classified as drought prone and those between 750 mm to 800 mm as vulnerable to drought. The Meteorological Department of the Government of India has defined an area as drought hit when annual rainfall is less than 75% of the normal. (i.e. the rainfall deficit is 25% to 50%) and severe drought when deficiency is above 59%. The Irrigation Commission assumes that the area is drought hit if the irrigated area falls short by 70% of the irrigable area. So, when the irrigated area is less than 30% of the irrigable area, the area are to be called as drought prone.

Study Area

The Vaippar basin is located in the southern part of Tamil Nadu. It lies between a latitude of 8° 59′ and 9°49N′ and a longitude of 77°15′E and 78°23′ E. It is bounded in the west by the western ghats on the east by the Gulf of Mannar (Bay of Bengal) on the south of Tambaraparani.

The river Vaippar orginates from the Echamalai Mottai, Neduntheri Mottai and Kiladiparai hill ranges of western ghats at an elevation of 15000m M.S.L. near Sivagiri in Thirunelveli district and flows generally and south easterly direction for a length of 140km and joins with the Gulf of Mannar. The total geographical area of Vaippar is 5423Km^2. The basin map of Vaippar basin is presented.

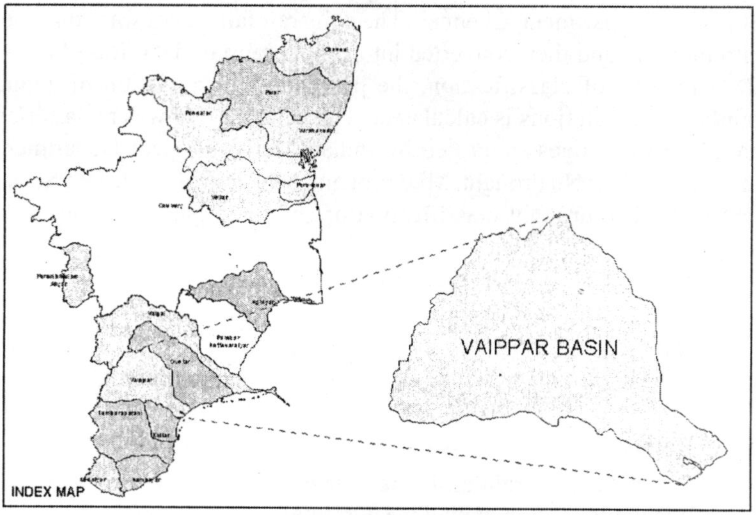

Figure-1 Vaippar Basin Index Map

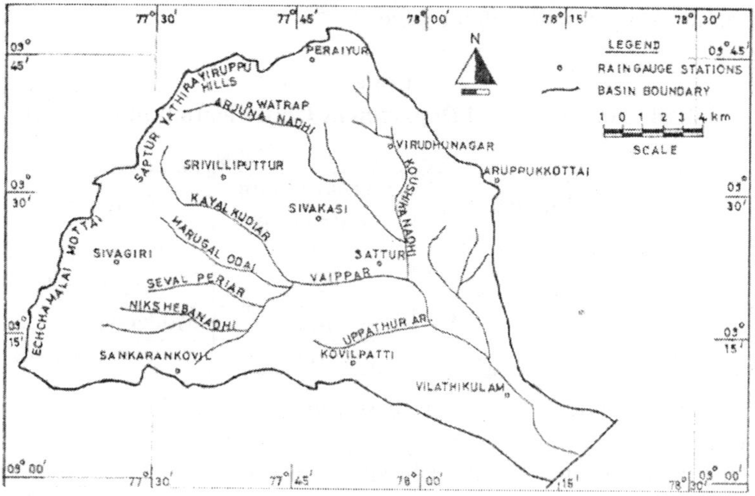

Figure 2 Vaippar Basin Map with Tributaries and Rain gauge Stations

Materials and Methods

In this study the IMD method is used which is a simple and widely used one that will give a preliminary idea about the drought condition of the area. In this method, drought is assessed on the basis of percentage deviation of rainfall from the long term rainfall. The daily rainfall values for various stations were collected for 26 years (1980-2006) from State Statistical Department, Chennai. The daily rainfall values was converted into monthly and then converted into annual rainfall data. Based on the IMD method of classification, the percentage of deviation of annual rainfall for all stations is calculated. The value of 'Di' was reclassified, based on the ranges suggested by Indian Meteorological Department, into four such as No drought, Mild drought, Moderate drought and Severe Drought. The drought classification of each rain gauge station in the Vaippar basin is calculated.

The percentage (Di) given by:

$$D_i = \frac{P_i - \overline{P}}{\overline{P}} \qquad \text{———} \qquad (1)$$

Where,

P_i is the annual rainfall in the year i:

P is the long term annual mean rainfall.

The percentage deviation of rainfall and the category of drought assessment is presented in the table -1

Table-1

Indian Meteorological Department Classification of Drought

S.No.	Range of D_i	Classification of drought	Category
1	>0	M0	No drought
2	0 to - 25	M1	Mild drought
3	-25 to -50	M2	Moderate drought
4	< -50	M3	Severe drought

Water deficit is the main reason for drought, which depends on rainfall; it is wise option to analyze long term rain fall data and to develop a risk index to view the drought situation in the region. Metrological drought risk index is a single number useful for decision making which is calculated based on frequency analysis that how many times the precipitation has deviated for a given period of time from historically established normal or mean rainfall value.

The drought severity classes were found out for each station on a yearly basis. The frequency of various classes of drought severity for each station was found out. The weight ages 1, 2, 3 and 4 are assigned to drought severity classes of no, mild, moderate and severe droughts respectively. The meteorological drought risk index of each station is found out by the corresponding weight age. The spatial distribution of drought risk was found out using the neibourhood analysis in GIS environment using vertical mapper, module of Map/Info software. The drought severity values for the study period are entered in GIS data base. The temporal variation of drought in 26 years is analyzed. The weight drought risk index is calculated and Map is developed using GIS.

Results and Discussion

(i) Drought Classification and Drought Severity Index of Raingauge Stations in Vaippar Basin

Based on the IMD method of classification, the percentage of deviation of annual rainfall for all stations is calculated. The value of 'Di' was reclassified, based on the ranges suggested by Indian Meteorological Department, into four such as No drought, Mild drought, Moderate drought and severe drought. Based on the percentage of deviation of annual rainfall the number of drought classes in each rain gauge station in the Vaippar basin is calculated. Drought severity index for each category is calculated by the number of occurrence divided by total year taken for analysis. This is carried out for the entire 12 stations and the variations of drought severity index in each station.

(ii) Temporal Drought Classification in Vaippar

The number of occurrence of each drought severity class in each station was determined and the temporal variation of drought over the basin was studied. During 1985-86 many part of Vaippar basin were experienced severe drought. In that year, nine stations falls under severe drought and three stations under moderate drought class. The occurrence of very less rainfall make that year falls into worst affected period. The spatial variation of maximum mild drought affected year (1990-91). During this period eight stations come under mild drought and four stations received good rainfall. The moderate drought classes occurred in six stations and rest of them fall in severe, mild and no drought classes.

Table 2 Drought Severity Index for each Raingauge Station

Sl.No.	Station	Drought Severity Index			
		M0	M1	M2	M3
1	Kavalur	0.42	0.23	0.31	0.04
2	Aruppukottai	0.46	0.35	0.15	0.04
3	Sivagiri	0.81	0.12	0.00	0.08
4	Villathikulam	0.23	0.38	0.23	0.15
5	Pilavakkal	0.58	0.23	0.15	0.04
6	Srivilliputhur	0.69	0.27	0.00	0.04
7	Watrap	0.62	0.35	0.00	0.04
8	Sankarankoil	0.31	0.42	0.19	0.08
9	Koilpatti	0.42	0.38	0.19	0.00
10	Sathur	0.38	0.38	0.19	0.04
11	Virudhunagar	0.50	0.31	0.12	0.08
12	Sivakasi	0.42	0.50	0.04	0.04

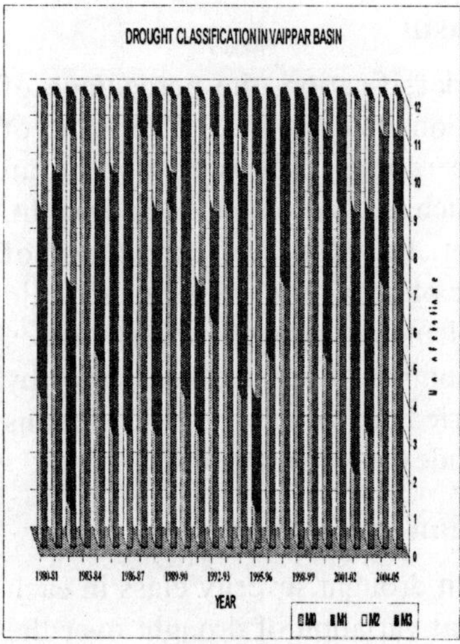

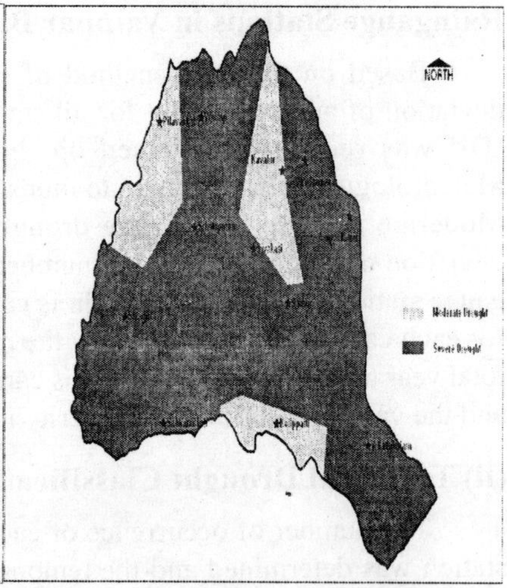

Figure- 4 Temporal Drought Classification in Vaippar Basin

Figure- 5 Maximum Droughts Year (1986-87)

(iii) Weighted Drought Risk Index

Drought risk analysis aims at improving techniques for drought prediction and management are based on the spatial variation of drought and are mainly focused on the magnitude, duration, intensity and spatial extent of droughts. The weighted drought index for each station is calculated from the drought index and presented in figure7. The spatial distribution of weighted drought index is used to produce drought severity map of Vaippar basin and which is presented in figure8. The drought is classified in mild, moderate and severe drought risk map of the Vaippar basin. Most of the area falls under moderate drought class and very less area under severe drought class in the basin. Srivilliputhur, Sivagiri region falls under mild drought class and Kavalur, Aruppukottai Villathikulam, Pilavakkal, Watrap, Sankarankoil, Sathur, Virudhunagar, Sivakasi region falls under moderate drought class and only Ramanathapuram area falls under severe drought class. Historical records also showed that the area affected by drought is also the same in the basin.

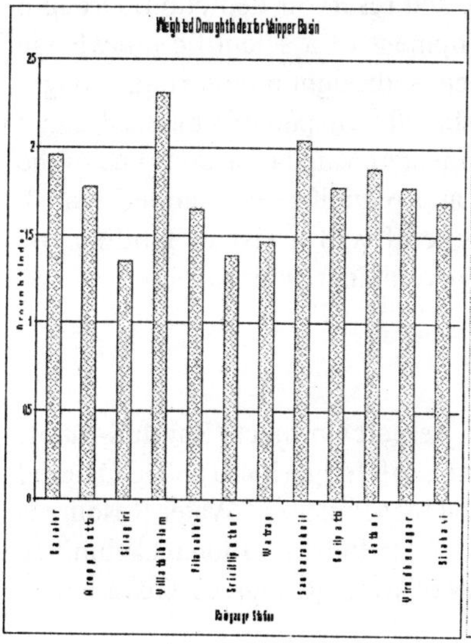

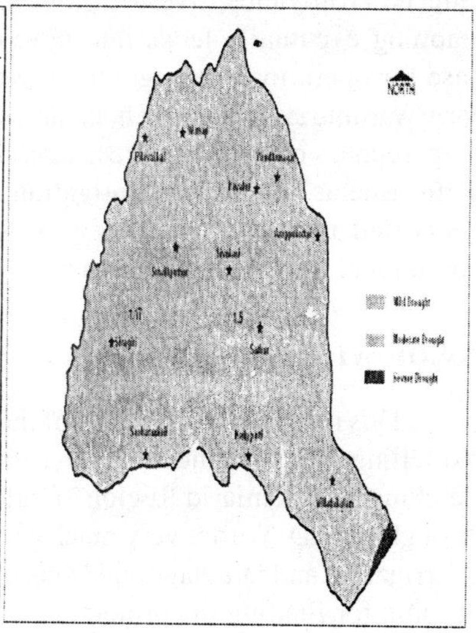

Figure- 7 Weighted Drought Index for Vaippar Basin

Figure -8 Drought Risk Map of Vaippar Basin

Conclusion

Mapping of the extreme events is one of the adaptation strategies to consequences of increasing climatic variability and climate changes. Drought is one of the short-term extreme events. In India, there is no operational practice to forecast the drought. One of the suggestions is to update mapping of drought prone areas for developmental planning. In the present study, the regions experienced drought has been demarcated to different years in the Vaippar basin. According to this, most of the areas are affected by severe drought during 1985-86, In 1990-91 major area of Vaippar basin experienced mild droughts, the moderate droughts were occurred over many parts of Vaippar during the year 1995-96, the drought risk map shows that major area is covered under moderate drought class, There by the Vaippar basin comes into the category of Drought Prone area. Such type of information would be a good tool for planning purposes and for input in modelling. Drought impact assessments can also be done by identifying direct consequences of the drought, such as reduced crop yields, livestock losses, and reservoir depletion. Drought mapping eventually feeds into development of a scientific knowledge base for operational drought tools such as drought monitoring, drought early warning systems, which, in turn, should form part of national drought preparedness plans. As the risk assessment of Vaippar basin is one of the main aspects of drought mitigation and planning, it is hoped that th generated maps for the study in general will assist in guiding the operational responses of the various government authorities.

Acknowledgement

This research work is part of the Research project "Rainfall-runoff modelling and Groundwater Dynamics of Irrigation Tank Clustered Catchment of Semiarid Region" funded by Ministry of Water Resources through INCID. We are very much grateful to Indian National Committee on Irrigation and Drainage and Ministry of Water Resources, Government of India for funding the project.

References

- Beard,L.R.(1972), "*Drought Severity and Water Supply Dependability*', ASCE Journal of Irrigation and Drainage Divition, Vol. 98, pp.443-442.

- Burrough, P.A.(1976), "*Principles of Geographic Information System for Land Assessment*",Clarandon Press, Oxford, 496p.
- Dracup,J.A., (1980), "*On the definition of droughts*", Water Resources Research, Vol.116,Drought from Rainfall Data', Journal of Hydrology, Vol.4, pp.264-272.
- Herbst,P.H., Bredenkamp, D.B. and Barker, H.M.G.(1966), "*A Technique for the Evaluation of Irrigation and Drainage Divition*", Vol. 98, pp.443-442.
- Kagon, F.N. (2000), Global drought detection and impact assessment form space, in *Drought: A Global Assessment*, Vol. 1, edited by D.A. Wilhite, pp 196-210, Routledge, London.
- Krishnaveni .M. (1993), "*Use of Geographic Information System for drought assessment*", M.E. Thesis, Centre for Water Resources, Anna University.
- Palmer, W.C.(1965), "*Meteorologic Drought*", US Weather Bureau Research Paper No.45.pp291-302.
- Ravikumar .G, Krishnaveni M. and Karmegam M. (2003), "*Remote sensing based agricultural drought information system for Palar and Thamiravaruni basins using GIS*", Centre for Water Resources, Anna University, Chennai-25.
- Shyam Sundar B., Jayaprakash H.P. & Rajkumar C., (2004), "*Droughts over Tamil Nadu including Pondicherry and Karaikkal*", Southern Eastern Costal Region, Chennai.
- Wilhite DA (2000) Drought as a natural hazard: concepts and definitions. In: Wilhite DA (ed) Drought: a global assessment. hazards and disasters: a series of definitive major works. Routledge Publishers,London.

Section-III
Sustainable Development and Land Management

18. Land Use Land Cover and Status of Development in Plain, Hilly and Coastal Watersheds of Gangavali River Basin of Central Western Part of Karnataka

Komala Bhat

The present study is related to the Gangavali river basin in Dharwad and Uttara Kannada districts of Kanranata state. All the rivers and streams of Dharwad district flow towards east and join the Bay of Bengal but Gangavalli alone flows towards the west and join Arabian Sea. . Administratively the study area stretches mainly in two districts and three taluks of Karnataka state and the watersheds are located in three different physical background of Gangavalli basin (Plain, hilly and coastal area)Number of maps has been prepared using SOI topographical maps and IRS IC-LISS AND PAN 2001 data of 1:50,000 scale. Socio-economic data collected during field survey have been used and ground water and surface water estimation have been done in the present evaluation. Relationship between land use land cover and its impact on socio-economic status in the watersheds have been highlighted in this paper.

Key Words: Land use, Land cover, Watershed, Ground water, Socio-economic status.

Introduction

Monitoring and evaluation are vital components of watershed development planning. Advanced technology of Remote Sensing and Geographic Information System provide a sound scientific platform in this endeavor. A case study is placed here to demonstrate the capability of the three watersheds of the one basin for evaluation purposes. To ascertain the impact of Watershed Development using Remote Sensing and GIS - by assessing the influence of land use / land cover condition on overall socio-economic conditions.

Materials and Methods

The investigation is based on primary and secondary data. Topographical maps and satellite imageries form the source for creating spatial database. The socio-economic, demographic, cropping pattern details have been collected through the field survey for the three watersheds separately by considering different economic groups and total agricultural area of the households. Villages have been selected on the basis of stratified method of sampling. For primary data collection, 5% of the villages and 10% of the households have been selected

Results and discussion

The study area has divergent physiographic features, climate and biodiversity and offers an excellent location for the study of interrelationship between land use land cover and socio-economic development.

Relief Features of the Watersheds

I. Bedti halla Watershed- The Bedti and the Shalmala rivers originate in the hill ranges in the vicinity of Dharwad city. In the initial catchment area dissected hills, structural hills, undulating topography and pediments are common. The height ranges from 600m to 780 m and undulating topography is prominent. Wide contour interval indicates more flat terrain or slopes with low angle.

II. Sondha halla Watershed- This part of the area consists of steeper slopes in the western half. Further, narrow and deep gorges, rapids and waterfalls are quite common. Apart from this, a widespread contour indicating more flat terrain or slopes with low angle is also evident in this region. The watershed comes under Western Ghat regions of Uttarakannada region.

III. Shirur Watershed- This area is almost homogeneous in elevation, except a small portion of low-lying area near the confluence of river with the Arabian Sea. In addition to this low land, there are several smaller lowland areas caused by uneven slit deposition and shifting courses of the river flowing in the area.

Regional Variations in Rainfall

As already mentioned in relief features, the watershed is located in three physiographic divisions and obviously the rainfall pattern is also differs. On the basis of the30 year's rainfall details rainfall distribution are classified into:

- Low rainfall area 600mm-900mm in Bedti halla watershed
- Moderate rainfall area 2500mm-3500mm in the Sondha halla watershed
- High rainfall area of more than 3500mm in the Shirur watershed

Under Ground Water

It is observed that depth to ground water level in the Bedti halla watershed depletes up to 25 meters below ground level, while it is 13.27 meters in Sondha halla region of the river basin. It is thus evident that in the Gangavali river basin, ground water utilization is certainly more in the Bedti hall watershed, compare to that in the Sondha halla and Shirur located in Central and Coastal Region respectively. Fig.1 shows the depth of ground water in the three watersheds.

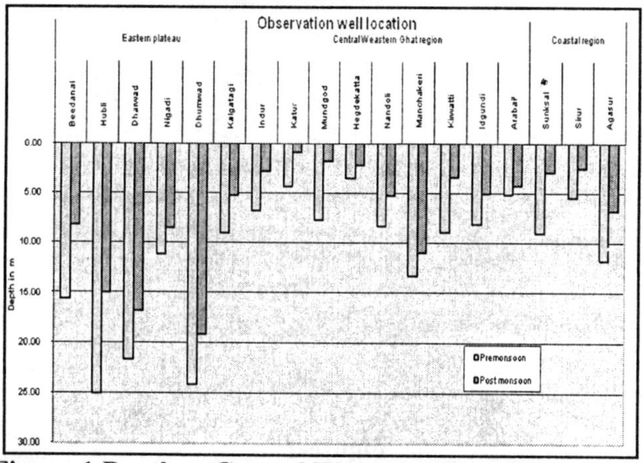

Figure-1 Depth to Ground Water, in Three Zones

II. Sondahalla Watershed -This watershed is named after a perennial stream locally called as Sonda River that is flowing in Sirsi and Yellapur taluks. The watershed covers an area of 613 sq km and is bounded by 14° 35' 13" and 14° 15' 18" North latitudes and 74° 37' 12" and 74° 56' 30" East longitudes and contained in SOI top sheet Nos.48J/ 9, J/10,J/13 and J/14. There are 73 revenue villages in the watershed Figure-3.

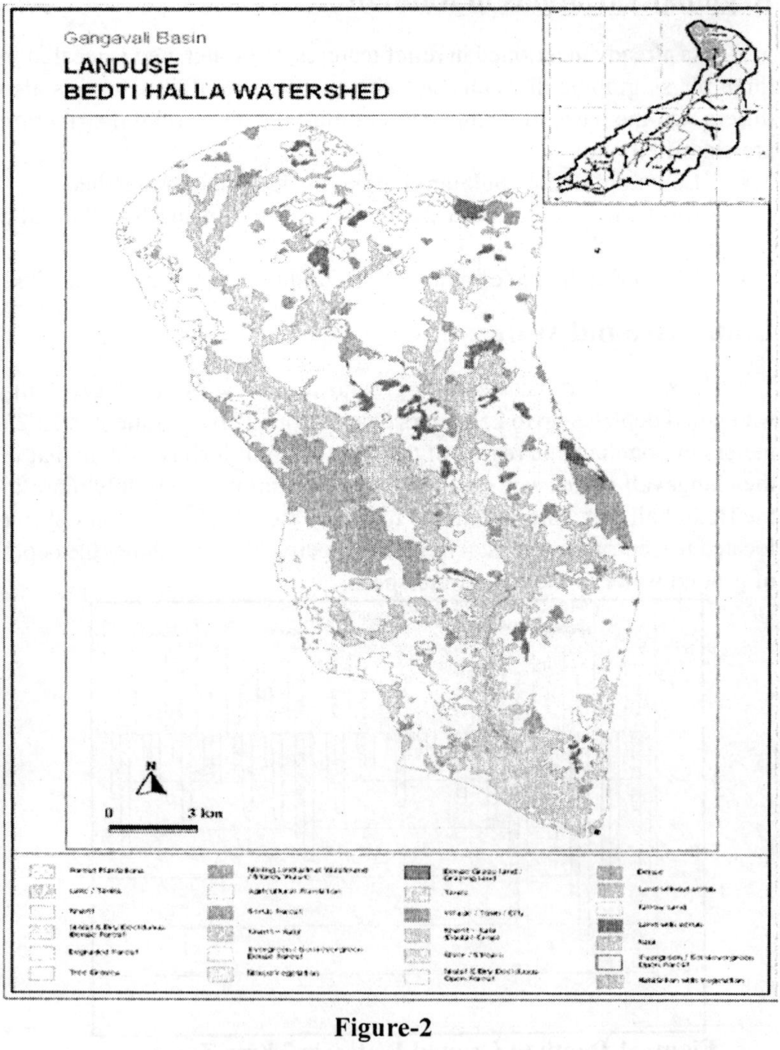

Figure-2

I. Bedti halla watershed -It is named after the Bedti Halla, which is flowing mainly due southeast, originated in the initial catchments area of the Gangavali river basin. It covers an area of 309 sq.km and is located between latitudes 15°11'10" and 15°28'30" North and, the longitudes between 74°52'45" and 75° 03' 07" East. Land use analyses based on the land use land cover map prepared 19 categories of land use are identified and given in the map in Figure-2.

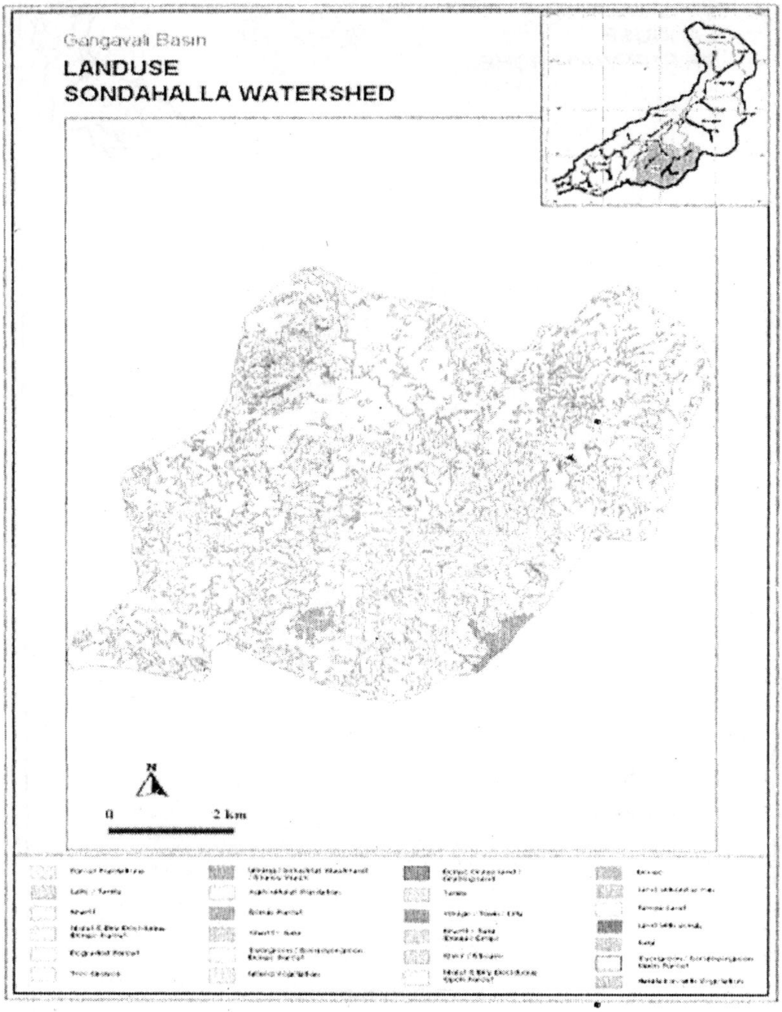

Figure-3

III. Shirur Watershed- This watershed is named after village Shirur in Ankola taluk, UK district and is located adjacent to Gangavali River in the southwestern part of the river basin. The watershed covers an area of 26 sq.km. The area lies between the longitudes 74Ú 19' 02" and 74Ú 39' 06" east, and the latitudes between 14Ú36'0" and 14Ú 39' 06" north and is contained in the SOI top sheet, no 48J/6. Figure-4

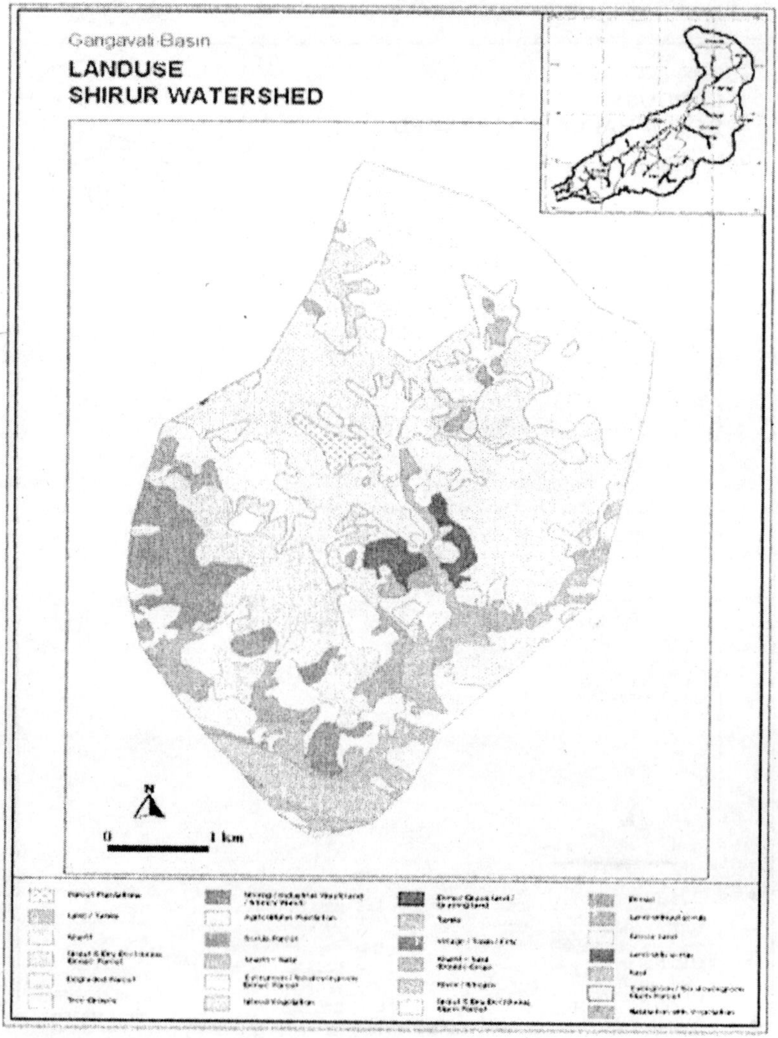

Figure-4

The table 1 shows the watershed wise land use land cover details in percent and Fig-5 clearly indicates the highest percentage of agricultural land in Bedtihalla and lowest in Sondahalla, where as forest is concerned the Sonda halla covers 85.80 percent of and lowest in Bedtihalla watershed.

Table 1
Watershed wise Land Use and Land Cover (values in %)

Watershed	Agriculture Land	Built-up land	Forest land	Waste land	Water bodies
I - Bedti halla	80.20	1.79	13.29	1.82	2.91
II - Sonda hall	12.36	0.97	85.80	0.48	0.38
III - Shirur	35.72	4.62	54.03	2.32	3.32

Fig. 5 Land use of the watersheds

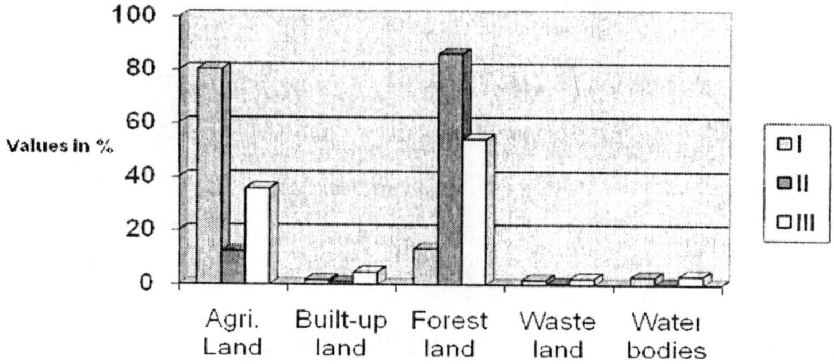

Important visible observations in the watersheds are summarized below:

- Land use picture of the watersheds clearly indicates that watershed located in plain terrain and in coastal area has adopted agricultural as the major occupation
- More than 80% of the land is used for agricultural purposes in watersheds I, it is 35.72 % in watershed III, whereas it is as low as only 12.36 % in watershed II.
- More than 80% of the area is covered by forestland in case of watershed II(comes under western ghat region) and more than half of the area in covered by forest land in coastal area whereas it is only 13% in the plain area.
- Wasteland is noticed more in watershed III because of back water area and salty, rocky area,

- In watershed I (in plain area) waste land is increasing due to intensive cultivation, overexploitation of groundwater and use of chemical fertilizer. At the same time surface water bodies in the form of tanks and reservoirs are also more in the watershed

Demographic Characteristics of the watersheds

- **Human resources and concentration of population**

For a better representation of the growth rate of population, is presented in a chart prepared and given in Fig. 6. The study of Fig. 6 shows that the growth rate is highest in Sonda halla (hilly watershed) Watershed II (11.51 %) and lowest in Shirur (Coastal) Watershed III (6.72%). The higher growth rate in Watershed II is due to the ideal climatic conditions and the overall development of the watershed.

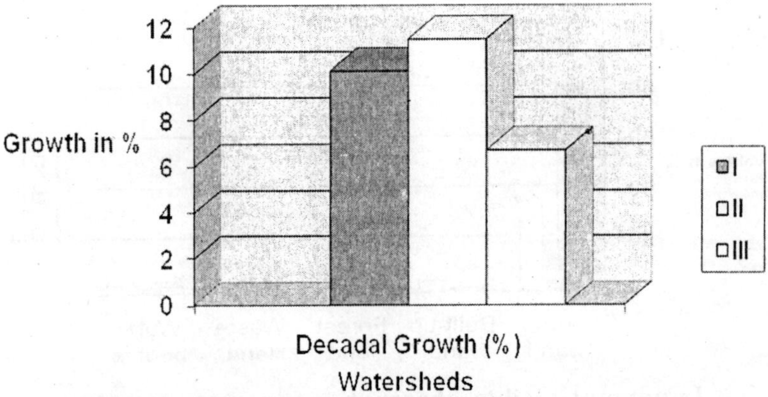

Figure- 6 Decadal Growth of Population (in Percentage)

Socio-economic conditions

The socio-economic conditions shows that watershed II(Sondha halla) located is hilly region of the Gangavalli basin is developed because the occupation adopted , literacy level, vehicles used, family system, female literacy, existing cropping pattern, house types , land holdings, sources of finance, market facilities, income structure and other infrastructure facilities are positively developed. (Table-2) Though land holdings are more in the watershed located in the plain region cropping system is dependent on underground water and the very often watershed is affected by the drought and famine due to erratic rainfall.

Table -2

Socio-economic conditions

Socio-economic conditions	**Watersheds**
Socially backward watersheds	**Bedtihalla**
Economically backward watersheds	**Shirur**
Watersheds developed socially and economically	**Sondahalla**

Developmental Profile and Resource Evaluation

The study intends to assess the regional imbalances in the development of land and human resources. The important parameters considering various attributes are demographic, social and economic indicators. In order to determine the regional disparities in the levels of development within and between the watersheds, said indicators have been used at different stages. The rank values are assigned to each indicator on the basis of percentage contribution. In the present study, higher rank value signifies the lower development. The ranks of all watersheds with respect to each selected indicator and the total rank order scores of levels of development are tabulated. In order to compare the developmental level between the watersheds, the **median rank value** has been taken to classify the basin into low and high developmental regions. Results of the developmental classification are set out in Table-3

Table-3 Level of development, Gangavali Basin

Watersheds	**Rank value**	**Level of Developmental**
II	116	High
I	133	Medium
III	190	Low

Median value=133

Source: Compiled by the Author

Highly developed area

Highly developed watershed is considered as watershed II. This watershed is highly developed because it is leading in many of the indicators like population, male and female literacy, total income and other facilities available.

Medium developed area

For the sake of comparison, watershed having the median total rank value (133) is considered as medium developed region in the Gangavali river basin. From the point of view of land holdings, basic facilities and female literacy the watershed is comparatively better than the low developed watersheds of the basin.

Low developed area

The study has indicated that watershed III located in coastal area is low developed. The region has not been provided with many basic infrastructure facilities. The demographic, social and economic ranks also support this observation.

The foregoing account of the status of development clearly demonstrates the regional imbalances in the existing socio-economic conditions in the basin. Cropping pattern adopted, female literacy and the demographic characteristics are supporting the status of development. Agriculture is not the only criteria to assess the status of development of a region. This gives more scope for further thinking about the availability of natural resources and its optimum utilization, generation of infrastructural facilities, integrated area development programmes and increase in literacy. .

- The analysis proves that the cropping pattern, the method adopted in agricultural activities, female literacy and the tertiary income are some of the important criteria influences the land use practices and the status of development
- The status of development is determined by the family system, occupation adopted, level of education, marketing facilities and the source of finance to the farmers
- The geographical factors like rainfall, availability of underground water, relief are also having positive relationship with the level of development.
- Though more than 80% of the area comes under agricultural land use(Bedtihalla) the medium status of development is found due to erratic rainfall and unsettled economy. Use of chemical fertilizer in the area , lack of forest cover, high dependence on underground water, less power supply to the farmers are some of the reason for the medicum development

- High development in Sondha halla watershed is due to adaptation of plantation crops, spices and perennial source of water supply, high rainfall, higher literacy rate.
- The low development is due to less available land for agriculture, high rainfall, less land holdings etc.

Conclusion

There are the regional imbalances in the development of land and human resources. High development in Sondha halla because of watershed is due to adaptation of plantation crops, spices and perennial source of water supply, high rainfall, and higher literacy rate. The low development is due to less available land for agriculture, high rainfall, less land holdings etc. The important parameters considering various attributes are demographic, social and economic indicators. In order to determine the regional disparities in the levels of development within and between the watersheds, said indicators have been used at different stages. The foregoing account of the status of development clearly demonstrates the regional imbalances in the existing socio-economic conditions in the basin. Cropping pattern adopted, female literacy and the demographic characteristics are supporting the status of development. Agriculture is not the only criteria to assess the status of development of a region. This gives more scope for further thinking about the availability of natural resources and its optimum utilization, generation of infrastructural facilities, integrated area development programmes and increase in literacy.

References

- Akiko Domoto (2003), International environmental governance, Its Impact on Social and Human Development. Article published in Human Development and Environment. Edited by Hans van Ginkel, Brendon Barrett Julius court and Jerry Velasquez. Rawat Publications.
- Alka Roy,Jain,S.K. and Shukla,R.K.(1994), Land Use Land Cover Changes in and around Roorkee using Multi data Remote Sensing data. Hydrology Journal vol.XVII No.3&4 July-December p.52
- Archana Mishra (2001), Watershed Management. ISBN 81 7273-055 E35/103 Delhi pp1-9 and 257-259
- Balak Ram. (2002) Resource Appraisal for Land Use Planning in Watershed, Edited Watershed Management for Sustainable Agriculture by M.A.Khan.

- Bhat, Komala M, Krishnamurthy (2006) Application of GIS and Remote Sensing in Land Use Land Cover and Its Impact on Socio Economic Pattern in Shalmala halla watershed, Dharwad-Karnataka Published in Population and Environment Issues and sustainable Development A research Publication volume Proceedings of II International Geography Conference, Goa January 2005.
- Bhat,L.S.(1989), Status Report on Land Use System Analysis for Evaluation of Resources at Micro-level. Discussion paper for the fifth review meeting for NRDMS project,DST..
- Department of Mines and Geology (2005), Report on Dynamic Ground Water Resources or Karnataka as on March-2004, DMG and CGWB, SWR, Bangalore, June 2005,132 p.
- Deresal Van, (1943), The American land, Its History and Its Uses. Oxford University press, London and Newyork.
- Dhruva Narayana,V.V.,Sastry,G. & Palnaik,V.S.(1997), Watershed Management. Publication and Information Division ICAR,New Delhi.
- Encyclopedia volume 23 Britannica Rao,V.L.S.P. and Bhat,L.S. (1959), The Role of Geographers in Land Use Planning, Bombay Geographical Magazine Vol. VII, p. 1.

19. Application of Remote Sensing and GIS in Wasteland Mapping:

A Case Study of Agra District, Uttar Pradesh, India

A. K. Rao,
Alia Yusuf
Ziaur Rehman

In a developing country like India, land is not only an important factor of production, but also the basic means of subsistence for majority of the people. About three-fourth of the total population draw their livelihood from agriculture and supports the vast population of India. But there is evidence to indicate that the land sector cannot bear the burden of growing population, not withstanding the untapped potentials for agricultural productivity growth in many regions. To make optimum use and to bring more and more areas under cultivation and forest etc. the satellite images (IRS – 1D LISS III on 1:50 000 scale) of Agra district of Uttar Pradesh have been visually analyzed to map the various wasteland categories. Wastelands have been analyzed for change detection over time and have reduced by 4.49% from 1993 (13.09%) to 2003 (8.06%), clearly indicating the effectiveness of the work carried out by various agencies regarding reclamation, soil conservation and afforestation. There is a major decrease of 1.5% in area dissected by ravenous land which is an effort of various national reclamation programs.

Keywords: Agra; Wasteland; Remote Sensing; GIS; Change detection

Introduction

India being the second most populous and seventh largest country in the world requires vast areas under cultivation and plantation. Due to increase in the population day by day there is a growing demand for food, fodder and fuel and over exploitation of natural resources, extensive degradation of forests and agricultural land leading to the substantial increase in area of wastelands. Therefore, there is an urgent need to plan substantial management of degraded forests and ecologically fragile areas. The increasing population and rapidly diminishing agricultural lands aggravated by the problems of land degradation and groundwater depletion have assumed serious proportions in many areas which threaten not only the sustainability of agriculture, but also the overall livelihood system of the people are facing ever-escalating problems day by day. As a result the land has extensively degraded, protective plant cover has eliminated and surface soils has exposed.

Information on existing wasteland, their categories and spatial distribution are the essential prerequisites for land management strategies. According to the Ministry of Agriculture, Govt. of India, degraded land is of the order of 92 million hectares, of which 73 million hectares are arable land and 19 million hectares are non-arable land. With the advent of remote sensing and GIS techniques, it has become possible to identify and delineate various wasteland (an uninhabited wilderness that is worthless for cultivation or land that is desolate, barren, or ravaged.) categories in order to prepare wasteland maps for their effective management and sustainable development (Gautam and Narayan 1988) in both spatial and statistical forms.

An attempt has been made in the present study to map wasteland information using remote sensing data (IRS- 1D LISS III) to ascertain the changes taken place in the wasteland pattern which are effective in preparing and maintaining database and to upgrade the information very frequently to bring more and more wastelands under productive use through afforestation, plantation, agro-forestry, horticulture and other productive uses.

Study Area

Agra district forms the extreme southwestern corner of the state of Uttar Pradesh, lies between 24°44': 27°25' north latitude and 77°26': 78°32' east longitude covering an area of about 4805 km^2 (Figure.1). the study area has an almost flat surface with few irregularities. The general slope is from northwest to southeast and elevation is 176.8 m above the msl with an average slope one meter per five km. Since overall the region has gentle slope the area is not prone to intense soil erosion where soil is under vegetation cover.

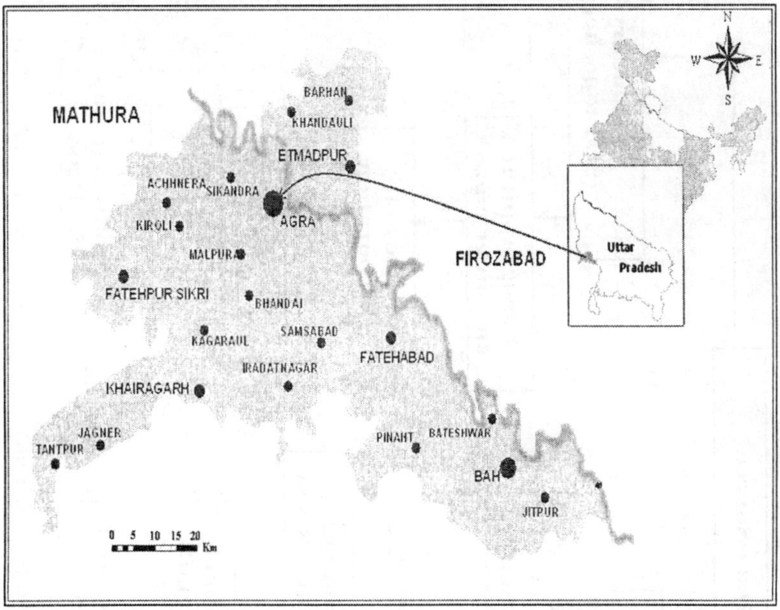

Fig.1 Location map of Agra District

The principal rivers of the region are Yamuna, Jherna, Sirsa and Sengar in the northeastern part whereas Chambal, Uttangan, Khari and Kavar rivers in the south and western part. The area around these rivers is severely affected by land degradation problems. Agra district includes 6 tehsils, 15 community development blocks, 15 towns and 940 villages (940 inhabited and 36 uninhabited). The climate of the area is semi-arid and temperature ranges from 45°C (May-June) to 0°C (January-February).

Database

The following data products have been used in the present study:

(i) Satellite image of IRS-1D LISS III acquired on 20 March 2003, Path 97/Row 52 on 1:50,000 scale.

(ii) Survey of India toposheets (1: 50,000 scale).

(iii) Secondary data such as Census Report and Wasteland Atlas of India.

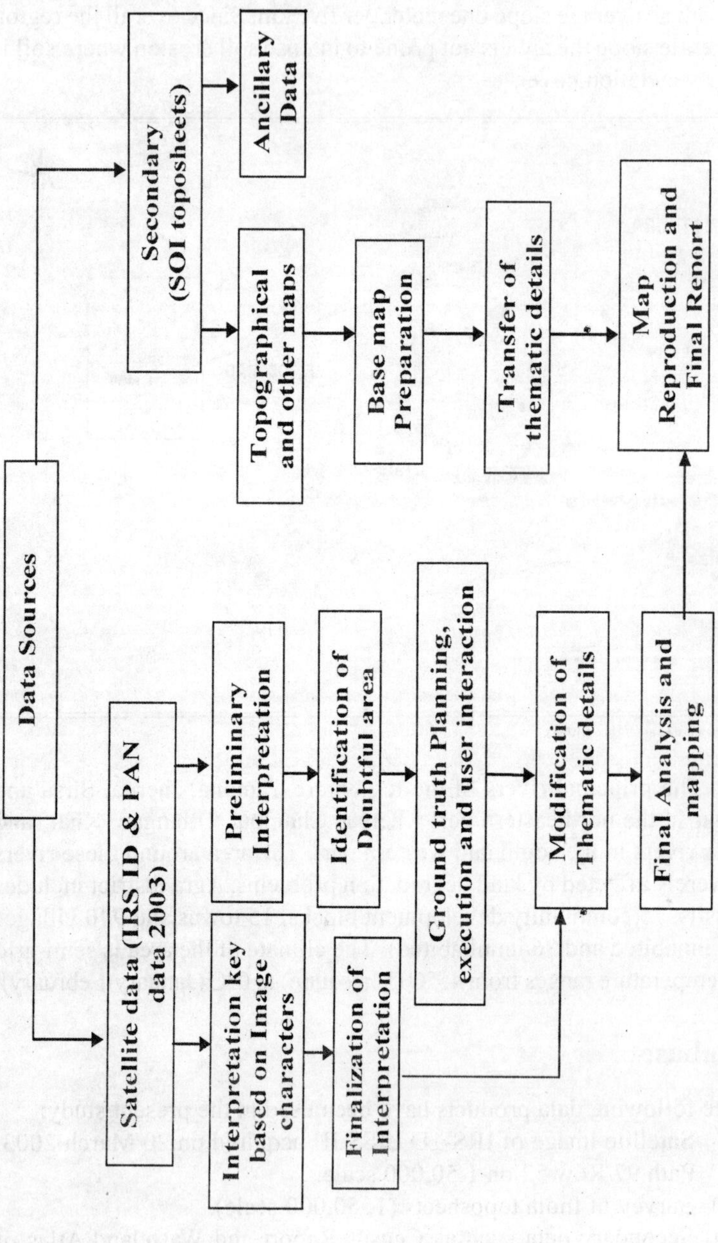

Fig 2 Methodology Adopted for the Study

Wasteland Mapping

The methodology essentially involved the visual interpretation of the standard FCC generated from the band 2(Green-0.52-0.59), 3(Red-0.62-0.68), 4(NIR-0.77-0.86) of IRS- 1D, LISS-III, for identification of different categories of wasteland using standard image interpretation techniques, based on tone, texture, pattern, color etc. in addition to field based knowledge of the interpreter. Other ancillary data included topographical maps for reference purpose to map the wasteland. The interpreted maps were checked for the ground truth verification and the wasteland boundaries were finalized.

Methodology

The Survey of India toposheets No. 54I/3, 54I/4, 54I/7 and 54I/8 (1:50,000 scale) were used for the preparation of base map. The IRS-1D LISS III data in the form of false color composites (FCC) generated from band 2, 3, and 4 on 1:50,000 scale was visually interpreted to prepare the wasteland map of Agra district. The variations in texture, shape, size, tone, pattern, association etc. were used to identify and delineate various wasteland categories and map was finalized after ground truth verification. The methodology and classification system was carried out using Manual of Instruction prepared by NRSA Hyderabad (1986). The GIS operation (Table-1) was used to derive the wasteland area. The methodology involves assimilation of GIS techniques by incorporating spatial (thematic map) and non spatial (attributes) data for effective utilization and management of wasteland and to create a wasteland information database (Figure 2).

Results and Discussion

The wasteland map on 1:50,000 scale of Agra district have been prepared based on visual interpretation of the FCC generated from IRS-1D LISS with required ground check. The various wasteland categories identified have been briefly discussed. The distribution of wasteland is shown in the study area (Figure 3).

The various wasteland categories in the study area (wasteland Atlas of India) are given in the Table 1. The areas of various wasteland categories in the Agra district have been computed and presented (Table 1). The change analyses between the year of 1993 and 2003 have been carried out in order to find out the temporal changes in the areas of wasteland and are given in Table 1.

The classification of wastelands as suggested by National Wasteland Development Board (1986) has been adopted for studying the distribution of various categories of wasteland. The different wasteland categories occurring in the study area and change analysis carried out from the years 1993 to 2003 are discussed briefly as follows:

1. Gullied and/or Ravenous land

Gullied and/or ravenous land are the result of the localized surface run off affecting the friable unconsolidated material resulting in the formation of perceptible channels resulting in undulating terrain. Networking of these gullies leads to the development of ravenous land. In Agra district, the area under gullied and/ or ravenous land has decreased from 5.4% to 3.9% from 1993 to 2003, which is a good indicator of reclamation of gullied land. Bushes/ grasses of fast growing varieties could be planted in such areas to stop further soil erosion and thus checking out the increment of such lands. However in some areas unplanned vegetations has developed (Figure.5) which has resulted in decrease in area of Ravines.

2. Upland with or without scrub

The land, which is generally prone to degradation, may or may not have scrub cover, occupies topographically high location. About 52.70

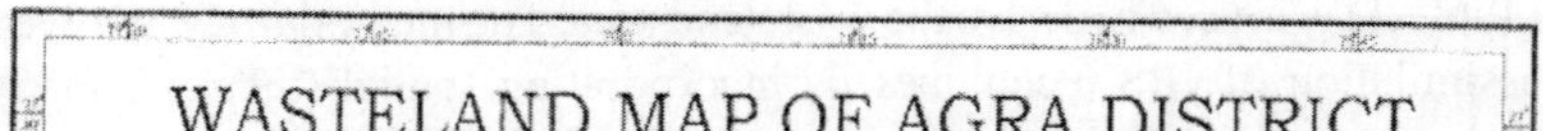

Fig.3 Wasteland Map of Agra District

km^2 area of Agra district is mainly covered by undulating uplands, which are mostly rocky with no soil cover. These upland lands (Figure.6) found in the south-western part of the district do not has any productive land usability as these are undulating terrain, low fertility and lack of water retention capacity.

3. Water logged and Marshy land

The land where the water is at/ or near the surface and water stands for most of the year. Waterlogged areas are also reduced from 2.43 km^2 to 1.60 km^2 during the time of ten years (1993-2003). It occupies small area but is a potential source of contamination for the population around (Figure.7) sub-surface drainage and shallow tube wells. These systems sink down the water table and reduce the water logging area. Afforestation with selective species of trees and plants help in lowering the water table on account of increased transpiration from the new vegetation.

4. Land affected by Salinity/ alkalinity

High rate of evaporation rate has caused concentration of salt of various types in the plain areas scattered in various part of the district (Figure.8). However the percentage of this category of wasteland has decreased from 0.46% to 0.07%. Most of the land is reclaimed under various scheme of USAR BHUMI SUDHAR PARIYOJNA (is a programme launched for the reclamation of land rendered unproductive due to excessive salinity and to increase the productivity of agriculture sector in parts of Uttar Pradesh) which has recommended for proper land leveling and banding to ensure uniform application of water and leaching. Application of amendments like gypsum and pyrite are quite effective.

5. Barren Rocky/stony

This wasteland category has decreased by 0.1% i.e. from 0.89% to 0.79% since 1993 to 2003. An area of 4.3 sq. km has decreased since 1993, from 38.3 sq. km in 1993 to 42.6 sq. km in 2003.

6 .Underutilized/ degraded notified forestland

Degraded forestlands have limited crown cover (about 10%). The relative concentration of scrubs, bushes and smaller trees are predominant and taller trees are limited (Figure.9)

In Agra district, about 230 sq. km of area were under degraded forests in 1993, which has been reduced to 109.6 sq. km in 2003 which shows a decrease of 2.5%. This clearly indicates the degradation of forest add the wasteland in the study area.

Fig.5 Gullied/Ravenous Land, Bah,Agra

Fig.6 Upland with and without scrub Jagner, Agra

Fig.7 Water logged and marshy land, Kindal, BrauliAheer, Agra

Fig.8 Land affected by salinity/ alkalinity, Achhnera, Agra

Fig.9 Under utilized/degraded forest l and, Mangatpur,Bah,Agra

Fig.10 Sandy Wasteland, Pinahat, Agra

Table 1 Wasteland Analysis of Agra District for the year

S.No.	Wasteland categories	1993		2003		Change Detected	
		Area (km^2)	Area (%)	Area (km)	Area (%)	Area (km)	Area (%)
1.	Gullied and/ or Ravenous Land	263.4	5.4	191.6	3.9	(-) 71.8	(-) 1.4
2.	Upland with or without scrub	67.8	1.4	52.7	1.1	(-) 15.1	(-) 0.3
3.	Waterlogged and marshy land	2.4	0.4	1.6	0.0	(-) 1.3	(-) 0.03
4.	Land affected by Salinity/ Alkalinity	22.1	0.4	18.9	0.3	(-) 3.2	(-) 0.07
5.	Barren Rocky/ stony	42.6	0.8	38.3	0.7	(-) 4.3	(-) 0.10
6.	Under utilized/ degraded notified forest land	230.0	4.7	230.0	2.2	(-) 120.4	(-) 20.5
7.	Sandy	-	-	0.8	0.02	(+) 0.8	(+) 0.02
	Total Wasteland	623.9	13.0	413.5	8.6%	(-) 215.4	(-) 4.4%

7. Sands-deserted land This category has stabilized accumulation of sand in reverie or inland areas, which was not traceable in 1993. About 0.8 sq. km of land had been turned sandy as interpreted from satellite images (IRS-1D, LISS-III) of 2003 and field checks at few places near Pinnahat (Figure.10).

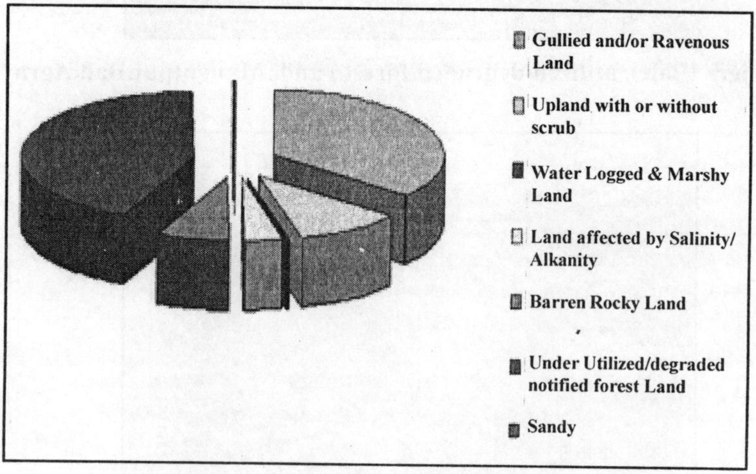

Fig. 3. 3D Pie-Diagram showing various categories of wasteland of Agra district, U.P.

Conclusion

Information obtained on wasteland of Agra district for the periods 1993 to 2003, and the trend of the changes in the areas of different categories of wastelands would undoubtly be of great values to various agencies, concerned with land reclamation, soil conservation and afforestation, for planning, selecting priority areas, allocating funds and executing reclamation measures on sound scientific basis.

The wasteland map of the district (Figure.3) indicates the areas along the rivers and central portion are severely affected by gully erosion; sandy area of the district is also encountered along the course of the river. Rocky barren land is found mainly in the south-western part of the district where Vindhyan sandstone is found as outcrop. Land affected by salinity/alkalinity is found scattered in north- eastern and in the western part. Category wise picture of wasteland as revealed by remotely sensed data indicates that the total agricultural land during 1993-2003 has been reduced and degradation of forest cover and sandy areas has been increased. The area under gullied and/or ravenous land has decreased from 5.4% in 1993 to 3.9% in 2003. Bushes/grasses of fast growing varieties could be planted to stop further erosion of topsoil. There has been a decrease of underutilized/degraded-notified forestland due to fast increasing population and growing demands for food, fodder and fuel etc. from 4.7% in 1993 to 2.2% in 2003.

In Agra district, the rocks belong to the Vindhyan Super Group covering an area of about 38.3 sq. km is categorized as barren/rocky stony and about 52.7 sq. km areas is occupied by uplands with or without scrub with a scanty soil cover. Efforts need to be made to reclaim by growing plants on rocky slopes and upland areas. It was also noted in the year of 2003 about 0.8 sq.km area of Agra under sand deserted land, which was not traceable in 1993. This land may require immediate attention to reclaim by a forestation under various governmental programmers to check the increase of desertification. Action plans also be generated on village and tehsil wise to initiate steps for reclamation. Availability of such details on wasteland helps to prepare effective micro level developmental plans for land uses.

Acknowledgment

The authors wish to thank Dr. M. Kudrat, Scientist, IIRS, Dehradun for this guidance any valuable suggestions. Financial assistance provides by UPCST is also acknowledge.

References

- Gautam N.C. and L.R.A.Narayan., 1988, Wasteland mapping in India. Pink Publishing House, Mathura.
- National Remote Sensing Agency., 1986, Manual of procedures of wasteland mapping, National Remote Sensing Agency, Hyderabad.
- National Remote Sensing Agency., 1991, Guidelines to use wasteland maps. National Remote Sensing Agency, Hyderabad.
- National Remote Sensing Agency., 1995, Wasteland Atlas of India, National Remote Sensing Agency, Hyderabad.
- National Wasteland Development Board., 1987, Wasteland definition and classification. Technical Task Group Report, Ministry of Environment and Forests, Govt. of India New Delhi.

20. An Analysis of Unsustainable Urbanization, Land Use and Urban Environment in the National Capital Territory Delhi (NCTD)

Rashmi Rani Anand

India is facing a serious crisis of urban growth at the present time. Whereas urbanisation has been an instrument of economic, social and political progress, it has led to serious socio-economic problem. The sheer magnitude of the urban population, haphazard and unplanned growth of urban areas, and a desperate lack of infrastructure are the main causes of such a situation. The rapid growth of urban population both natural and through migration, has put heavy pressure on public utilities like housing, sanitation, transport, water, electricity, health, education and so on. Poverty, unemployment and under employment among the rural immigrants, beggary, thefts, decoities, burglaries and other social evils are on rampage. Urban sprawl is rapidly encroaching the precious agricultural land. A significant facet of Indian urbanization has been greater concentration of population in metropoliam cities, especially in the recent years. Delhi is one of the fastest growing cities in the country. The Delhi is growing at unprecedented manner in terms of demography and space and has given rise of various problems related to urbanization.

Key words :- Urbanization, Migration, Land Uses, Growth Rate Sustainable

Introduction

Sustainable urbanization promotes the concepts of access and equity of resources and safety/ security to all its inhabitants[1]. Unprecedented growth of Indian cities in terms of demography and space has given rise to various problems related to urbanization. Almost all cities and towns here are facing acute problems of rural-urban migration, lack of urban governance, transport crisis, urban poverty, increasing rate of crime, high density living, pollution of air and water, insufficient job opportunities etc. with varying scale and intensity. So it can be said that current urban form is unsustainable and economically unproductive and there is an immediate need to have integrated ecosystem with proper urban land use planning in order to have equilibrium and to make Delhi as a sustainable city. A significant facet of Indian urbanization has been greater concentration of population in metropolitan cities, especially in the recent years[2]. Delhi is one of the fastest growing cities in the country. It is considered to be the second largest metropolitan city next to Mumbai with approximately 1,38,50,507 people dwelling in New Delhi, according to Census 2001. Delhi, the National Capital Territory (NCT)[3], has been emerged as a gigantic magnet, attracting people not only from the proximate states of Haryana, Rajasthan and U.P but also from the far-flung regions of the country by offering diverse scope for employment and opportunities.

Study area

The Delhi Capital Territory covers total area of 1483 Sq kms. As per Population Census 1991, its rural and urban composition is as following Rural Area: 797.66 Sq. Kms, Urban Area: 685.34 Sq. Kms, total = 1483 Sq. Kms. It includes areas administered in tandem by the Government of Delhi and three Municipal Corporations i.e MCD, NDMC and Cantonment Board. The largest of these is the Metropolitan Corporation of Delhi (MCD). The National Capital of Territory of Delhi is divided into 15 zones as per the Master Plan of Delhi-2001, notified by the Central Govt. on 1.9.90, under the Delhi Development Act 1957. Out of these 8 Zones (A to H) are in Urban Delhi and 6 Zones J to N & P fall in Urban Extension /Rural Area. Zone-O covers the area of River Yamuna (River Front Area). Delhi was considered as a single district for Population Census 1991. In 1996, government of N.C.T. of Delhi created 9 districts.

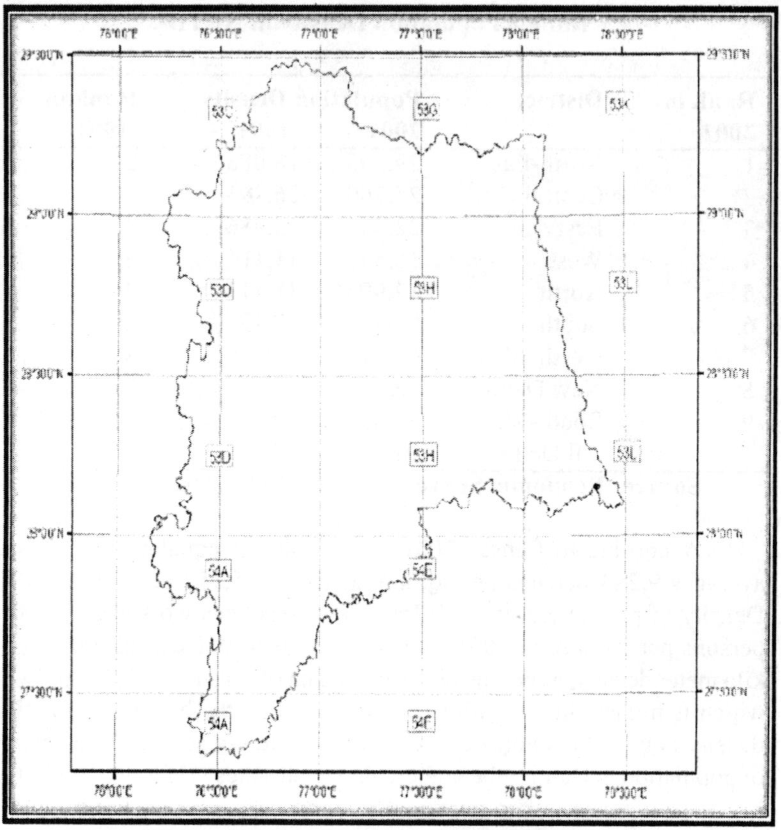

Figure-1 Location Map of the Study Area

Source: Survey of India Toposheet, 1:25000

Urbanization in Delhi

The urbanization of Delhi dates back to the beginning of the 20th Century. In 1901, 52.76% of the population of Delhi was urban. The urban area in Delhi territory has increased from 22% in 1961 to 62.5% of the total area in 2001.

The rapid urbanization of Delhi has resulted in sharp increase in the density of population. In 1901, the density was 274 persons per square km., this increased to 1176 persons per sq. km. in 1951 and 9294 persons per sq. km. in 2001. The total density of Delhi is 9340 per sq. km. The density of population in Delhi is also the highest among all States and Union Territories.

Table 1 Population Density in NCTD

Rank in 2001	District	Population Density 2001	1991	Rank in 1991
1	North-East	29,395	18,088	2
2	Central	25,760	26,261	1
3	East	22,637	15,986	3
4	West	16,431	11,116	5
5	North	12,996	11,471	4
6	South	9,033	6,012	6
7	North-West	6,471	4,042	8
8	New Delhi	4,909	4,791	7
9	South-West	4,165	2,583	9
	All Delhi	9294	6352	

Source: Economic Survey of Delhi, 2001-2002

According to Census 2001, the density of population, in Delhi, averages 9,294 persons per sq. km. as against 6,352 persons in 1991. Density of population at All-India level has been worked out at 324 persons per sq. km. in 2001. According to the 2011 census per square kilo meter density, as per the provisional data of census, is 11,297 persons which is highest in the entire country. In India, the highest population density is in Delhi's north-east district (37,346 per sq km). The density of population in Delhi is the highest among all states / UT's in the country.

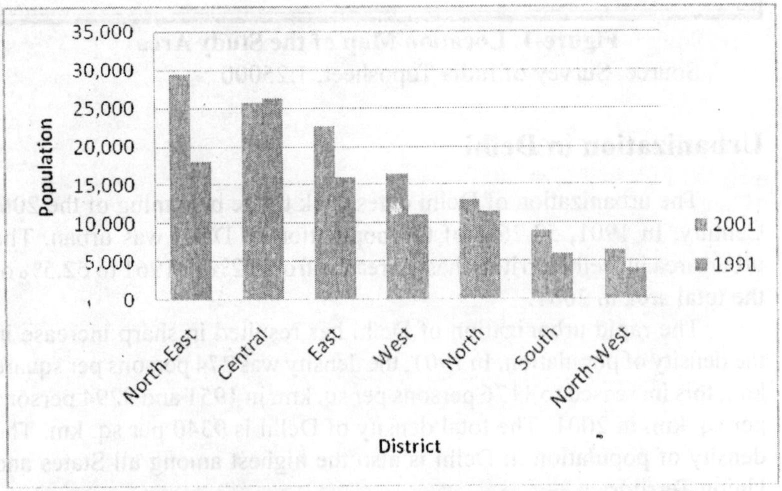

Figure 1: Population Density of NCTD, 1991 and 2001

As per Census report 2011, the population of Delhi is 1.67 crores with the increase of 29 lakh in last 10 years. The 2001 Census recorded 137.83 lakh population of Delhi with 3.81% annual growth rate and 46.31% decennial growth rate during 1991-2001. More than the 45% of Delhi's population resides in slums, unauthorized colonies and other unplanned settlements[4]

Table 2

Rank in 2001	**District**	**Population 2001**	**% of total Population of District**	**Population 1991**	**% to total Population of District**	**Rank in 1991**
1	North-West	2,847,395	20.66	1,778,268	18.88	1
2	South	2,258,367	16.38	1,502,878	15.95	2
3	West	2,119,641	15.38	1,434,008	15.22	3
4	North-East	1,763,712	12.80	1,085,250	11.52	4
5	South-West	1,749,492	12.69	1,084,705	11.51	5
6	East	1,448,770	10.51	1,023,078	10.86	6
7	North	779,788	5.66	688,252	7.31	7
8	Central	644,005	4.67	656,533	6.97	8
9	New Delhi	171,806	1.25	167,672	1.78	9
	Total	13782976	100.00	9420644	100.00	

District-wise Population of NCT

Source: District Census Handbook, 2001 – 2002.

Note: The population of nine districts for 1991 is derived by recasting the Census data of 1991 Census according to the present jurisdiction of the districts.

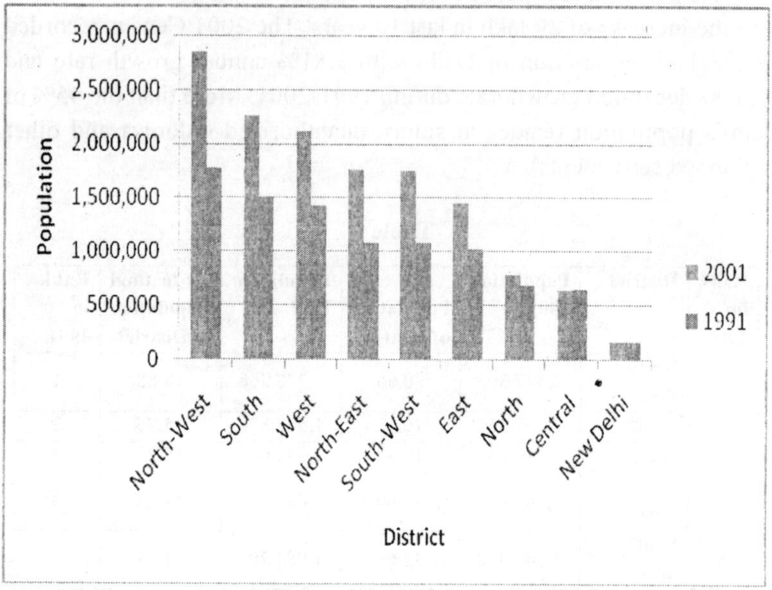

Figure 2: District-Wise Population of NCTD, 1991 and 2001

Migration

Migration has been a major driver in the population increase in the city. The Delhi being the vibrant centre of trade, commerce with excellent employment opportunities attracted people from all over the country. It was a small town in 1901 with a population 0.4 million. Its population started increasing after it became the capital of British India in 1911. During the Partition of the country, a large number of people migrated from Pakistan and settled in Delhi, and in the decade 1941-51 the population growth rate was 90%. The annual average exponential growth rate of population of Delhi was the higest (6.42%) during 1941-1951 due to large-scale migration from Pakistan to India post partition in 1947. The migration process continued to the city ever since partition[5]. The following table shows the trend of migration in Delhi from 1991 to 1999.

Table 3
Trend of Migration in Delhi between 1991 and 1999

Year	Population as on 1st July (Lakhs)	Increase in population over previous year (Lakhs)	Total Birth	Total Deaths	Natural increase (4-5)	Increase due to migration (3-6)	Ratio Natural to migration
1	2	3	4	5	6	7	8
1991	95.50	3.89	2.72	0.61	2.11	1.78	1:0.84
1992	99.37	3.87	2.74	0.62	2.12	1.75	1:1.02
1993	103.38	4.01	2.70	0.64	2.06	1.95	1:1.18
1994	107.50	4.12	2.62	0.68	1.94	2.18	1:1.42
1995	111.74	4.24	2.75	0.69	2.06	2.18	1:1.38
1996	116.10	4.36	2.83	0.76	2.07	2.29	1:1.50
1997	120.57	4.47	2.89	0.71	2.18	2.29	1:1.55
1998	125.14	4.57	2.84	0.80	2.04	2.52	1:1.79
1999	129.82	4.68	2.88	0.79	2.09	2.59	1:1.24

Source: Economic Survey of Delhi, 2001-02

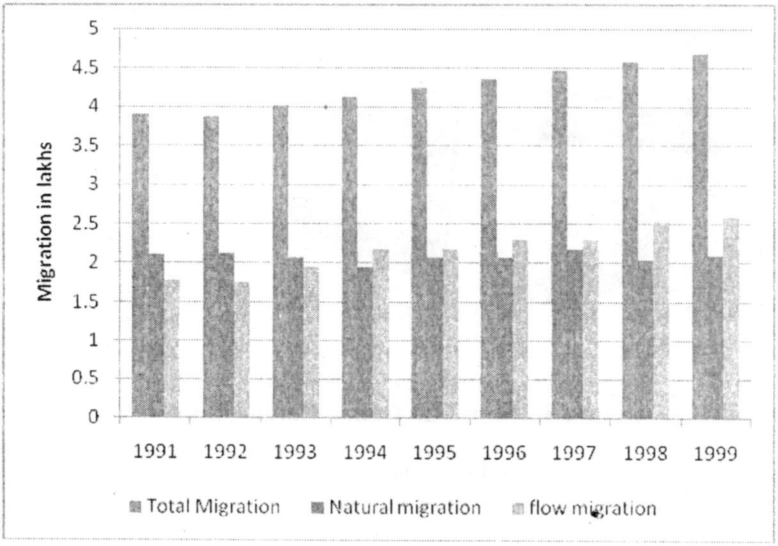

Figure 3: Trend of migration during period of 1991 - 1999

Meanwhile, the table given below shows the percentage of migration taken place from various States in Delhi during Census period 1981 to 1991:

Table 4
Migration from different states

States	% of migration taken place
Uttar Pradesh	49.61%
Haryana	11.82%
Bihar	10.99%
Rajasthan	6.17%
Punjab	5.43%
West Bengal	2.79%
Madhya Pradesh	2.71%
Other States	10.48%

Source: Economic Survey of Delhi, 2001-2002

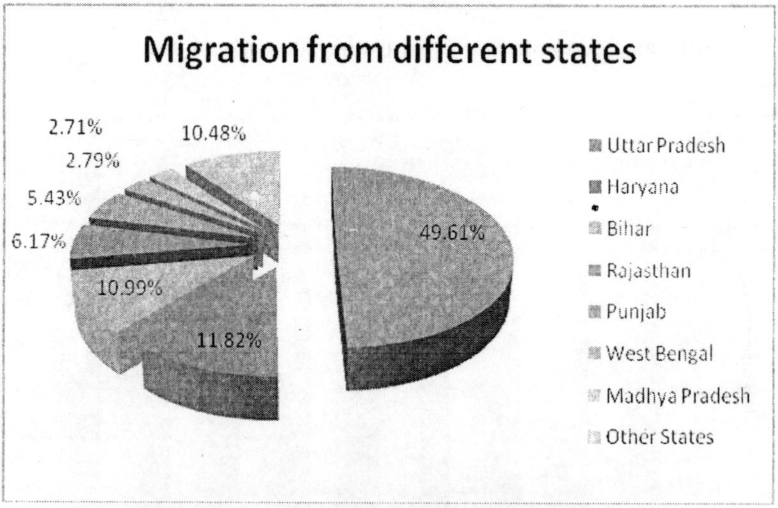

Figure 4: Migration from different states

Migration has resulted into long term deterioration of urban open space and rapidly causing adverse impact on urban ecology. Unskilled migrants often living in marginalised waste-lands and open space, cause to urban sprawl and mushrooming of informal settlements. These settlements have lack of basic infrastructural facilities like water supply, drainage system, organised refuse disposal etc. and act as breeding point of several diseases, crime and delinquency. The census data also reported that almost 15.72 percent population of the NCT of Delhi resides in slum clusters while Delhi Government declared that 52 percent of its population is residing in JJ Clusters, Slum Designated Area, Unauthorized Colonies, Resettled Colonies, etc; may be considered as slums[6] .

So, these unskilled migrants, to a great extent, are responsible for urban environmental and urban ecology degradation and at the same time victims of the same. Unplanned urbanisation puts strain on available resources and creates conditions even more conducive to crime. High population density leads to competition for limited resources. On the other hand, state's inability to provide the necessary infrastructure such as houses, schools, hospitals etc., leads to increase in haphazard and unplanned land use in the city.

The brief history of land use urban planning of NCT

In Delhi —the Delhi Capital territory—the history of development and planning has experienced subsequent changes over time since its formulation. The era of urban planning in Delhi commenced in 1824 when a Town Duties Committee was constituted for the development of the Cantonment, Khyber Pass, the Ridge area and the Civil Lines area by the Britishers. Therefore, urban development picked up in Delhi with the composition of the first Municipal Committee of Delhi in 1883. The Municipal Building byelaws were first made operational in Sadar Baazar, Subzi Mandi and other suburbs.

In order to address the uncontrolled haphazard growth of Delhi, Urban planning approach was adopted by the planners. Post-independence Delhi was immediately overwhelmed by deluge of refugee immigrants from West Pakistan as a result of partition. Many colonies were set up in the south and west like Kailash colony, Kalkaji, Malaviya Nagar, Moti Nagar and Kirti Nagar to provide for this huge Punjabi population influx[7]

It was further realised that in order to have proper land use of Delhi, there is a necessity of Planning Body. In 1957, the Delhi Development authority (DDA) was constituted through an act of parliament. Under DDA guideline, the Land use Plan for Delhi i.e. Delhi Master Plan (1962) was prepared. The plan was basically land use which took projected population into consideration and tried to develop the infrastructure. For that reason DDA undertook large scale acquisition of land in order to control future development of Delhi.

There have been two previous master plans in the city, the MPD 62 and the MPD 2001. The first master plan was developed in consultation with the Ford Foundation. The second plan was to be formulated 20 years hence, that is in 1981 but with the advent of the 1982 ASIAD games it was put at hold and the new plan came out only in 1991. These were mainly land development plans, the strategy being that DDA was to acquire new lands and develop them for housing and commercial purposes with a subsequent auctioning of flats/shops/plots. Throughout, planning followed a top down approach with the formulation of a master plan and keeping in mind its aims, subsequent zonal and sub zonal plans to be made[8].

The present study is the analysis of urban land use change pattern NCT (National Capital Territory). The urban planning is basically an efficient form of land use in order to cope up with uncontrolled growth of the city.

Land Uses

Land use is the utility of land for various purposes by human being like agricultural, industrial, residential, recreational and other purposes. In other words, it is use of land by human beings. Land use has received considerable attention since the 1968 and 1971 under Town and Country Planning Acts, and local authorities are now required to make annual returns of land use change in their areas. To ensure comparability, a National Land Use Classification (NLUC) has been developed and some authorities have also developed their schemes.

Table 5

Area used /to be used for Urbanization in NCT

S.N.	Land Use	Area (ha)	% age of total area (ha)
1.	Total area	148,300.00	100.00
2.	Built up area (As per IRS IC LISS III Statelite data 1999)	70162.0	47.31
3.	Natural Features (Forest, Wild Life Sanctuary, Ridge,River Yamuna and Other Water Bodies/Drains)	19509.1	13.16
4.	Sub – Total (Built-Up + Natural Features)	89671.1	60.47
5.	Balanced land available in NCT-Delhi (1-4)	58628.9	39.53

6.	Land to be kept reserved for:		
I	Disposal of Solid Waste generated up to 2051 (sanitary landfill & statutory green belts)	10,000.0	6.74
II	Metro Services/Utilities e.g. power plant, grid stationwater and sewerage treatment plant, etc.	10,000.0	6.74
III	Agriculture zone in NCT Delhi including dairy farming, horticulture, greenbelts etc.	11,000.00	7.42
7.	Sub-Total-6	31,000.00	20.90
8.	Proposed/Actual land available for urbanization	27,629.00	18.60
9.	Total Urbanisable area 2021 (including built up area1999) (2+8)	97790.9	65.94
10.	Population, which can be accommodated in 97790.90 hect. @ 225 PPH = 220 lakh		

Source: IRS, IC, LISS III, 1999 Remote Sensing Data.

The use of land in an attempt to avoid land use conflicts. Land use plans are implemented through land division and use ordinances and regulations, such as zoning regulations.

As Albert Guttenberg (1959) wrote many years ago, 'Land use' is a key term in the language of city planning[9]. Commonly, political jurisdictions will undertake land use planning and regulate Land use and land management have become important for the urban ecosystem due to inevitable process of urbanization in NCT. Change in land use takes place through two ways:

- Natural areas like water bodies and forest land etc are converted for habitation and industrial activities.
- Agricultural areas are converted for the township development and habitation.

In terms of availability of urbanisable land in NCT-Delhi 2021, there are clearly severe limitations with respect to urbanisable land in Delhi. In the words of the Base Paper for Preparation of Regional Plan 2021, there is specifically a "significant shortage of land to accommodate the 2021 projected population within NCT-Delhi". In order to accommodate this projected population – estimated at 230 lakhs by the draft MPD 2021 – in NCTD, a review of available land for urbanization must be made. Table 1 details land availability figures drawn up by the Draft NCR Plan 2021.LAND USE GIVEN IN MPD-2021 AD[10].

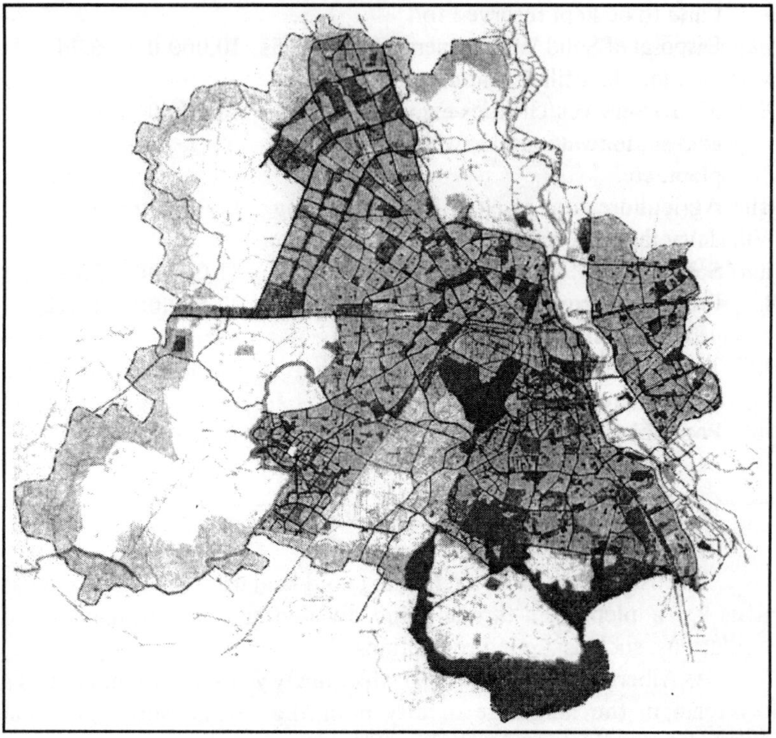

MAP 2: Land Use Map 2001-21

Source: Satellite Imagery 2007, 1 :25000

The above figures propose that only 220 lakh of population can be accommodated in the 97790.90 hectare of land. Therefore, this much of land is urbanisable land in the NCT. With the rapid pace of urbanization, the Delhi is growing at unprecedented manner in terms of demography and space and has given rise of various problems related to urbanization. The above available figures suggest that there is great pressure over NCT in terms of sustainable urbanization. The sustainable urbanization of NCT means appropriate shelter, food, employment etc for everyone. The city which is free from the acute problems of rural-urban migration, lack of urban governance, transport crisis, urban poverty, increasing rate of crime, high density living, pollution of air and water, insufficient job opportunities etc. with varying scale and intensity.

Therefore, the major issues regarding urbanization of NCT are the lack of urban planning in terms of urban environment and land use. The haphazard growth of Delhi in recent years resulted in demand of land for various uses and significant environment hazards. According to Rahman the Delhi ridge, which was once considered the lungs of Delhi is fast degrading. There was a considerable decrease in the ridge area, from 6.7 % in 1992 to 5.5 % in 2004, because of continuous illegal tree cutting, quarrying and construction activity especially in the S-E around Vasant Vihar, Vasant Kunj and other areas. Also there is a reduction (16. 8 %) in agricultural land because of urban expansion in the fringing areas affecting air, water, land and overall urban environment quality significantly.

In order to cope up with the problem of unsustainable urbanization of Delhi, good governance and urban land use and environment planning is required. Else this will lead to the decline in investment and business activities. Tourist may put off causing hamper to the tourism industry. Fear of crime, income inequality, leads to fragmentation and polarisation in cities with gated community. Therefore, unsustainable urbanization and urban planning at NCT crime has emerged as major challenge of urbanization.

Foot Notes

1 Anand , Rashmi. Rani (2010), "Unsustainable Urbanisation in South Africa: A Historical Appraisal", *Newsletter of African Studies Association of India*, vol III (3) pg:

2 National Capital Region Growth and Development, (1996), *National Capital Planning Region*, Har-Anand Publication Private Limited.

3 The NCT is a recent name: before 1994, the same area carried the name Delhi Union Territory (DUT). The DUT was founded in 1911, when Delhi became the capital of British India. Later, in 1947, Punjab and Haryana were split up, and Delhi was given a considerable rural area around the city.

4 Ministry of India (2002), Economic Survey of Delhi 2001-2002, Ministry of Finance, New Delhi.

5 Ministry of India (2001), Economic Survey of Delhi 2001-2021, Ministry of Finance, New Delhi.

6 Shtiyaq and Kumar (2010), "Typology of Informal Settlements and Distribution of Slums in the NCT of Delhi", Institute of Town Planners, vol (7 – 3): Pg:

7 http://home.iitk.ac.in/~skandoi/Urban%20Planning%20in%20India%282%29.pdf

8 *http://www.ccsindia.org/ccsindia/interns2006/Delhi%20Master%20Plan%20-%20Snigdha.pdf*

9 *http://www.ireference.ca/search/Land%20use/*

10 http://www.ccsindia.org/ccsindia/interns2006/Delhi%20Master%20Plan%20-%20Snigdha.pdf

References

- Anand , Rashmi. Rani (2010), "Unsustainable Urbanization in South Africa: A Historical Appraisal", *Newsletter of African Studies Association of India*, vol III (3) pg: 26-32.
- Berry, B.J.L and Spodek, H, (1971), "Comparative Ecologies of Large Indian Cities", *Economic Geography,* vol 47,pg 266-85.
- Guttenberg, Albert Z, (1959), "A Multiple Land Use Classification System", *Journal of the American Planning Association,* vol 25: 3,pg: 143 — 150.
- Ishtiyaq and Kumar (2010), "Typology of Informal Settlements and Distribution of Slums in the NCT of Delhi", *Institute of Town Planners,* vol (7 – 3).
- Jain, A.K. (1990), *The Making of a Metropolis: Planning and Growth of Delhi,* National Book Organisation, Delhi.
- Master Plan for Delhi 2021, DDA, Delhi.
- Ministry of India (2001), *Economic Survey of Delhi 2001-2021*, Ministry of Finance, New Delhi.
- Ministry of India (2002), *Economic Survey of Delhi 2001-2002*, Ministry of Finance, New Delhi.
- National Capital Region Growth and Development, (1996), *National Capital Planning Region*, Har-Anand Publication Private Limited.
- Rahman, Atiqur et al (2007) *An Assessment of Urban Environmental Issues Using Remote Sensing and GIS Techniques: An Integrated Approach: A Case Study: Delhi, India,* Paper Presented to the Pripode workshop on Urban Population, Development and Environment Dynamics in Developing Countries Organised by CICRED, PERN and CIESIS, Nairobi (11- 13 June), available at websitehttp://www.cicred.org/Eng/Seminars/Details/Seminars/PDE2007/Papers/Atiqur-RAHMAN_paperNairobi2007.pdf

Web Sources

- http://www.ccsindia.org/ccsindia/interns2006/Delhi%20Master%20Plan%20-%20Snigdha.pdf.
- http://www.ireference.ca/search/Land%20use/
- http://home.iitk.ac.in/~skandoi Urban %20 Planning% 20in% 20India % 28 2%29.pdf.
- http://www.real-estate-india.com/delhi-city.html

Section-IV
Sustainable Development and Energy

21. Energy-Environment: Interaction and Linkages

Vinita Katiyar
S.K. Verma

Energy plays a crucial role in economic development. Energy is an essential input for all productive economic activities, economic development and human resources development. As the growth rate of economy is increased, the consumption of energy inevitable increases high level. India is a poor country, where about 30 percent of population is either below the population the poverty line or very close to it, so per capita consumption of energy in India is one of the lowest in the world. The level of per capita energy supply is a good indicator of the level of economic Development. The per capita consumption in India is far below that in other countries. India is suffering energy shortage and poor quality of supply of electricity. Energy generation and end-use adversely effected the environment via producing externalities, so to achieving sustainability in the energy chain it is important to identify, measure, value and integrate the impact of energy sector on environment, so for sustainable economic development alternative energy sources are required. It usually also includes technologies that improve energy efficiency. Energy efficiency and renewable energy are said to be the twin pillars of sustainable energy.

Key words: Renewable energy, Sustainability, Energy efficiency, Sustainable energy.

Introduction

Energy plays an important role in economic development. There is positive correlation between economic growth and per capita income and per capita consumption of energy. Since energy is an essential input for all productive economic activities, the process of economic activities, the process of economic development inevitable increasing high level energy. Consumption of energy in domestic uses and public lighting also increases with growing affluence. E.A.G.Robinson has denoted that a 3 percent rise in industrial production in world accompanied by a 2 percent increase in energy consumption. India is a poor country, where about 30 percent of population is either below the population the poverty line or very close to it, well over half of the population does not have the purchasing power to buy commercial energy. Population of this section involve mostly rural population survive on non commercial energy. India faces challenges in meeting its energy needs and providing adequate and varied energy of desired quality to users in a sustainable manner and at reasonable costs. India needs economic growth for human development, which in turn requires access to clean, convenient and reliable energy for all. As India near the 8-10% growth rate that we aspire for, the quantity & quality of energy we need, will increase substantially. Thus the energy challenge is of fundamental importance.

Methodology

The present study is entirely based on secondary data and data were obtained from the various reports of Reserve Bank of India (RBI), Central Statistical Organization (CSO), National Accounts Statistics, Various publications of Indian Government, National Bank for Agriculture and Rural Development (NABARD), Various Economic Surveys and Internet. The data collected, tabulated was subjected to statistical analysis as per the objectives of the study. Tabular analysis and graphical representation have been used for analysis the data.

Objectives of Study

Paper mainly focus linkage of energy and environment in term of effects of energy production and end –use on environment. Paper also includes Indian approach about climate change related to energy. The main objectives of this paper are -

- Evaluate the trends and pattern supply and consumption of energy in India.
- Focus on effect of energy generation on environment.
- Focus on effect of energy end –use on environment

- Highlight the energy sustainability in India

Supply and Consumption of Energy in India

Per capita consumption of energy in India is one of the lowest in the world. India consumed 439 kg of oil equivalent (kgoe) per person of primary energy in 2003 compared to 1090 in China, 7835 in the U.S. and the world average of 1688. India's energy use efficiency for generating Gross Domestic Product (GDP) in Purchasing Power Parity (PPP) terms is better than the world average; China, U S and Germany as shown in Table 1. Clearly, significant reduction in the energy intensity of growth can be achieved based on existing technologies

Table-1
Selected Energy Indicators for 2003

Country	Total Primary Energy Supply Per Capita [Kilogram of Oil Equivalent(Kgoe)]	Total Primary Energy Supply/GDP (K geo/$2000PPP)	Electricity Consumption Per Capita (K wh)*	K wh/$-2000PPP
China	1090	0.23	1379	0.29
Australia	5630	0.20	10640	0.38
Germany	4210	0.17	6898	0.27
India*	**439**	**0.16**	**553**	**0.20**
Saudi Arabia	5805	0.46	6481	0.52
United Kingdom	3906	0.14	6231	0.23
United States	7835	0.22	13066	0.37
Japan	4052	0.15	7816	0.29
World	**1688**	**0.21**	**2429**	**0.31**

**Data for India are corrected for actual consumption and the difference in actual and IEA assumed calorie content of Indian coal.*

Source: *IEA (2005), Key World Energy Statistics 2005, International Energy Agency, Paris, http://www.iea.org*

The level of per capita energy supply is a good indicator of the level of economic development. The per capita consumption in India is far below that in other countries. Moreover, access to electricity is very uneven. Even though 85 percent of villages are considered electrified, around 57 percent of the rural households and 12 percent of the urban households i.e. 84 million households (over 44.2% of total) in the country did not have electricity in 2000. Improvement in human development is also strongly associated with access to electricity. Even those who have access to electricity suffer from shortages and poor quality of supply. Unscheduled outages, load shedding, fluctuating voltage and erratic frequency are common. Consumers and the economy bear a large burden of the consequences of this poor quality of supply. The extent of power shortage varies from state to state. In 2004-05, the peak shortage varied from 0 to 25.4% with an all-India average of shortage 11.7%. Similarly, energy shortage also varied from 0 to 20.1% with an all-India average of 7.3%.

Table-2

Power supply, demand and Shortage in India (in million units)

Year	Energy Availability	Energy Requirement	Energy Deficit /Surplus	%Energy Deficit /Surplus
2002-03	4,97,890	5,45,983	-48,093	-8.8
2003-04	5,19,398	5,59,264	-39,866	-7.1
2004-05	5,48,115	4,91,373	-43,258	-7.3
2005-06	5,78,819	6,31,757	-52,938	-8.4
2006-07	6,24,495	6,90,587	-66,093	-9.6

Source: *Central Electricity Authority (CEA)- Power Scenario at a Glance(2007-08)*

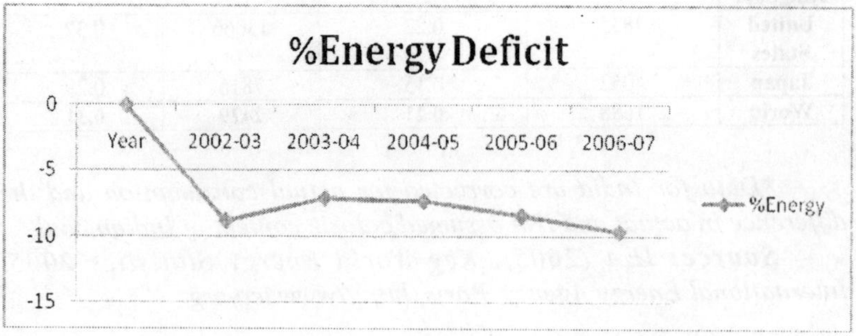

Figure –1 Percent Energy Deficit /Surplus

Consumption pattern of energy via different energy source is not similar. Contribution of renewable sources of energy in total energy consumption in India is only 27.4 percent, rest 72.8 percent energy is used non- renewable sources including coal, oil, natural gases, hydropower, and nuclear energy. Coal is largest energy source of India, it contribute 40.6 percent in total energy consume in India. Consumption of natural gases energy only 5.6 percent, oil 23.7 percent, hydro 1.8 percent. Least energy consumes is nuclear energy obtained by nuclear resources. Total energy consumption by different type sources in India (2007) is shown in Figure –2.

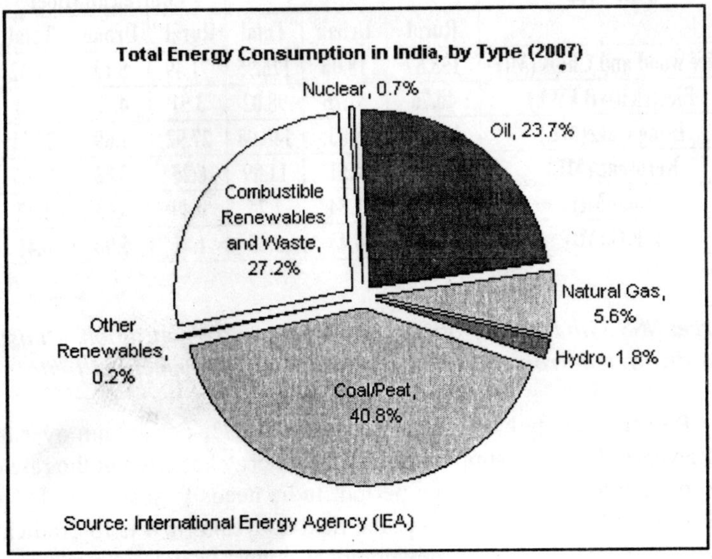

Figure -2 Percent Energy consumption in India by different sources in 2007

A majority of India's people use traditional fuels such as dung, agricultural wastes and firewood for cooking food. These fuels cause indoor pollution. The 1999-2000 55th rounds of the National Sample Survey (NSS) revealed that for 86% of rural households the primary source of cooking energy was firewood and woodchips or dung cakes. In urban areas as well, more than 20% of all households relied mainly on firewood and chips. Only 5% of the households in rural areas and 44% in urban areas used LPG. Kerosene was used by 22% of urban households and only 2.7% of rural households. Other primary sources of cooking energy used by urban and rural households include coke, charcoal, gobar gas (cow dung gas), electricity and other fuels. The use of traditional fuels for cooking with the attendant pollution and the cost of gathering the imposes a heavy burden on people, particularly women and girls. Household energy consumption pattern in India shown in table-3

Table - 3
Household Energy Consumption in India (July99 – June 00)

Fuel Type	Physical Units			Million Tonnes of Oil Equivalent(Mtoe)		
	Rural	Urban	Total	Rural	Urban	Total
Fire wood and Chips (Mt)	158.87	18.08	176.95	71.49	8.13	79.62
Electricity(B kWh)	40.76	57.26	98.02	3.51	4.92	8.43
Dung Cake(ML)	132.95	8.03	140.98	27.92	1.69	29.61
Kerosene (ML)	7.38	4.51	11.89	6.25	3.82	10.07
Coal(Mt)	1.20	1.54	2.74	0.49	0.63	1.12
L.P.G.(Mt)	1.25	4.43	5.68	1.41	5.00	6.41

Source: *NSS 55th Round, National Sample Survey Organization, Ministry of Statistics and Programme Implementation, Government of India*

Power capacity has risen at the rate of 5.87% per annum over the last 25 years. The total supply of electrical energy has risen at the rate of 7.2% per annum over the same period. India needs to sustain an 8% to 10% economic growth rate, over the next 25 years, if it is to eradicate poverty and meet its human development goals. To deliver a sustained growth rate of 8% through 2031-32 and to meet the lifeline energy needs of all citizens, India needs, at the very least, to increase its primary energy supply by 3 to 4 times and, its electricity generation capacity/supply by 5 to 6 times of their 2003-04 levels. With 2003- 04 as the base, India's commercial energy supply would need to grow from 5.2% to 6.1% per annum while its total primary energy supply would need to grow at 4.3% to 5.1% annually. By 2031-32 power generation capacity must increase to nearly 8, 00,000 MW from the current capacity of around 1, 60,000 MW inclusive of all captive plants. Similarly requirement of coal, the dominant fuel in India's energy mix will need to expand to over 2 billion tonnes/annum based on domestic quality of coal. Meeting the energy challenge is of fundamental importance to India's economic growth imperatives and its efforts to raise its level of human development.

Energy and Environmental Linkage

Energy sector effects environment in many dimensions. Energy end-use not only adversely effected the environment but the production of energy also effects environment via producing externalities, so to achieving sustainability in the energy chain it is important to identify, measure, value and integrate the impact of energy sector on environment. Environmental concerns are associated with all forms of energy including fossil fuels, nuclear energy, and renewable sources of energy, throughout the energy chain from exploration or mining, transportation, generation and end-use. However, the nature, intensity, and impacts on environment vary according the forms of energy. These effects can occur at the different level including household, local, regional, national or global. To achieving the goal of poverty alleviation, India, needs to use and to develop all forms of energy sources optimally and sustainably.

Effects of Energy Generation on Environment

Energy generation involves social and economic externalities, such as production of bio- diesel programmes provide employment to people and improve their lifestyle by increasing their income. Energy generation also involves environmental externalities and emits Green House Gases. Thus production of energy also damage environment, but different forms of energy generation impact different ways. The environmental impacts of various energy options are of growing concern owing to increasing and widespread use of energy.

- **Fossil Fuels**

Fossil fuels are the main source of energy. The total estimated resource coal in India is placed around 148,790 milliontonnes but the mineable reserve may account to only about 60,000 million tones. Coal mining, exploration and production of oil and gas (both on and off shore) have a wide range of adverse effect on environment. Some of these impacts may be reversible, while most of these effects are irreversible. Open cast mining of coal polluted the environment and also the main cause of loss of biodiversity due to loss of or severe disturbance the habitats of plant and animals. Advanced technologies and better management minimized the impact on natural ecosystem and environment. Transportation of these fossil fuels coal and petroleum products by rail, road and sea also causes environmental damage. The main pollutants associated with the combustion of fossil fuels are Suspended Particulate Matter (SPM),

Sulphur Oxide, oxide of nitrogen (NO_2, NO_3), and CO_2. Either in transformation activities (to petroleum products or power and heat combustion) or in end-uses pose a major threat to ecological, artificial resources, and human health. Fly ash generated in thermal power plants is another major problem for environment and sustainability. Effects associated with energy based on fossil fuels are shown in the table 4

Table-4
Environmental Impacts Associated with Energy Transformation Based on Fossil Fuels

Stages of Fuel Cycle	Natural gases	Oil	Coal
Exploration/ Mining	Drilling accidents, Sludge	Drilling accidents, Sludge, SO_2	Mining injuries, Land degradation, SO_2, NO_x
Processing/Refining	Refinery accidents, Waste disposal	Refinery accidents, Waste disposal, SO_2	SO_2, SPM
Transport/Distribution	Pipeline accident, Liquefied natural gas explosion	Pipeline accident, Oil spills, SO_2	SO_2, SPM, NO_x CO, CO_2
Conservation/Electricity generation		Ash disposal, SO_2	Fly ash, SO_2, NO_x, CO_2, Sludge
All Stages/all fuels	SPM, NO_x CO, CO_2 CH_4, Relative Organic gases, Hydrocarbons, and Tharmal pollution.		

• Nuclear Power

Nuclear power is associated with risks of radioactive release. Generation of nuclear power adversely effects environment at various stages of the nuclear fuel cycle –

✓ **Mining** - Accidents, release of radon gas and radioactive dust from Uranium mines, radioactive sewage from waste and land degradation,

✓ **Processing** - Accidents,

✓ **Transport** - Accidents, risk of proliferation, and

✓ **Electricity generation** - risk of catastrophic accidents, low and high level Radioactive wastes.

Nuclear power plants entail the disposal of radioactive wastes. While significant technological has been developed to dispose radioactive wastes, but there is always risk. However, despite these risks, global data suggests that of all the conventional energy options, nuclear energy has posed the least risks in terms of mortality per billion megawatt hours of generation.

- **Hydro power**

Large hydro power projects are rather site specific, and in particular, centre on submergence of land and displacement of local population and endemic species from their ancestral domains. Such displacement inflicts a high social cost. In addition, loss of natural forests and its associated unique biodiversity may also be lost due to submergence. Green House Gas Methane may also be emitted in large quantities if proper precautions are not taken to remove vegetative matter. Submergence of large areas may also increase the risk of seismic events, for which dams need to be designed, for if they lead to failure of the dam structure, it may be catastrophic for downstream populations. Other adverse impacts include the spread of water-borne disease, such as schistosomiasis. Impact of the major forms of conventional energy options on environment shown below in the table- 5

Table -5

Energy production Impact on Environment

Effects	SO_2 NOx, TSPs	Solid waste	Liquid effluent	Thermal impact on receiving wastes	Forest loss	Involuntary resettlement
Thermal power (including Mining)	Yes	Yes	Yes	Yes	Yes	Yes
Hydro power	No	No	No	No	Yes	Yes
Nuclear power (including Mining)	No	Yes	Yes	Yes	Yes	Yes
Petroleum (including Mining)	Yes	No	Yes	Yes	Yes	Yes
Transmission & pipelines	No	No	No	No	Yes	Yes

Effect of Renewable Energy Generation on Environment

Energy from renewable sources is generally viewed as involving lower impacts on environment than conventional energy and large hydropower. The main environmental benefits of renewable energy sources are that they avoid the air pollution emissions from fossil fuels and the catastrophic risks associated with nuclear plants. Small hydro projects may also help conserve water supply. Biomass based systems promote the cultivation of energy crops in wastelands and help to reduce land degradation. Nevertheless, renewable energy sources may also cause environmental degradation of a different kind. Unsustainable use of biomass leads to depletion of forests, generation of wind energy produce noise, which may cause bird mortalities, and despoil the aesthetics of landscapes. Large arrays of solar photovoltaic panels cover large land areas and impair aesthetics. Use of chemicals in the manufacture of solar panels and use of lead acid batteries cause several adverse environmental impacts. While bio-energy is generally considered carbon neutral over the vegetation life cycle, the potential environmental impacts of such projects can include impacts on soil and water resources in addition to an increased competition for land use. Since net carbon emissions can be prevent only to net deforestation, sustainable harvested biomass fuels do not add to green house gases.

Effect of Energy End –use on Environment

The end-use of energy may also impose severe environmental costs. Industrial and Vehicle emit large amount of emissions in urban areas. Petrol-driven vehicles are the major source of CO emissions, contributing over 85% of total emissions, while diesel-driven vehicles are the major source of NOx, contributing over 90% of total emissions. This emission may cause many respiratory diseases. Improper use of energy in the agricultural sector has resulted in the depletion of groundwater in several parts of the country. Domestic consumption of both traditional and commercial fuels is main cause of indoor air pollution. Indoor air pollution is the second largest source of disease, particularly among women and children, in the India. To understand end-use energy-environment interactions, we can begin with its effect on environment, indicators, and level of impact.

Energy Sustainability in India

Sustainable energy is the provision of energy that meets the needs of the present without compromising the ability of future generations to meet their needs. Sustainable energy sources are most often regarded as including all renewable sources, such as hydroelectricity, solar energy, wind energy, wave power, bio energy, geothermal energy, tidal energy, and wave power. It usually also includes technologies that improve energy efficiency. Energy efficiency and renewable energy are said to be the twin pillars of sustainable energy. Some ways in which sustainable energy has been defined are:

- "Effectively, the provision of energy such that it meets the needs of the future without compromising the ability of future generations to meet their own needs. ...Sustainable Energy has two key components: renewable energy and energy efficiency." – Renewable Energy and Efficiency Partnership, British.
- "Energy which is replenishable within a human lifetime and causes no long-term damage to the environment." – Jamaica Sustainable Development Network

This sets sustainable energy apart from other renewable energy terminology such as alternative energy and green energy, by focusing on the ability of an energy source to continue providing energy. Sustainable energy can produce some pollution of the environment, as long as it is not sufficient to prohibit heavy use of the source for an indefinite amount of time. Sustainable energy is also distinct from low carbon energy, which is sustainable only in the sense that it does not add to the CO_2 in the atmosphere. Green energy is energy that can be extracted, generated, and/or consumed without any significant negative impact to the environment. The planet has a natural capability to recover which means pollution that does not go beyond that capability can still be termed green. Green power is a subset of renewable energy and represents those and those renewable energy resources, technologies that provide the highest environmental benefit. The Green power is define as green power as electricity produced from solar, wind, geothermal, biogas, biomass, and low-impact small hydroelectric sources. Customers often buy green power for avoided environmental impacts and its green house gases.

Sustainable Development has been made possible by a range of factors, including India's sustainable patterns of consumption, enhanced competitiveness, proactive policies to promote energy efficiency, and more recently, the use of the Clean Development Mechanism to accelerate the adoption of clean energy technologies.

(i) Patterns of Consumption

Food habits and recycling processes have mitigated growth in energy demand and GHG emissions. The specific GHG emissions from food production and processing are much lower in India than in developed countries. India emits only 0.1 tonnes CO_2/ million Kcal of food Energy. In other hand UK, Germany, Netherland, Australia, USA emit respectively 1.7, 1.8, 1.9, 2.0 and 2.2 tonnes CO_2/ million Kcal of food Energy. As seen in Figure-3

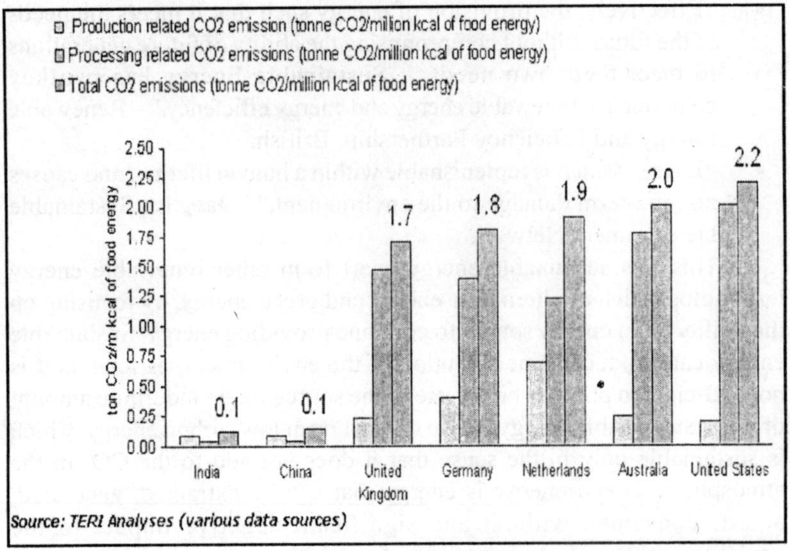

Figure-3 CO_2 Emission in Production and Processing in Major Economies

The high ratio of recycling in India, compared to that of other major economies has also limited the growth in energy use, and GHG emissions, because of the lower demand for virgin material such as steel, aluminum and copper. Recycling ratio is in US only 30 in Germany and Japan respectively 47.3 and 53 comparison India recycling ratio 70.

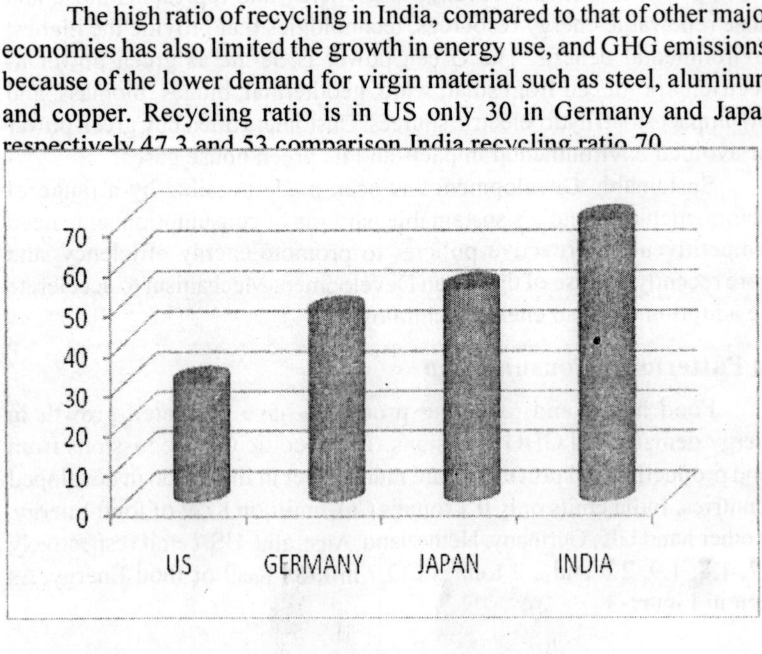

Figure-4 Recycling Ratios in Major Economies

(ii) Increased Industrial Energy Efficiency

Over the past decade, energy efficiency in Indian industry has increased steadily. In the major energy-consuming industrial sectors, such as steel, cement, aluminum, fertilizers etc., average specific energy consumption has been declining because of energy conservation in existing units, due to new capacity. The specific energy consumption of Indian cement plants and of Indian iron & steel plants has been declining rapidly. In the cement sector, the specific energy consumption of the most efficient plants is now comparable to that of the most efficient plants in the world.

(iii) Policies to Promote Energy Efficiency and Renewable Energy

Several affords has been adopted for energy efficiency and energy security-

- **Energy from Renewable sources:** The Electricity Act, 2003, requires State Electricity Regulatory Commissions to specify a percentage of electricity that the electricity distribution companies must procure from renewable sources. Several Commissions have already in almost every industrial sector; some of the world's most energy-efficient units are located India operational zed this mandate, and also notified preferential prices for electricity from renewable. This has contributed to acceleration in renewable-electricity capacity. Estimated Potential capacity from renewable energy sources shown in table – 6

Table -6

Estimated Potential Capacity from Renewable Energy Sources

Sources	Approx. Potential
Biomass Energy	19500
Solar Energy	20000
Wind Energy	47000
Small Hydropower	15000
Ocean Energy	50000

Source: Annual Report -2008, Ministry New and renewable Energy, India

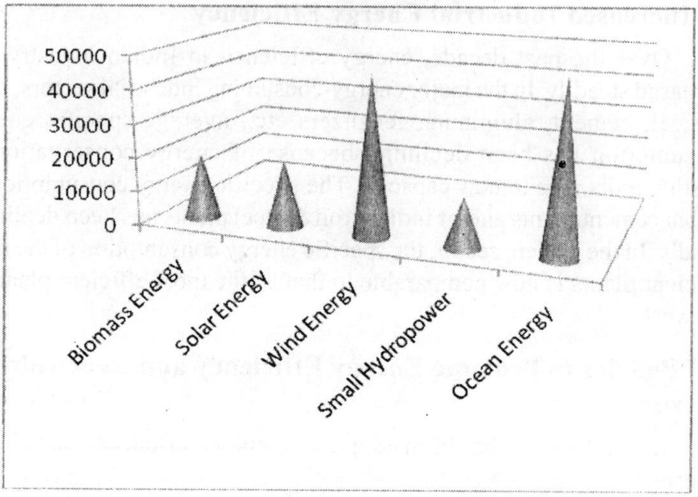

Figure-5 Estimated Potential Capacity from Renewable Energy Sources

- **Enhancing Efficiency of Power Plants:** Efficiency can be increased in energy extraction, conversion, transportation as well as in consumption. Coal is the Main stay of India's energy economy, and coal-based power plants account for about two-thirds of the total electric generation installed capacity of about 135,000 MW. In addition, the Electricity Regulatory Comhissions are also linking tariffs to efficiency enhancement, thus providing an incentive for renovation and modernization. New plants are being encouraged to adopt more efficient and clean coal technologies, and four new plants under construction have adopted the more-efficient supercritical technology for power generation.
- **Introduction of Labeling Programme for Appliances:** An energy labeling programme for appliances was launched in 2006, and comparative star based labeling has been introduced for fluorescent tube lights, air conditioners, and distribution transformers. Figure 6 shows labels for refrigerators and tube lights. The labels provide information about the energy consumption of an appliance, and thus enable consumers to make informed decisions. Almost all fluorescent tube lights sold in India, and about two-thirds of the refrigerators and air conditioners, are now covered by the labeling programme. India has over 7,000 MW of wind energy capacity – the fourth largest in world. The growth of electricity from renewable has been accelerated by the legislative mandate for its procurement.

- **Energy Conservation Building Code:** An Energy Conservation Building Code (ECBC) was launched in May, 2007, which addresses the design of new, large commercial buildings to optimize the building's energy demand. Commercial buildings are one of the fastest growing sectors of the Indian economy, reflecting the increasing share of the services sector in the economy. Nearly one hundred buildings are already following the code, and compliance with it has also been incorporated into the Environmental Impact Assessment requirements for large buildings.
- **Energy Audits of Large Industrial Consumers:** In March 2007, the conduct of energy audits was made mandatory in large energy-consuming units in nine industrial sectors. These units, notified as "designated consumers" are also required to employ "certified energy managers", and report energy consumption and energy conservation data annually.

(iv) Accelerated Introduction of Clean Development Mechanism

Over 700 CDM projects have been approved by Clean Development Mechanism (CDM) National Designated Authority, and about 300 of these have been registered by the CDM Executive Board. The registered projects have already resulted in over 27 million tones of certified CO_2 emissions reductions, and directed investment in renewable energy and energy projects by reducing the perceived risks and uncertainties of these new technologies, thereby accelerating their adoption. The Clean Development Mechanism has accelerated the diffusion of renewable energy and energy efficient technologies by mitigating some of the risk associated with the adoption of new technologies.

Indian Approach about Climate Change Related to Energy

Climate change is recent phenomena; it is subsequent effect of global warming. Due to increases the concentration of Green House Gases (GHG) in the atmosphere, rise the global temperature. This is called global warming. GHGs mainly due to fossil fuel use, certain agricultural and industrial activities, and deforestation, has the potential, over the next few generations, to significantly alter global climatic conditions. This would result in large changes in ecosystems, leading to possibly catastrophic disruptions of livelihoods, economic activity, living conditions, and human health. On the other hand, the abatement of GHGs involves significant economic costs.

While climate change is a global environmental issue, different countries bear different levels of responsibility for increases in atmospheric GHG concentrations. Further, the adverse impacts of climate change will fall disproportionately on those who have the least responsibility for causing the problem, in particular, developing countries including India. India's GHG emissions in 1994 were 1228 million tons (Mt) CO_2 equivalent, which is below 3% of global GHG emissions. In per-capita terms, it is 23 percent of the global average, and 4 percent of USA, 8 percent of Germany, 9 percent of UK and 10 percent of Japan's per-capita emissions in the same year. In terms of the GHG intensity of the economy in Purchasing Power Parity terms, India emitted a little above 0.4 tone CO_2 equivalent per 1000 US dollars in 2002, which is significantly lower than those of the USA and the global average. In terms of primary energy use, India's share of renewable energy (with zero net GHG emission) at 36 percent is far higher than industrialized countries can hope to reach in many decades. Since GHG emissions are directly linked to economic activity, India's economic growth will necessarily involve increase in GHG emissions from the current extremely low levels. Any constraints on the emissions of GHG by India, whether direct, by way of emissions targets, or indirect, will reduce growth rates, and impair pollution abatement efforts. Though India is a signatory to the United Nations Framework Convention on Climate Change (UNFCC), she is not required to contain her GHG emissions. India's policies for sustainable development, by way of promoting energy efficiency and the use of renewable energy, changing the fuel mix to cleaner sources, energy pricing, pollution abatement, afforestation, mass transport in and of themselves result in a relatively GHG benign growth path. Anthropogenic climate change, significant responsibility for which clearly does not lie with India, may, on the other hand, have severe adverse impacts on India's precipitation patterns, ecosystems, agricultural potential, forests, water resources, coastal and marine resources, besides the increase in range of several disease vectors. Large-scale resources would clearly be required for adaptation measures for climate change impacts, if catastrophic human misery is to be avoided. As the global concern vis-a-vis climate change grows and as the threat of climate change is perceived to be real, pressures on India to contain GHG emissions will rise. India should be willing to contain her GHG emissions as long as she is compensated for the additional cost involved. India acceded to the Kyoto Protocol on 26

August 2002. The Government of India established National CDM Authority (NCDMA) in December 2003, with its office in Ministry of Environment and Forests. As of 8th May 2006, a total of 297 projects have been approved by the NCDMA. These projects expect emissions reduction of 236 Million CERs till 2012 at a potential total investment of more than Rs.190 Billion. India recognizes the possibilities of selling GHG emission reductions. We should choose options that permit doing so at a later date when it is found to be attractive. In any case, for us the imperative is to push energy efficiency, promote modern renewable, develop new technologies that augment our energy supply such as in-situ coal gasification that also provides scope for carbon capture, and emphasis nuclear power. All of these will automatically help reduce the GHG emissions

Conclusion

India is very rich with vast resources of renewable energy in solar, wind, biomass and small hydro. However, there are many technical problems related with the exploitation of these sources such as problem in identifying places where a good amount of wind blows, regular sun heat comes, continuous and sufficient flow of water, and lack of cost effective technology etc. Hence, more and more investment should be directed in research and development of cost effective technology. It would be better to develop area specific technology to exploit these resources, because they are most eco-friendly and safe. Further, they would be helpful in reducing pace of global warming also. Tapping India's wind, solar, biomass, and hydro could bring quality job from a domestic resources. Only there is need to put sustained effort and strong commitment to realize this potential.

Energy is key factor of economic development. It plays an important role in human resources development in term of education, health and income. Human resources are axis of any development. But generation and utilization of energy adversely effects environment and create hinder in sustainable development. Sustainable Development can be made by sustainable patterns of consumption, enhanced competitiveness, proactive policies to promote energy efficiency, adopt renewable sources for energy generation and more recently, the use of the Clean Development Mechanism to accelerate the adoption of clean energy technologies.

References

- Unlocking Finance for Clean Energy: The Need for 'Investment Grade' Policy, Kirsty Hamilton, Chatham House, 2010.
- Clean Energy Investment Policymaker's Summary, Aaron Cosbey, Jennifer Ellis, Mahnaz Malik and Howard Mann, International Institute for Sustainable Development, 2008.
- Economic survey-2007-2008, Oxford University Publication, Ministry of Finance, Government of India. 'Energy and Resource – India', Earth Trends-2003.
- Meisen, Peter and Eleonore Queneudec (2006), "Overveiw of Renewable Energy potential of India," Global Energy Network Institute, October, 2006.
- Annual Report-2008, Ministry of New Renewable Energy, Government of India.
- 'India Energy Data, Statistics and Analysis -- Oil, Gas, Electricity, Coal' (2007), Energy Information Centre, Retrieved from www.eic.doe.gov.
- World Development Indicators- 2005, The World Bank.
- World Development Report- 2005, The World Bank.
- Sell, Malena (2006) "Trade, climate Change and the Transition to a Sustainable Energy Future: Farming the Debate", ICSTD Project on Trade and Sustainable Energy, Nov.2006, Pg1-2
- Vaishampayan, J.V., "Energy and Economic Development: An Analysis of Energy Economy Interactions".
- Ruhl, Christi (June, 2007), BP Statistical Reviewof World Energy.

22. Renewable Energy:
Eco-Friendly Measures of Energy

Shikha,
Rakhi Sharma
Ambica Prasad

India has 15% of world's population and fast developing country, as economic growth rate increases the aspiration of its people to better quality of life; India has a voracious appetite for energy. But the country lacks sufficient domestic energy resources, and growing its requirement due to fast development. This growing requirement of energy has led to increasing dependence on fossil fuels such as coal, petroleum and natural gas for the supply of electricity. These fossil fuels emit carbon emission and causes global warming, which ultimately leads to climate change. The resulting environmental pollution and health problems caused by the use of these fossil fuels brought about the need to develop and utilize alternative sources of clean energy.

Keywords: Renewable energy, Eco-Friendly, Global warming, Bio-mass, Hydro- Power, Bio-Gas, and Thermal power.

Introduction

Energy is one of the basic requisites for economic development. Every section of society, whether agriculture, industry, transport, businesses or households consume energy. Over the years, as the country progresses, increased amounts of energy are needed by these sectors.

This growing consumption of energy has led to increasing dependence on fossil fuels such as coal, petroleum and natural gas for the supply of electricity. These fossil fuels emit carbon emission and causes global warming, which ultimately leads to climate change. The resulting environmental pollution and health problems caused by the use of these fossil fuels brought about the need to develop and utilize alternative sources of clean energy. Also, the rising prices of fossil fuels and fears about future shortages in energy supply called for the creation of a sustainable path of energy development. This is best done through two avenues-the promotion of energy conservation and the use of environmentally friendly renewable sources of energy.

Renewable energy is energy created from natural sources such as sunlight, wind, water, waste products and other sources that can be naturally replenished. India is fortunate to be blessed with an abundance of such sources. These sources of energy are available locally throughout the year and do not need elaborate arrangements for distribution. This makes them well suited to decentralized applications for use in remote areas. Other advantages of renewable energy sources are its environment friendliness and low operation costs.

Characteristics of Renewable Energy

- Large and inexhaustible source of energy like solar energy.
- Clean and eco-friendly source of energy.
- Large collection area and high cost.
- Availability varies with time like in summer solar energy produce large amount but in rainy season production of solar energy reduce.
- Additional cost is required for storage.

Sources of Renewable Energy

Renewable energy is created with the natural resources sunlight, wind, water, waste products. They are eco friendly energy sources, and not emit carbon in production of energy. There are some renewable sources given in flow chart-

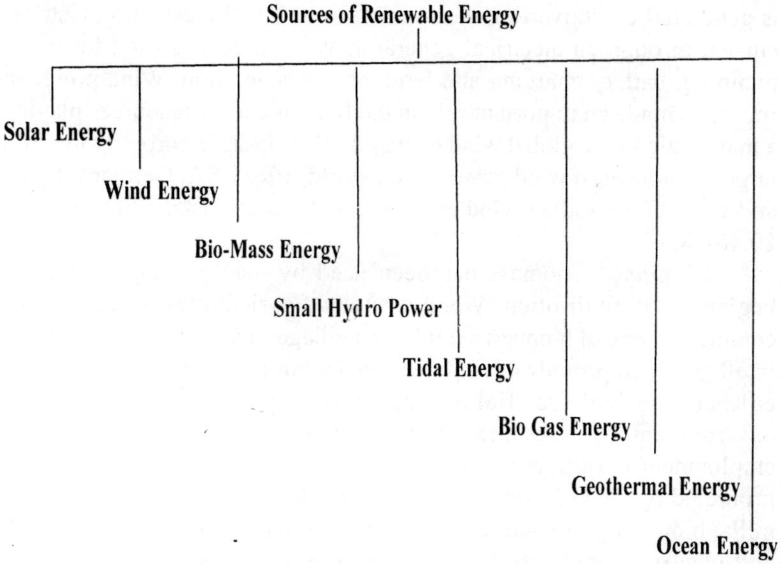

Figure - 1

Here are some details about the types of alternative sources of energy being produced in India.

Solar Energy - India is a sunny country with most parts receiving about 4 to 7 kilowatt hours of solar radiation per square metre per day with 250 - 300 sunny days in a year. This makes solar energy a very attractive option for generating both power and heat. The thermal route uses the heat from the sun for water heating, cooking, drying, water purification, power generation and much more. This alternative energy can be harnessed using a photovoltaic cell that converts sunlight directly into electricity which can then be used for lighting, pumping, communications and power supply in non electrified areas. Solar power is a clean renewable resource with zero carbon emissions and is one of the prime renewable sources of green energy in India. More than 700,000 PV systems of capacity over 44 MW are installed all over India for diverse solar applications such as home lighting, street lighting, solar lanterns and water pumping for irrigation.

Wind Energy -Wind power in India has made an important role in the field of energy resources playing a major role in the global wind energy market. The development of wind power in India began in the 1990s, and has significantly increased in the last few years. Wind power is generated by converting the rotation of turbine blades into electrical current through an electrical generator. Wind energy is used for water pumping, battery charging and large power generation. Wind power in India has made an important role in the field of energy resources playing a major role in the global wind energy market. India is currently the fifth largest producer of wind power in the world, after USA, Germany, Spain and China. The Indian wind energy sector has an installed capacity of 10,464 MW.

Biomass - Biomass has been used by mankind ever since the beginning of civilization. Wood waste and agricultural wastes are the common source of biomass in all Indian villages and it needs to be tapped intelligently to provide not only electricity but also water to irrigate and cultivate agricultural fields for steady generation of power. It is environment friendly and has the potential to provide significant employment in rural areas. The three main biomass technologies being promoted by the Government are biogases -based cogeneration in sugar mills, biomass power generation, and biomass gasification for thermal and electrical applications. Biomass power has become an industry attracting annual investments of over Rs.1000 crores while generating more than 9 billion units of electricity per year. Increased usage of biomass-based fuels will help in modernizing the agricultural economy at lower investment costs. More than 2,000 biomass units are estimated to have been established with a capacity in excess of 22 MW in India.

Biogas - Biogas is derived from organic products, primarily cattle dung, kitchen waste and agro forestry wastes and is used mainly in rural areas. The Government promotes the use of biogas through the National Biogas and Manure Management Programme. Biogas can be used for cooking, heating, lighting, motive power in certain gas engines and generation of electricity through attached alternators. There is an estimated potential for the installation of 12 million family type biogas plants in the country. Currently, India ranks second in the world in biogas production.

Small Hydro Power - Hydro power is the largest source of renewable energy for generating electricity. Hydropower is obtained from the movement of water in rivers, oceans or other energy sources to generate electricity using turbines, or can be used mechanically to do useful work. In India, hydro power capacity is up to 25 megawatts (MW) are categorized as small hydro power projects. The major advantages of hydro power includes saving scarce fuel reserves, cost of generation, operation and maintenance is far lower than the other sources of energy.

Impact of Non- Renewable Energy Sources on Environment and its Regulation

Carbon-dioxide emissions in the country have been gradually increasing and are expected to increase sharply in the next 20 years as India enters a period of sustained economic growth and higher consumption of energy. Over 50% of the total CO_2 emissions are from the power sector. Cement, shipping, and iron and steel are other industries that have experienced annual growth rates above 4% between 1985 and 2005. In fact many industries have great potential for improving energy efficiency through use of combined heat and power. The primary driver for combined heat and power is economic, but it is also an effective means of reducing CO_2 emissions. Over the past two decades, the Government of India has been very active in participating in international efforts for climate change mitigation:

- Vienna Convention – ratified in 1991
- Montreal Protocol – ratified in 1992
- UN Framework Convention on Climate Change – ratified in 1993
- Kyoto Protocol – ratified in 2002
- Establishment of National Clean Development Mechanism Authority (CDM) in 2003
- 337 CDM projects are currently registered with the UNFCCC with estimated annual emission reduction of 31.62 million tonnes of CO_2 per year

Still, India is facing enormous international pressure to reduce its greenhouse-gas emissions under a successor agreement to the Kyoto Protocol that is now in development. In June 2008, the National Action Plan on Climate Change (NAPCC) was announced as a major move toward reducing carbon intensity, but its real effect is still unclear, What is obvious is that India will have to choose between increasing its fossil-fuel use and reducing its emissions; it cannot do both. Investors should keep this uncertainty in mind and closely watch developments in Indian climate-change policy to make the wisest use of their investment capital.

Suggestions for Eco-friendly Energy Use

- It is not a mission impossible to reduce these carbon dioxide emissions. Developed countries have started taking various measures to reduce their emissions at all levels right from schools, universities and colleges to houses, supermarkets, industries, production units etc.

- Try to recycle items that could be recycled so in order to save energy, water and natural resources. Things like batteries and electronic items when thrown away emit dangerous chemicals that leak through the landfill and pollute the groundwater, and also contaminate the soil which is important for our crops which in turn causes health risks to humans.
- Government and other officials should spring forward in their efforts to look for ways to expand recycling and bring awareness amongst the masses on waste reduction and its harmful effects. It would be best to encourage and motivate manufacturers and companies to be environmentally compliant and energy efficient.

Renewable Energy and India

India has 15% of world's population and fast developing country, as economic growth rate increases the aspiration of its people to better quality of life; India has a voracious appetite for energy. But the country lacks sufficient domestic energy resources, particularly of petroleum and natural gas, and must import much of its growing requirements. Currently, about 35% of India's commercial energy needs are imported. Table 1 indicates the primary commercial energy consumption in India.

Table-1
Commercial Energy Consumption

Source	Unit	2007-08
Petroleum products	MMT	140.7
Natural Gas (Net)	BCM	31.5
Coal	MMT	457.1
Lignite	MMT	34.0
Electricity	Billion kWh	813.1

Source: Basic Statistics on Indian Petroleum and Natural Gas, 2008-09, Ministry of Petroleum and Natural Gas, New Delhi

At present most of power generated in India through thermal power, which emits large amount carbon, responsible for environment pollution and global warming. Indian Government does some eco-friendly measure in term of renewable energy. The Ministry of New and Renewable Energy is responsible for putting into action comprehensive programmes for the development and utilization of renewable energy. It is promoting many new technologies and devices that are now commercially available. These include biogas plants, solar water heaters, solar cookers, street lights, pumps, wind electric generators, water-pumping wind mills, biomass gasifiers and small hydro-electric generators. Currently, renewable sources contribute around 9 per cent of the total power installed capacity in the country. Approximate 2.96 BN US $ the amount of financial investment in India's clean energy sector in 2009. The 75 percentage of India's renewable energy that power installed capacity comes from wind power. Since 2002, the percentage contribution of various sources has generally increased, with the large increase being with renewable sources. Thermal nuclear sources have decreased by 0.5- 1 percent since 2007.

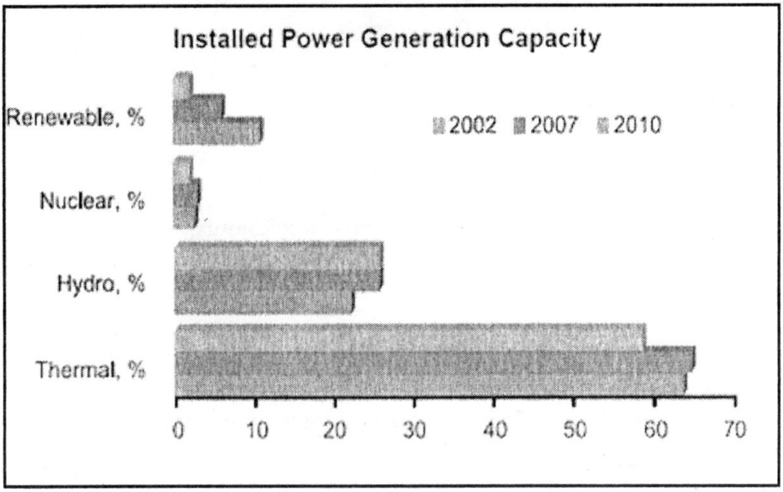

Figure- 2

Source: Ministry of New and Renewable Energy, India

The Government is also working to develop advanced technologies of renewable energy such as geothermal energy, ocean energy, fuel cells, bio fuels and tidal energy to satisfy future energy requirements. Tapping of renewable sources of energy and energy conservation are two steps in the direction of ensuring energy security and bequeathing a green planet to our children. The Government of India has aims to increase the contributed capacity of alternative energy sources to the country's energy mix over the next decade. Clean energy sources would ideally become large part of country's energy sources. The goal to increase overall installed capacity by 15.9 percent by 2022.

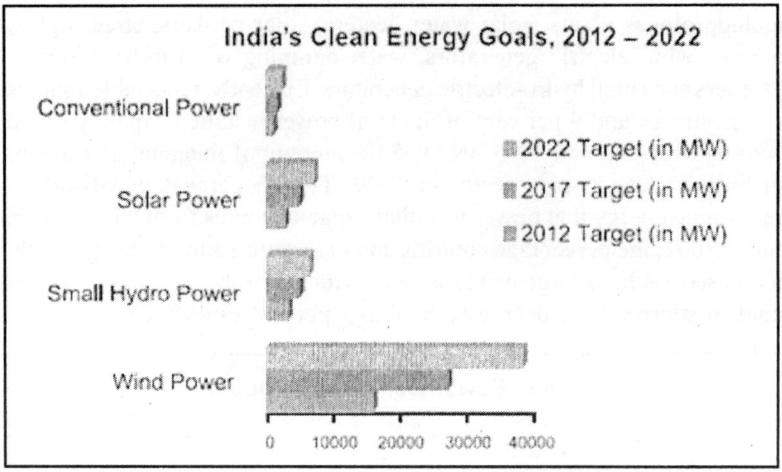

Figure-3

Source: Ministry of New and Renewable Energy, India.

Conclusion

Over decades, the depletion of ozone layer and its preservation had been the high priority of environmentalists and developed nations and campaigns about educating the masses of the importance of reducing carbon emissions thereby helping the preservation of the ozone layer. Now, there are a lot more issues along with this that has been worrying the environmentalists, one of this being depletion of natural resources and increase in CO2 emissions, pollution, wastage of resources, and so on. The conventional energy production challenges towards environment. As renewable energy is obtained from natural sources, adopting such technologies would protect our environment by reducing pollution. Thus, it is important to promote the benefits of conserving energy and adopting such 'clean' technologies through public awareness campaigns. As citizens learn and adopt energy saving behaviour and technologies in their homes and businesses, the nation will also progress. Let's GO to GREEN, and reduce emissions and wastage, encourage recycling and pave way for pure and clean surroundings for our future generations.

References

- Amulya KN Reddy (2003), 'Energy and Social Issues', World Energy Assessment: *Energy and the Challenges of Sustainability',* India.
- Advisory Board on Energy (ABE) (1985), *Towards a Perspective on Energy Demand and Supply in India in 2004-05,* Report of the Advisory Board on Energy, Government of India, New Delhi.
- Batliwala S and AKN Reedy (1994) 'Energy Consumption and Population', In *Population: the complex Realty,* London: The Royal Society.
- CMIE (2001), 'India's Energy Sector', *Centre for Monitoring Indian Fconomy*, New Delhi.
- UNDP, UNDESA, WEC, (2000), *'World Energy Assessment: Energy and the Challenges of Sustainability',* United Nations Development Programme, New York.
- Viklund, Viklud E (ed.) (1989), *'Nordic Seminar on Domestic Energy in Developing Countries; A Seminar Report',* Publication 31, Lund University, Lund Centre for Habit Studies, Sweden.
- WB (2003), *'Household Energy Use in Developing Countries – A Multicountry Study',* ESMAP Report, The World Bank, Washington D.C.
- WB (2000), *Energy Development Report 2000: Energy Services for the World's Poor',* World Bank Publication.

23. The Problems in Development of Bio-gas Energy:
A Study of Mysore District

Y. P. Chandrashekara
H.Kamala

With the spread of new agricultural technology there is an increased use of conventional energy sources such as diesel, petrol and electricity especially in rural areas. Paradoxically at the same time, the supply of sources of energy is becoming scarce and it threatens to grow into crisis of disproportion. This has resulted in the high cost of production of crops with its chain reaction alternately affecting the consumers. A solution to this problem is to exploit the sources of supply of biogas energy at a faster rate. The other way out is to discover and explore other possible non-conventional energy, such as solar energy, wind power, atomic power, thermal power and gobar gas. This study aims at evaluating the present status in biogas energy use in household and agricultural sectors, options available to them and the economics of these options. The results of this study would be useful for the development of potential energy resources for the integrated development of rural areas. Therefore a further attempt in the present study is to analyze the present problem and future prospects of energy use in the Mysore District.

Key words: Biogas energy, Manure, Biogas plants, Cattle population

Introduction

Energy is an essential input for any kind of economic development and social advancement, as well as in making the basic necessities available to mankind. Whatever the development strategy may be, one of the prime requirements is adequate and growing energy production, distribution and efficient utilization.

Karnataka state in general and Mysore district in particular is facing the energy crisis. In Mysore district because of various reasons, people still heavily depend on traditional sources of energy, especially on fuel wood. Looking into alternatives of this traditional energy, the biogas program was started in 1982 with the aim of maximizing the number of plants at rural households and as such is still in the preliminary stage. With the growing population and increased demand for firewood exerts lot of pressure on the forest wealth, which ultimately becomes the contributing factor for ecological imbalance which furthers the need for non-conventional sources? The biogas generated by using cow dung is supposed to be the best way to solve energy crisis to some extent and it helps to check the environmental degradation to certain extent. In this paper an attempt is made to high light the problems and prospects of biogas production in the Mysore district.

Objectived of the study

The present research is an analytical study of installation and functioning of biogas plants in Mysore district. It evaluates spatio-temporal pattern, eco climatic conditions, and levels of development of the biogas energy. Keeping in views of this, the study is focused on its impact on cultural landscape. The main objectives of the study are

1) To explore the possibilities for improving the biogas and assess the potentiality of biogas energy resources and its utilizations.
2) To analyze the impact of environmental factors on the spatio-temporal distribution of energy resource
3) To evaluate and identify the problems and prospect of diffusion of biogas energy in the rural community.

Study area

Mysore is one of the 28 districts of the state lying in the southern most part of Karnataka state with an area of 6269 sq.km, which constitutes 3.27 percent of the total area of the state. It ranks 13th in size and 4th

(4.97%)in terms of population (2001).Latitudinal it extends from 11°-45' to 12°-50'North and longitudinally 75°-45' to 77°-10' East. On the north is bounded by the Hassan, Mandya and Bangalore districts and Chamarajnagar district, Wynad district of Kerala state on the south and east, on the west by Kodagu. (Fig.1.1)

It consists of seven taluks namely Mysore, Thirumakudala Narasipura (TNP), Hunsur (HUN), Krishnarajnagar (KRN), Nanjanagud (NGD), Periyapatna (PPT), and Heggadadevana Kote (HDK). These taluks are grouped into two subdivisions namely Mysore and Hunsur for administrative purposes.

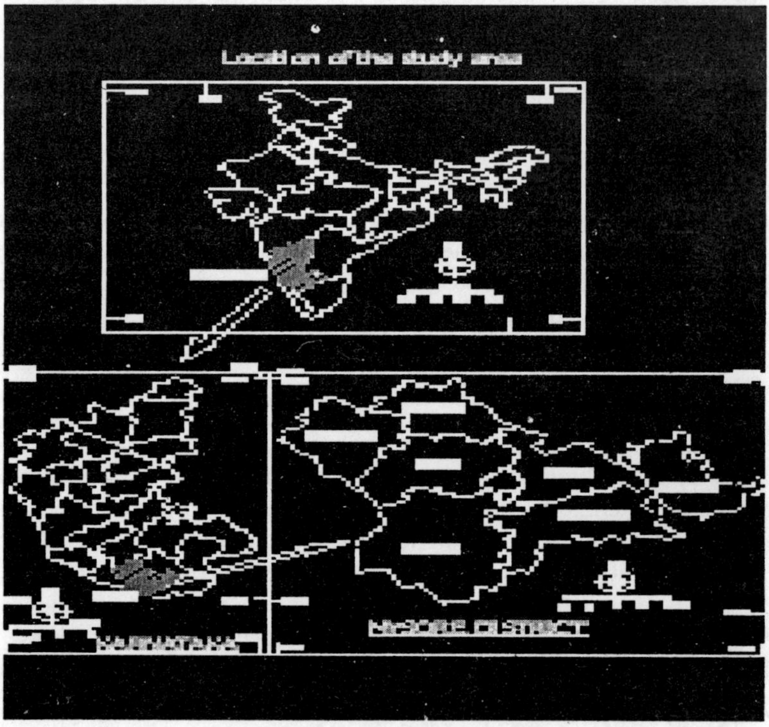

Figure-1

Problems

The major existing problems in installation of biogas plants in the study area are social, economic and technology oriented. These problems have been categorized with respect to construction of biogas plants and consumption of biogas energy in the rural area and suggestions also have been made to overcome these problems.

1. Choice of Energy Technology: - Faced with world wide energy crisis, scientists are searching for new sources of energy, such as solar, wind, tides, biomass and other means. Animals are ready sources of energy and most appropriate, lying within the economic and technological capacities of the millions of farmers. Biogas energy is more appropriate and essential in many counts and will be depended upon for decades to come. Nevertheless, bio energy has been, neglected by the policy makers, scientists and professionals. This gap has to be redressed.

There is abundant opportunity to develop biogas energy in the study area. As biogas energy is available at about one third of the cost of electrical energy and cheaper even than petroleum energy, it will be wise to utilize and improve the bio energy. And more importantly biogas energy is an inexhaustible source compared to petroleum energy, which is estimated to be exhausted in the near future.

2. Potentials of Biogas Energy: - Biogas energy from animals is the principal source of motive power in rural economy. Many attempts are being made to assess the energy output which is depend upon bovine animals, fodder, feeding, nutrition method etc., The total cow dung produced every day in the study area is about 7.81 million tones per day from 650445 Bovine population. Out of this, 89 percent of cow dung remains unused; only 11 percent of cow dung is being used for the production of biogas energy. Of the total cow dung it is possible to install at least 50000 biogas plants with 2 m^3 capacity. However, so far only 13942 plants have been installed (2004), which is 28 percent of the total rural potential. Hence effort should be made to make use of the total potential biogas energy and to apply more intensively, productively and efficiently.. If the potential productivity of biogas energy could be fully exploited, the district could meet her rural power needs with lower or cheaper than electric energy.

3. Conservation of Manure: - Farmyard manure from bovine constitutes one fourth of the income from cattle breeding. Cow dung compost applied to soil enables the latter's humus contents and maintains its fertility. The biogas plants digest not only the cow dung but also urine of cattle and human excreta. After these materials pass through the gas plant, the end product, which remains in the digester, is manure. This manure is not only rich in its nutrient but is also innocuous and odor free from the flies and other sources of infection and rich in NPK. The qualitative and quantitative nature of dung and urine were characterized as shown in table below.

The table reveals that about 2848.95million tones of cow dung and 3413 million tones of urine are being produced per Annum in the district. The value of cattle manures has been credited to urine, unfortunately, much of the urine is being wasted and only a negligible portion of it is used as manure.

Table: 1
Cattle Population and Existing Biogas Plants

1	Total number of villages	1257
2	Bovine population	64.04 Lakh
3	Total cattle dung available	0.781 million tones/day
4	Total cattle dung available	23.42 million tones/month
5	Total cattle dung available	285.06million tones/year
6	Total urine available	10.9-million liters/day
7	Total urine available	341.3 million liters/year
8	Total number of existing biogas plants	13942 in numbers
9	a. 4 cubic meter capacity	4142
10	b. 3 cubic meter capacity	2800
11	c. 2 cubic meter capacity	6900
12	Total cattle dung being used in biogas	749775 kg [11%]
13	Unused cattle dung in potential	11232225kg[89%]
14	Biogas used	89136.41 cum
15	Potential of biogas	653666.99 cum
16	Equivalent energy in Kw	13 lakh

Source: Zilla Panchayath Office

Biogas produced by the fermentation of organic contains methane [55-65%] and carbon dioxide [45-35%] when fed into special burners gives a blue flame. Gobar gas is used for lighting with specially designed lamps. It can also be used for running oil engines for irrigation purposes with special attachments to the engine. The requirements of gas for different purpose are as follows.

- For cooking 0.227 m^3 (8 cft) per person per day.
- For lighting 0.127 m^3 (4.5cft) per hundred candle bulb
- For motive engine (15 cft) per HP per hours.

Cow shed flour should have very gentle slope in such a way that urine runs along a drain and to collect in a pit. Therefore some better means should be found for conserving the valuable urine, which is practically as valuable as dung. In the study area around 3413 million liters of urine is produced from a bovine population of 65.04 lakh [2002.] Of this only 10 percent is being utilized and the rest of the urine is wasted or unused. The total value of the liquid manure at a cost of Rs.550 Per tone works out to Rs.153.6 lakhs.

The method of conserving cattle excreta followed in rural parts is to pile up the dung and other refuse in the form of heaps on the ground. This method is not desirable, as it results in a great deal of loss of plant nutrient. The proper method has to be devised to store manure in a pit with top cover sealed in order to prevent leaching of urine and the liquid manure should be properly preserved. This will also improve village sanitation by eliminating flies and mosquitoes. The area is facing a chronic shortage of fertilizer on one hand and a tremendous wastage of useful and valuable manure of animal origin on the other, which looks paradoxical. This has to be urgently solved.

4. The Problems Associated with Biogas Installation:

a) Status of biogas energy use:- This indicates the normal pattern of biogas development. About 21 percent of the plants are dysfunctional, which is comprised of KVIC model. As in the normal case, concentration of such dysfunctional plants is more in the dry land. Another notable phenomenon is the ubiquitous nature of dung use. The distance decay effect has been almost clearly marked on the northeastern and central part of Krishnarajanagar. In the case of Hebbal, Varanchi, the presence of facilitates are responsible for a higher density of plants.

b) Changes in Biogas Resources: - The major changes brought in the biogas plants are here under:

- Biogas use has increased its share considerably
- There has been only a slight increase in the biogas energy consumption.
- The share of Deenabandu model increased than that of KVIC model.
- The biogas values also shows marked variation during 1981-2002. The major increase in the provision of biogas plants is observed in older villages. However, about 10-15 percent of additional areas in the rural villages have brought under the biogas energy.

c) Perception and Participation of People in the Biogas Development Process: -As mentioned in previous sections, to facilitate a viable biogas development, it is highly indispensable to ensure the full co-operation and participation of the people in the biogas installation process. To assess the perception of the people in rural environment, and the nature and magnitude of biogas development process and their problems, associated with local environment, therefore conduction of a primary survey is highly essential. To achieve the target, primary survey has been conducted in the district to understand the perception and participation of people in the process of development. Along with this, questionnaires and interviews have also been conducted to understand the nature and level of facilitates made available, magnitude of problems and the perception of the people in the rural environment.

d) Characteristics of Samples: -The general characteristics of the samples are given in the table. It is noticed from the table that, medium sized families (4-6 person) dominate the rural area accounting for 50 percent of the total share, followed by low income sized families (38 percent) and the remaining is constituted by the big families (12 percent). Regarding the annual income of inhabitants, it has been noted that more than 40 percent of the families have an annual income ranging between Rs.10000 and Rs.20000. and 47 percent of families ranging between above Rs. 5000-10000. About 13 percent of families have annual income below Rs. 5000. Similar trend has been noticed in all the taluks of the district.

Table: 2

General Characteristics of Samples*

Size of family				Total	Annual income				Biogas plants	
zone	Less than 4	4 to 6	above 8		below 5000	5000-10000	10000-20000	20000-above	number	percent
Inner	38	54	8	100	17	8	50	25	2000	14.34
Mediate	20	50	330	100	20	12	38	30	9042	64
Outer	40	35	25	100	15	27	32	26	2900	20.8

Source: Field survey, Computed by Author

e) Details of Plants: - Regarding the ownership of biogas plants, the medium farmers own more than 45 percent of biogas plants in the study area. Having seen the general socio-economic conditions of the district, it is apt to extend the investigation on to the major problems related to installation and facilities available in the rural area.

f) Interview Method: - From the interview of the concerned persons at Zilla panchayath in the district, it has been noted that almost all the persons concerned are actively assisting the local bodies which consists of nearly 40% of the total population and independently undertaking activities to minimize rural energy problems, to improve the conditions of the rural energy crisis, the supply of biogas plant installation and its environment in the study area.

The Main Problems

The most urgent problem fermenting with the farmers is the fuel and fertilizers, firewood becoming very scarce and prices are exuberating. In this context dried cattle dung becomes the sources of the fuel. However, the problems involved in the wide spread dissemination of biogas technology are not usually highlighted and analyzed systematically by the TERI [Tata Energy Research Institute] which has been working for the design, construction and monitoring of biogas plants in the study area. Problems associated with the implementation of biogas technology and the potential prospect for the development of biogas plants in study area has been discussed such as.

- The problems in the installation of biogas plants.
- Brief reference to various drawbacks in promoting the renewable technology such as biogas energy and the potential of biogas plants are revealed.
- Since there is a chronic deficiency of fertilizer in one hand and wastage of useful manure of animal origin on the other

The problems associated with the biogas plants are summarized here under:

- Motivational
- Technical
- Financial problem/ Economic
- Operational
- Organizational

Motivational Problems: Among many exiting problems associated with biogas, the motivational problem is the most important aspects. The associated problems in it are:

- Livestock population in the district is very less
- Availability of electricity free of cost under Bhagya jyothi scheme.
- Introduction of Modern mechanized cultivation
- Non-suitability of Soil for erecting small sized biogas plants
- Higher percentage of illiteracy

- Firewood dominated energy for cooking
- Low income due to dependency on rain fed agriculture
- Technical ignorance about the management of the biogas plants

MYSORE DISTRICT

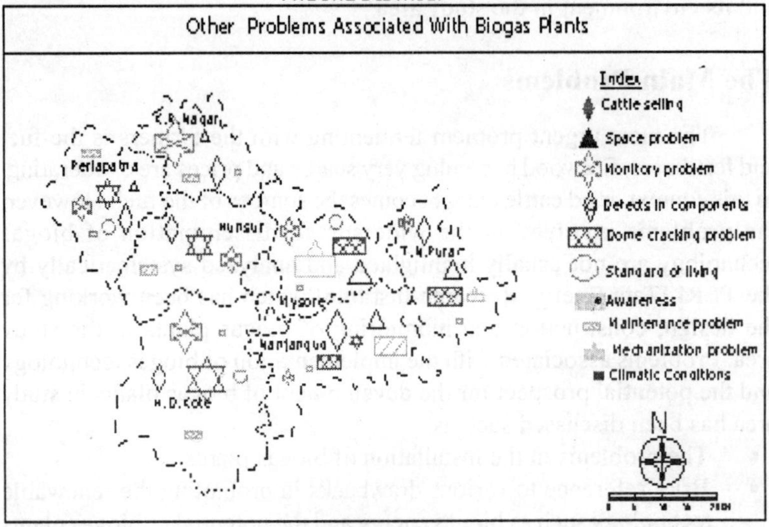

Figure-2

The main importance of biogas plants is as the source of fertilizer. The amount of fertilizer obtained from a 2cm³ per day in a year is equivalent to 109 kg of urea, 275 kg of super phosphate, and 50 kg of potash. This can meet the fertilizer requirements for one acre of land for 2crops in a year. The financial gain from the fertilizer amounts to about Rs.540 per year.

The financial benefit offered by the government varies from place to place. For example, places where fuel wood is purchased at a high cost that might be higher motivation for installing biogas plants. A farmer may not be making calculations as above, but it is easy to see that biogas plant is not a direct or immediate gain system. The motivation for using biogas engines for irrigation pumping could, in principle, be quite strong. But it can be calculated that the minimum size of the biogas plant needed to support a 5 hp pump for 5 hours is about 18cum/day which would be tough for an individual farmer. It might be concluded that the incentives provided by the government for installing biogas plants are not adequate. The only way to encourage farmers to install biogas plants is to offer an additional subsidy. It is found from the field survey that the motivation problem is found to be more among the women folk, since they are involved in using gobar gas for cooking purpose.

The degree of motivation also depends up on the educational level of the individual. Illiteracy blinds the people to the advantages of using biogas. Poverty makes even those willing to use biogas unable to set up plants. Illiteracy coupled with poverty is one of the main reasons for the present status .Quite contradicts to this the rich and literate don't use biogas as they consider LPG to be an economic standard along with the hesitation to use cow dung.

Table: 3

Sample Biogas Installed Villages and Educational Statues of Households

Name of the village	No. of house holds	Educational qualification			No.of biogas plants
		Below 7th	10th	Above 10th	
Hittavalli village (TN)	1240	60%	30%	10%	1
Hosahalli (TN)	800	70%	28%	2%	2
Talakadu(TN)	970	65%	20%	15%	5
Kaniyunahundi(HDK)	700	85%	12%	3%	Nil
Madahalli(MT)	300	80%	18%	2%	2
Turuganure (TN)	1500	50%	30%	20%	75
Beerihundi (MT)	1540	58%	30%	12%	30
Hebbal (KRN)	1650	40%	35%	25%	110
Varanchi (HUN)	900	42%	38%	20%	85

Source: Field Survey, Computed by Author

Technical Problems: Even though the biogas development in the study area dates back to 1960's, owing to lack of research efforts in this field has created many gaps and technical problem. It is evident from the field survey that the amount allocated for incentives in the form of subsidy was about Rs.12 Lakh in 2002 where as the allocation for research and development in the same year is very insignificant. There seems to be a large failure of KVIC drums type biogas plants than the fixed dome Jonatha model of plants. No subsidy is granted for KVIC drum type plants since 1996 affecting installation largely. The Deenabandu model with fixed dome is becoming more popular in the entire study area, except in Nanjangud, T.Narasipur and Mysore taluks. The fixed dome design was introduced in 1994 and is an accepted one. Innovation pertaining to the existing designs, quality, efficiency and performance are rather poor. The agency is commercial minded rather than the quality ones, particular about installing more number of plants without any improvement in the quality of plant. In Mysore district Deena bandu model plants are working in good condition with 75 to 85 percent of functional status. While KVIC type plants functional statues is shown a poorer performance, till 1996 KVIC plants have performed well in the district and had subsidy facilities. Later on owing to its technical problems subsidy has been completely stopped.

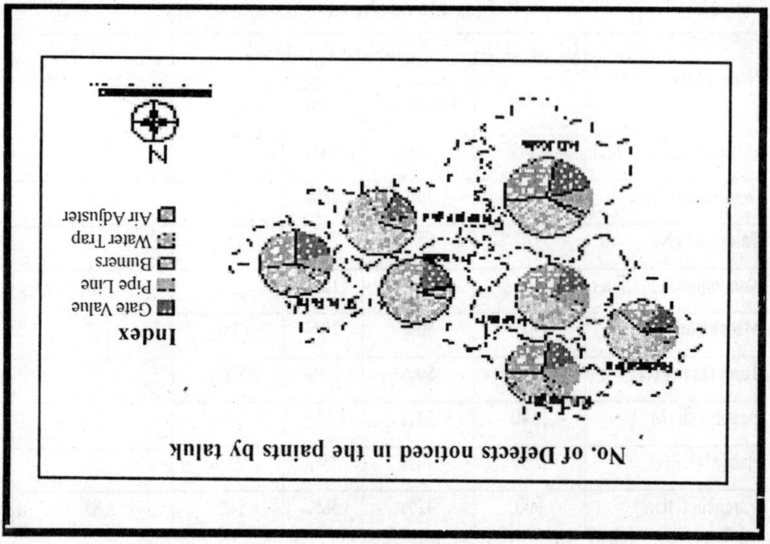

Figure-3

An analysis was also made of the functional status of plants installed in different years. As expected a large number of plants installed in the years have become dysfunctional and were found to be abandoned at the time of study. Only 40 to 45 percent of the plants which were constructed before 1995 are working. The functionality rate of plants installed between1998-2000 was as high as 90.5percent while those installed between1995-98 was 60% figure shows the functionality rates for biogas plants installed in different years. Construction of spherical brick dome involves considerable skill and training, technical problems from bad construction due to use of the unskilled labourers, consequently leakage of gas, cracking of domes, resulting in dysfunction of plants.

There is a problem of non-availability of proper material too. Since most of the drums are made of Iron, consequently there is a problem of rusting of the drums leading to leakage of gas. It is observed that most of the dome type plants have become a defunct due to poor quality of material used.

The most neglected task while operating a biogas plant is, perhaps, the removal of accumulated water from the pipeline using a water trap. It is interesting to note that many implementing agencies do not install a water trap in the biogas plant. In cases where water traps were installed, most of them were found buried under the ground with the beneficiary unaware of its existence. Among the various compound problems leakages in kitchen-valve, broken or jammed regulators, and choked burner holes. There is a lack of awareness among the users on the function and importance of different components of the stove.

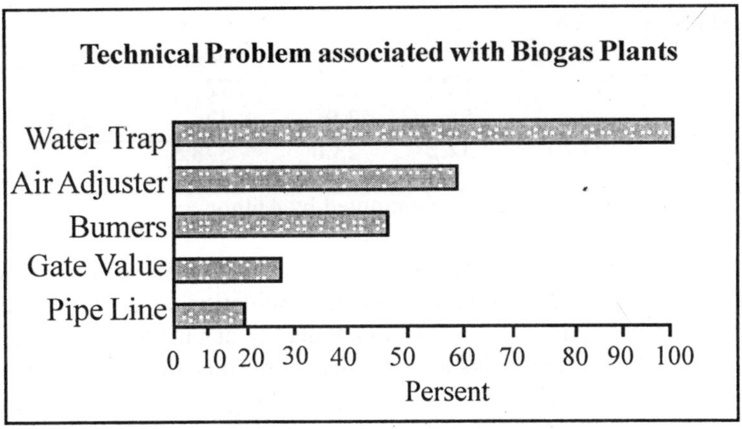

Figure -4 Computed by Author

The percentage of defective components varies from taluk to taluk. Nanjanagud and Periyapatna have shown more than 90% of defects in the water trap. Whereas the least defects of water trap is found in T.Narasipur. The highest defects in gate valve have noticed in the taluks of T.Narasipura (43%) Periyapatna (35.6%) and rest less than 20% in other taluks. In case of air adjuster problems is found in Hunasur (80%), K.R.Nagar (72%). T.Narasipura (63%), Periyapatna (62%), H.D.Kote (70%) and least defects of 32% and 15% are found in Nanjanagud and Mysore taluks. The defects in pipe line and burner comprise with less than 10% is most significant among all the taluks. This increase in the number of defective biogas plants is owing to non-availability of technicians within the villages. In fact, technicians are to be drawn by paying higher cost from the districts head quarters.

Table:4
Type of Defects Noticed in the Plants by Taluks.
(Defects in Different Parts of the Plants in percent)

Taluk	Gate Valve	Pipe Line	Burners Burners	Water Trap	Air Adjuster
HDK	15.4	7.6	1.7	48	30.3
HNS	20.2	5.1	2.1	52.6	22
KRN	20.8	13	2.9	34.1	30
MYS	20.8	0.2	4	61	14
NGD	12.9	0.1	6	68.8	12.2
PPT	8	1.2	3	72	15.8
TNP	21	5	3.9	42	28.1
Total	**119.1**	**32.2**	**23.6**	**378.5**	**152.4**

Source: Field Survey, Computed by Author

In spite of these problems, overall functionality of the biogas plants in the study area is good. However, formulation of comprehensive planning is required to improve in the quality of functioning of biogas plants. Most of the plants are suffering from some defect or the other. Further, the awareness level among the biogas users is low which contributes to the sub optimal performance of plants. Thus, is a need to focus on these aspects for improving the over all performance of the biogas programme. The other problem mainly found was that the distance between the digester and the kitchen wall, necessitating long pipeline and supply of gas from digester to the store is not optimal. Hence extra care has to be taken to install the digester.

Environmental Problem: -Another problem faced is sudden drop in temperature during winter, reduces the quantity in production of gas as there is no simple means available to keep digester warm in winter. Methane producing bacteria work best in temperature between 35° to 38° c. There is a fall in gas production at temperatures below 20 C and in fact, reaction stops completely at temperatures of 10° c and below. This sometimes imposes a serious limitation on the usefulness of biogas plants. In the Periyapatna, where cattle populations are generally found to be quite high, occurrence of low temperatures during the winter season reduces the possibility of using biogas plants for energy production.lack of large spaces is another problem to the establishment of plants.

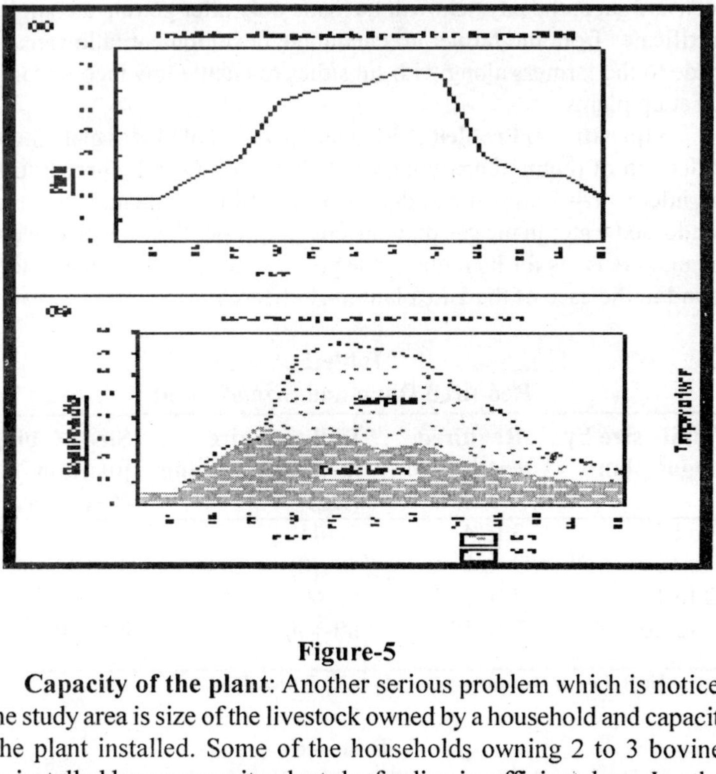

Figure-5

Capacity of the plant: Another serious problem which is noticed in the study area is size of the livestock owned by a household and capacity of the plant installed. Some of the households owning 2 to 3 bovines have installed larger capacity plants by feeding insufficient dung. In other cases, larger size of bovine population owned by a household installed a smaller capacity biogas plant. Table shows proportion of Bovine and number of existing plants in the sample villages. It is observed that there is under feeding of dung and over feeding of water in most of the plants.

Economic Problem: -The biggest problem is implementation and development of biogas technology in the rural community is associated with financial problems. Biogas plant is an expensive one, depending upon local costs of material and labor, a 2 m^3 plant costs about Rs.4000-

6000.The government offers loans and attractive subsidies for installation. The cost of constructing a 2 m^3 fixed dome plant in the study area is about Rs.5500 and subsidy available is Rs.3500 for scheduled castes, small and marginal farmers as well as landless laborers, and Rs.2500 for others. Non payment and misuse of loans has led to the stopping of loans by the government.

At present the Zilla Panchayanth has taken the responsibility of constructing the biogas plants. For this purpose private contractors have been assigned, the payment will be made only after getting a completion certificates from the farmer. to encourage development, allowances are made to the farmers along with subsidies to enable low income families to set up plants.

Operational Problem: - Regular operation of biogas plant involves collection of required amount of cow dung, followed by preparation of the inlet slurry, removal and drying of the split out manure, removal of condensed water in the gas pipeline etc., Improper feeding may be one of the main reasons for its failure. The problem of under-feeding is acutely found in the case of the 2 m^3 plant in study area.

Table:5

Required Dung and Size of Plant

Family size by biogas plants	Required cattle Size	Daily require ment of wet dung (Kg)	Size of biogas plant, in m^3 (Cft) / day
5 to 8	3 to 4	30-45	2 (70)
8 to 12	4 to 5	40-50	3 (105)
12 to 16	5 to 7	55-60	4 (140)
16 to 20	7 to 10	80-100	8 (210)

Source: Voluntary Hand Book, 2004-05

Table - 6

Required Quantity of Dung and Actual Dung Input

Plant Size in m^3	Daily Dung Required(Kg)	Average Daily Dung fed (Kg)
2	40	23
3	55	40
4	75	60
6	120	100
8	150	140

Source: Field Survey, Computed by Author

Table shows proportion of dung and water ratio's to maintain 2 m³ biogas plants. As it is already mentioned in the table, the daily input required for 2 m³ gas plant is 50 kg dung fed by biogas plant owners stands at 23 kg with 33.7 liters of water.

Table:7
Dung and Water Ratio for A 2 m³ Biogas Plant

Average Cattle holding Size	Daily Dung Fed (in Kgs)	Daily Water Fed (in Liters)
< 3	23	33.7
3 to 5	27	36
5 to 8	34	46
> 8	40	52

Source: Field Survey, Computed by Author

More problems are confined to the reduced gas production and reduced input rates are found with the fixed dome type plants. The fixed dome plants are designed in such a way that the spent slurry is thrown out only if there is sufficient gas pressure inside the digester. When gas production is decreased and if all the gas being produced is used up, the gas pressure cannot build up to the design value. As a result, the slurry level inside the digester will keep increasing and the space in the dome originally provided for gas is filled up with slurry, and makes the plant becomes inoperative. Hence, when the gas production goes down, it is advisable to remove the spent slurry manually from the outlet tank.

Selection of a Model and Suitable Size: - First a model suitable for a given area should be decided upon. Then careful selection of the size of a biogas plant has to be made to derive the maximum benefit by the user. The unit should be constructed as per the given specifications only which should be based on distance between kitchen and cattle shed, availability of dung and water, preference of the people model and family size.

During the time of field survey it is found that most of the plants installed by the households are the size suitable in relation to the number of cattle held by the owner or more than the number of cattle, it is the quantity of dung, which is important for the working of the plant. If a user has adequate number of cattle and if, cattle are taken out for grazing, a substantial quantity of dung gets lost, the available dung would be, therefore inadequate for the purpose of running the plant

Organizational Problem:- The planning, construction, and operation of biogas plants in an effective way calls for a well organized network providing; -

- Trained masons
- Supervising technicians
- Easy access to loans,
- Availability of materials for construction, etc.,

Under the existing set up, with all its bureaucratic loopholes, it is difficult to envisage the formation and proper functioning of such a network. When an individual takes up the construction of biogas plants procuring the required materials is the greatest problem. Proper monitoring of biogas plants is not being done and becomes greatest hindrance in the development of biogas energy in rural area.

On one hand there are regional changes in cattle holding sizes due to various problems while on the other there are regions where cattle are very less in number because of modern mechanization and urbanization. Other reasons for the biogas plants getting into a state of disrepair are components getting damaged and lack of maintenance of plants by users. From the questionnaire survey and the interview, the real dimension of Biogas energy distribution in rural environment together with the perception of people to it has been evolved.

Following are major rural energy problem: -

- Percentage of literacy should be increased to mitigate the diffusion of biogas plants.
- The provision of dung supply should be improved by proper management of collection.
- The condition of plants should be improved and yearly maintenance should be ensured. Additional services should be operated in the most of the villages

Prospects of Biogas Energy

The benefits perception by gender is analyzed on the basis of individual's matrix performance scores at sample household level. For this purpose five important components are fuel saving, monetary saving, cleanliness, time factor and smoke removal which were examined with the help of schedule based questionnaire. Among the various benefits from biogas technology, the users perceived wood saving as the important benefit. It was assessed during the filed survey about 40% of the biogas plant owners were more concerned about saving of wood. From a strategic point of view, people's perceptions on fuel saving are of special relevance in areas where: [a] fuel wood scarcity is a serious problem and a majority of the households have to purchase wood; and [b] fuel wood is the primary source of energy for cooking due to lack of availability of fossil fuel or poor purchasing power of the people.

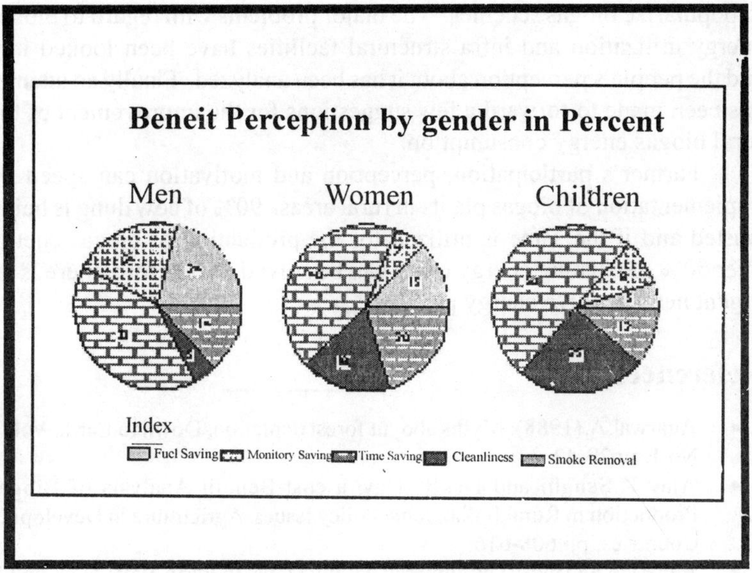

Figure-6

Biogas energy and its promotion have bright prospects when implemented among the rural community in developing countries Some of the identified areas to be addressed for rapid progress in establishing biogas technology among the rural folk are-

- Identifying the size of the plant on specific need
- Proper management of subsidy on biogas plants.
- Adequate and timely supply of good quality construction material particularly cements.
- Post construction care of the plants to cope-up with the frustration of farmers who have defective or failed plants.
- Careful and quality construction of the plants.
- At the same time an intensive effort to each of the villagers by organizing audio-visual demonstration may be quite helpful.

Conclusion

Biogas is one of the cheapest and easiest sources of energy most suited for the rural areas. The use of biogas saves up to 40% fuel expenditure. Central and state government should give more subsidies to popularize biogas schemes. The major problems with regard to biogas energy utilization and infra structural facilities have been looked into and the people's perception about it has been analyzed. Finally an attempt has been made to forward a few suggestions for the improvement of the rural biogas energy consumption.

Farmer's participation, perception and motivation can speed up implementation of biogas plants in rural areas. 90% of cow dung is being wasted and if the samc is utilized for the production of biogas energy over 50% of existing energy crisis can be solved. Therefore, there is an urgent need to solve energy problem

References

- Agarwal,A,(1988): Myths aboyut forest depletion, Down to Earth, Vol.7, No.1. pp29-42
- Ajay K.Sanghi and Dekly .Day, a cost-Benefit Analysis of Biogas Production in Rural India: Some Policy Issues. Agriculture in Developing Countries, pp.608-616.
- Banerjee,S,(1996):The enmergy within. Down to Earth, Vol5. No4, pp27-32
- Biogas Technology Handbook of Voluntaries.(1992) Mittal publications, P.26-38
- Colleran E.etal: "Anaerobic Digestion of High-Suphate-Content Waste water from the Industrial Production of Citric Acid", Wat. Sci.Tech.Vol30, No12, PP263-273, 1994
- Fanal.T.Etal (1985), energy and available energy, contents of cattle manure and digestor sludge. Aggricultural waste, pp 239-249.
- Ishwar and C Dingra,(1975), "Natural resources-Indian Economy", sultan Chad and sane publications, pp 97-108.
- Klass D.L. and Ghosh. [1981], In biomass as a non- fossile fuel sources, Arbor Science ann. pp.231.
- Kshethrabivrudhi adhikhari,(2001), "Karnataka",Jaivika Anila, Department of Graminabivrudhi and Panchayanthraj. pp 17.
- Lobo:(1998), "oil Sector, Emerging Scenario". FE, July 21, 1997. Also see K.L.N.Rao, "The Final Solution", Kurukshetra, March 13.1998

Section-V
Sustainable Development and Environment Pollution

24. Impact of E-Waste Recycling on the Environment and the Health and its Future Perspective in India

Ramesh Chandra Bharti
Subodh Kumar

E-Scrap or E-Waste or Waste Electrical and Electronic Equipment (WEEE) are the "Fastest Growing waste stream in the world". WEEE is one of the rapidly growing Challenge of India as well as the other countries in the world. E-waste can have an adverse impact on human life and the global environment due to the improper handling and the poor government legislation. In India, toxic-waste management assumes greater significance not only due to the generation of its own e-waste but also because of the dumping of e-waste from developed countries. Because of poverty and inequality in India, lack of appropriate infrastructure and procedures for its disposal and recycling, India becoming the global dumping ground for the African and Asian counties. The aim of this paper is to provide a concise overview of the current scenario of E-Scrap in India, like magnitude of the problem, health and environmental hazards, current methodology for disposal and recycling, existing legal framework, involved organizations and the future perspective.

Keywords: E-Scrap or E-Waste or Toxic-Waste, Health and Environmental hazard, E-Waste Legal Framework

Introduction

Electronic Scrap, "E-Waste" or "Waste Electrical and Electronic Equipment" (WEEE) is a waste consisting of any broken or unwanted electrical or electronic appliance. The E-Waste is one of the extremely major issues and challenge across the world. During the last decade, it has been assumed that the role of providing a forceful leverage to the socio - economic and technological growth of a developing countries. The consequence of its consumer oriented growth combined with rapid product obsolescence and technological advances are a new environmental challenge. It is an emerging problem as well as a great business opportunity for the developing countries, increasing significantly and produces both the toxic and valuable materials in them.

The ratio of the contents of the E-waste including iron, copper, aluminum, gold and other is over 60%, while plastics account for about 30% and the hazardous pollutants comprise only about 2.70% (Widmer et al., 2005). The Toxic – Waste packed and shipped from developed countries into the developing countries in the name of free trade (i.e. Donations, Toxics Link, 2004) as well as by paying high price. The recycling of this waste creates disaster long-term health consequences of constantly being fuels and toxic chemicals. As per the latest World Bank and census 2011 report (Times of India - 19th May 2011) 41.8% of total population is still below the poverty line due to the inequality and most of them are involving themselves in such a risky and dangerous recycling operations of the E-Waste because that is the way to make their living. The paper highlights the associated issues and strategies to address this emerging problem, in the light of initiatives in India.

What contributes to E-Waste?

The growing convergence of information, communication and entertainment has given a new impetus to the Electronics Hardware Sector which comprises mainly of four sub-sectors namely: Industrial Electronics, Computers and Peripherals, Communication and Broadcast Equipment and Strategic Electronics and Components. E-waste includes all types of electronic equipments which have become obsolete or have been discarded due to: Advancement in technology, changes in fashion, style, status or perception, nearing the end of their useful life. Figure 1 shows the different sources of the E-Waste:

Figure- 1: E-Waste Sources

Sources of E-Waste	
Household Appliances	Washing machines, Dryers, Refrigerators, Air-conditioners, Vacuum cleaners, Coffee Machines, Toasters, Irons etc.
Office, Information & Communication Equipment	PCs, Latops, Mobiles, Telephones, Fax Machines, Copiers, Printers etc.
Entertainment & Consumer Electronics	Televisions, VCR/DVD/CD players, Hi-Fi sets, Radios, etc
Lighting Equipment	Fluorescent tubes, sodium lamps etc. (Except: Bulbs, Halogen Bulbs)
Electric and Electronic Tools	Drills, Electric saws, Sewing Machines, Lawn Mowers etc. (Except: large stationary tools/machines)
Toys, Leisure, Sports and Recreational Equipment	Electric train sets, coin slot machines, treadmills etc.

Magnitude of E-Waste

An estimate, According to a report by UNEP titled, "Recycling - from E-Waste to Resources," 50 million tons of E-waste is produced each year and the amount of e-waste being produced - including mobile phones and computers - could rise by as much as 500 percent over the next decade in some countries, such as India.

1. Magnitude of E-Waste: in World

The United States is the world leader in producing electronic waste, tossing away about 3 million tonnes each year. China already produces about 2.3 million tonnes (2010 estimate) domestically, second only to the United States. Despite having banned e-waste imports, India and China remains a major e-waste dumping ground for developed countries. Environmental Protection Agency estimates that only 15-20% of e-waste is recycled, the rest of these electronics go directly into landfills and it can lead to the leaching of lead into the ground water can cause problems to the humans. Electrical waste contains hazardous but also valuable and scarce materials. Up to 60 elements can be found in complex electronics. In the United States, an estimated 70% of heavy metals in landfills come from discarded electronics.

2. Magnitude of E-Waste: in India

Studies so far reveal that the total e-waste generation in India is approximately 1,46,000 tonnes to 3.3 lakh tonnes a year and is expected to touch 5 lakh tonnes by 2011. 'The projected growth for e-waste generation for India is about 34% year on year' says Dr. Sinha (Associate Director of Toxics Link). Of the total e-waste generated in the country, western India accounts for the largest population at 35%, while the southern, northern and eastern regions account for 30, 21 and 14%, respectively. The top states generate 70% of e-waste in India in order of highest contribution to waste electrical and electronic equipment (WEEE) include Maharashtra, Andhra Pradesh, Tamil Nadu, Uttar Pradesh, West Bengal, Delhi, Karnataka, Gujarat, Madhya Pradesh and Punjab. The city-wise ranking of the largest WEEE generators is Mumbai - 50,000 tonnes, Delhi - 35,000 tonnes, Bangalore - 30,000 tonnes, Chennai - 25,000 tonnes , Kolkatta - 19,000 tonnes, Ahmedabad - 14,000 tonnes, Hyderabad - 13,000 tonnes, Pune - 10,000 tonnes, Indore - 8,000 tonnes. And also some amount at Surat and Nagpur.

Health and Environmental hazard in India

Electronic wastes can cause widespread health and environmental damage due to the emissions of toxic materials and poisonous gases. It contains both valuable as well as harmful components: Valuable components include precious metals such as gold, silver, copper, palladium, etc and harmful substances like lead, cadmium, mercury, beryllium, BFR, polyvinyl chloride and phosphorous compounds and hexavalent chromium in one form or the other are present in such wastes that produces several kind of diseases and pollution. According to N.M. Taphani, Regional Officer, Pollution Control Board, this Electronic Waste is highly dangerous for environment and hence should not be disposed with the regular garbage. The Ministry of Environment and Forests (MoEF) of the Government of India has developed a special guide for these issues and problems created by E-Waste. However, the workgroup for these divisions are currently functional only in Bangalore, Mumbai and New Delhi. While the density of e-waste in the country is on the rise, at present there is no well equipped e-waste recycling plant in India, only the informal methods has been used to disposed e-wastes. Figure 2 explain the various toxic constituents available in E-waste and their effect on the health in the form of various diseases.

Toxic Elements in Electronic Devices and its Impact health		
Components	**Toxic Constituents**	**Effect**
Cathode ray Tubes (CRTs) /TV	Lead Oxide and Cadmium And Barium (Be)	Disabilities, behavioral problems, seizures, coma and even death , central and peripheral nervous systems, hemopoietic system, genitourinary system, reproductive systems (male and female). Heart, spleen and Muscle weakness
Batteries (Automotive/Alkaline/ Button /Nickel-Cadmium Rechargeable/ Lithium/ computer)	Lead-Acid, Mercury, Sulfuric acid, Cadmium, Poly Vinyl Chloride(PVC)	Skin burn, skin allergy, inhaled or ingested, Cancer , fetus, neural damage, blood system and kidney damage, liver
Old Refrigerators, Heat Pumps and Air Conditioners	Freon, a chemical known as a Chlorinated Fluorocarbon	Each molecule of a CFC can destroy over 100,000 molecules of the earth's protective ozone coating , leading increased risk of and sunburn, cataracts and skin cancer
Motor Oil	Toxic Oxide	One quart of oil can kill fish in thousands of gallons of water
Printed Circuit Board	Lead and Cadmium	Skin burn, blood system and kidney damage
Capacitors and Transformers	Poly Chlorinated Biphenyls	Inhaled or ingested
Wires Burning	Dioxins, Bromine	Inhaled or ingested, air pollution, poisonous CO_2

Source: http://www.ces.iisc.ernet.in/energy/paper/ewaste/ewaste. html

Impacts of Informal Recycling

Due to lack of awareness, the workers are risking their life and the environment as well. The valuable fractions are processed to directly reusable components and to secondary raw materials in a variety of refining and conditioning processes. No sophisticated machinery or personal protective equipment like Masks or Glove is used for the extraction of different materials. All the work is done by bare hands and only with the help of hammers and screwdrivers. Not only the men, women and children are also routinely involved in these operations. Waste components which does not have any resale or reuse value are openly burnt or disposed off in open dumps. Pollution problems associated with such backyard smelting using crude processes are resulting in fugitive emissions and slag containing heavy metals of health diseases.

E-Scrap Strategies: in India

Recycling and reuse of material are the next level of potential options to reduce e-waste (Ramachandra and Saira, 2004). Recovery of metals, plastic, glass and other materials reduces the magnitude of e-waste. It is high time the manufactures, consumers, regulators, municipal authorities, state governments, and policy makers take up the matter seriously so that the different critical elements depicted in Figure 3 is addressed in an integrated manner. Sustainability of e-waste management systems has to be ensured by improving the effectiveness of collection and recycling systems.

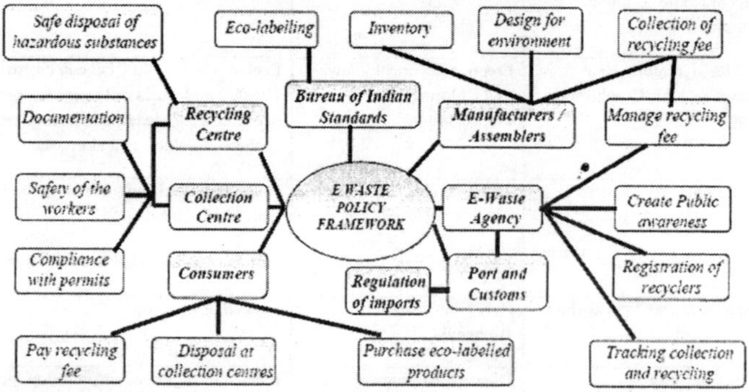

Source: http://www.swlf.ait.ac.th/UpdData/International/NRIs/Electronic%20waste%20management%20in%20India.pdf

The Policy shall address all issues ranging from production and trade to final disposal, including technology transfers for the recycling of electronic waste. Clear regulatory instruments, adequate to control both legal and illegal exports and imports of e-wastes and ensuring their environmentally sound management should be in place. The regulations should prohibit the disposal of e-wastes in municipal landfills and encourage owners and generators of e-wastes to properly recycle the wastes. Manufactures of products must be made financially, physically and legally responsible for their products. A public-private participatory forum (E- Waste Agency) of decision making and problem resolution in E-waste management must be developed. This could be a working group comprising Regulatory Agencies, NGOs, Industry Associations, experts etc. to keep pace with the temporal and spatial changes in structure and content of E-waste. This working group can be the feedback providing mechanism to the government that will periodically review the existing rules, plans and strategies for E-waste management. The efforts to improve the situation through regulations, though an important step are usually only modestly effective because of the lack of enforcement. While there has been some progress made in this direction under the pressure of NGO's like Green Piece and with the support of agencies such as GTZ, the government in the process of functioning and drafting the strategies and polices which make more difficult to bring all the E-Wastes. The enforcement of regulations is often weak due to lack of resources and underdeveloped legal systems. Penalties for non- compliance and targets for collection or recycling are often used to ensure compliance.

I. Controlling Authorities, Duties and Corresponding Rules

E-waste Rules are notified by the Ministry of Environment and Forests (MoEF), Government of India, on the May 12, 2011. The Rules will come into effect from the May 1, 2012. The rules shall apply to every producer, consumer or bulk consumer involved in the manufacture, sale, and purchase and processing of electrical and electronic equipment or components, collection centre, dismantler and recyclers of e-waste. These rules shall apply to every collection centre, dismantler, and recycler of e-waste and shall not apply to –

i). batteries as covered under the batteries rules 2001
ii). Micro, small and medium enterprises act 2006 (27 of 2006) and
iii). Radioactive waste as covered under the provision of atomic energy act 1962(33 of 1962)

Figure 4. Explain authorities and their duties regarding E-Waste:

Ministry of Environment and Forests, under the Environment (protection) Act, 1986	i. Identification of hazardous wastes as per Rule 3 ii. Permission to exporters as per rule 14(3) iii. Permission to importers as per Rule 13(3)
Central Pollution Control Board constituted under the Water (Prevention and Control of Pollution) Act, 1974	i. Coordinate activities of the State Pollution Control Boards and ensure implementations of the conditions of imports ii. Monitor the compliance of the conditions of authorization, import and export. iii. Conduct training courses for authorities dealing with management of hazardous wastes iv. Recommend standards for treatment, disposal of waste, leachate and specifications of materials v. Recommend procedures for characterisation of hazardous wastes
State Pollution Control Boards constituted under the Water (Prevention and Control of Pollution) Act, 1974	i. Grant and renew authorisation under rule 5(4) and rule 8 ii. Monitor the compliance of the various provisions and conditions of authorisation iii. Forward the application for imports by importers as per rule 13(1) iv. To review matters pertaining to identification and notification of disposal sites
Directorate General of Foreign Trade constituted under the Foreign Trade (Development & regulation) Act 1992	i. Grant licence as per rule 13(5) ii. Refuse licence for hazardous wastes prohibited for imports under the Environment (protection) Act, 1986
Port Authorities and Customs Authorities under the customs Act, 1962	i. Verify the documents as per rule 13(6) ii. Inform the ministry of Environment and Forests, Govt. of India of any illegal traffic as per rule 15 iii. Analyse wastes permitted for imports and exports iv. Train officials on the provisions of the Hazardous Wastes Rules and in analysis of hazardous wastes

II. Organizational Involvement in India

To tackle the big challenge and convert it into the employment opportunity, Government, NGO's and various private organizations looking forward and take initiatives in this direction. The **Cerebra Integrated Technology Ltd. (BSE: 532413),** India's leading total It solution provider announced on 31 August 2010 in Mumbai to setup its unit at Bangalore as India's largest recycling unit with Cemelia Resource Recovery Pvt. Ltd offering environmental restoration solution for the electronic, electrical and equip firm. The company will be investing Rs. 50 Crore for the proposal unit and aim to achieve a company turnover of Rs. 250 Crore in FY 2013 with a profit margin of 25%. **Wipro:** The environmentalist organization Green Peace International last September dumped some 500 kilograms of electronic waste outside of Wipro's Bangalore headquarters. The activists said that they had collected the scrapped Wipro-branded computers from recycling yards in Delhi, Bangalore and Chennai. ***Trishyiraya Recycling India Private Limited*:TPL** is a wholly owned Indian subsidiary of SIMS Recycling Solutions. SIMS group is a USD 8.0 billion corporation having facilities in over 200 locations across the globe. Radha Industry (world's largest and fastest growing manufacturing industry - Radha, 2002; DIT, 2003) HCL, Tata Teleservices, HP and Nokia and many other organization are also planning.

E-Scrap Future perspective in India

E-waste, with the rapid scientific progress has been made and new frontiers reached with the advent of the twenty - first century. Progress has brought with it some problems, viz. large quantities of Non-Decomposable Electronic Dump/Wastage; this is acquiring monstrous proportions. According to a study by the government of India, it has been found that the amount of E waste in the country is growing at 10% every year and 95% of it will end up in urban slums – 434000 metric tonnes. According to a new UN report released in Bali, Indonesia by 2020, India's e-waste from old computers will jump 500% from 2007 levels, whereas South Africa and China will witness a 200-400% rise in computer-related waste. The rapidly growing mobile telephony in India will take its toll by 2020 when e-waste from discarded phones will grow 18 times from 2007 levels, a period during which China is estimated to see a seven-fold rise in the electronic waste from mobile phones and as result India becoming a global dumping ground.

Conclusion

India can play a leadership role on the global stage and emerge as the third largest economy in the world after the US and China. Despite a booming economy India ranks poorly on poverty. Therefore India need to catch-up in various areas and also need an urgent approach to manage the e-waste hazard by technical and policy-level interventions, implementation and capacity building and increase in public awareness such that it can convert this challenge into an opportunity to show the world that India is ready to deal with future problems and can set global credible standards concerning environmental and health.

References

- Sardinia 2007, Eleventh International Waste Management and Landfill Symposium, Last accessed on June 17th, 2011 Available from: *http://www.swlf.ait.ac.th/UpdData/International/NRIs/Electronic%20waste%20management%20in%20India.pdf*
- Available from: *http://www.ncbi.nlm.nih.gov/pmc/articles/PMC2796756/* [Last accessed on June 17th, 2011]
- Available from: *http://en.wikipedia.org/wiki/Electronic_waste* [Last accessed on June 17th, 2011]
- Available from: *http://articles.timesofindia.indiatimes.com/2009-11-11/pune/28074435_1_e-waste-weee-maharashtra-pollution-control-board* [Last accessed on June 18th, 2011]
- Available from: *http://www.livemint.com/2010/02/22163815/By-2020-India8217s-ewaste.html* [Last accessed on June 18th, 2011]

25. Environmental Management: Internal and External Dimensions and Response of Defense Installations for Sustainability

Sharad Chandra Dwivedi

The defense sector in India has generally not been explicitly included within the purview of environmental regulation, perhaps out of deference to the military mission. But this is not likely to remain so into the future. This article presents how procedures for addressing environmental concerns, which are based on developments in ecosystems science and environmental management, can be integrated into the military mission in India.

Key words: Environmental conservation, Defense sector, Natural resources, Environmental Policy

Introduction

The Government of India has promulgated a number of acts, rules, and notifications for the protection and preservation of the environment that are addressed at various sectors – agriculture, industry, forestry, energy, mining, tourism, transportation and human settlement[1]. The defense sector has not been explicitly included within their purview, perhaps out of deference to the military mission. In recent times however, the environmental dimension is being increasingly introduced more directly into the military role. The International Association of Retired Generals and Admirals specifically resolved in 1993 to include the military role in environmental protection and restoration. There are several examples of micro level initiatives at reforestation and greening of the military areas, use of renewable sources of energy, recycling of wastes and other pollution control measures. The Indian military forces have also frequently extended logistics support to the civilian sector agencies in several environmental conservation and documentation efforts in difficult terrain. Many of these efforts are based on individual interest and initiative of the commanding officers. The Army Environmental Cell organizes annual Environmental Meetings, with representatives from the Army Commands, Navy, Air Force, Territorial Army, Border Roads and Environmental Ministry of the Government of India, as well as other non- governmental environmental organizations to discuss and present the progress on environmental concerns. All of these measures recognize the importance of environmental conservation and the role of the military. These are largely voluntary measures and there is no institutionalization of activities related to environmental conservation and protection of military lands through formal policies and procedures which mandate compliance with national environmental policies and laws, while furthering the military mission.

The need for such institutionalization will become increasingly relevant and urgent as available lands for military purposes decline with rising population, demand for land from other sectors rises, and as training and testing activities are modernized and intensified in existing defense controlled areas. Environmental groups and the government will also apply increasing pressure for compliance with environmental quality standards. This has happened in developed countries like USA, where following the enactment of a series of acts related to the environment conservation and protection, the Department of Defense adopted a policy that commits itself to act responsibly in public interest in managing its lands and resources. It directed its decision-makers and commanders to introduce integrated natural resources management plans while ensuring that they support the primary military mission. The defense forces in India too will have to be concerned, sooner than later, about long-term sustainability of military lands for testing, training and residential purposes, and conformity with national environmental policies and regulations which aim at sustainable development of the regions of which their lands are a part.

Objective

The objective of this article is to look ahead and understand how environmental concerns can be anticipated and procedures for addressing them integrated into the military mission in India. In doing so, three factors are kept in view. The first is that land use is vital to the sustainability of the defense mission, but land management is not the primary mission of the military forces. The second factor is that the procedures for protecting and improving environmental quality must be science-based. The third factor is that for effective implementation of an environmental policy, there is a need to have in place a standard management framework that will permit planning, monitoring and evaluating the performance of the system on a 'corporate wide' scale.

Methodology

Accordingly, to the objective this article is developed in three parts. The first part presents an international defense forces perspective on ecosystems and environmental management as applied to the defense lands. The purpose of this is to understand and learn about the processes of integrating environmental concerns into defense planning, so that similar processes can be initiated in India. The question of implementation of environmental management across the defense installations on a 'corporate-wide' process by employing appropriate environmental management systems is addressed in part two. In the final part, the implications of these developments for integrating environmental concerns into defense planning in India are presented.

(I.)The International Defense Forces Perspective on Environmental Management

Military activities and environmental impacts-

Field training in peace time is essential to keep the defense forces in wartime readiness. Most large defense training areas/installations/ cantonments typically consist of a small city or township with housing, commercial, maintenance and light industrial components; a road, rail and air transportation infrastructure; and large relatively undeveloped area for training, testing or storage. They come in a wide range of activities, sizes and locations. The impacts on the ecosystem of administrative and light industrial activities in the area may not be significant. But, environmental management may prove critical in the undeveloped land, which is used primarily to support field training requirements for combat and supporting units. In these areas the defense forces have to balance training and readiness requirements with the conservation of natural resources.

Training by armed units generates cumulative impacts on natural resources. Tanks, mechanized infantry, self-propelled artillery and combat engineer units are extremely mobile because they are tracked vehicles. They are heavy, fast and capable of producing significant physical damage to the natural environment and habitat in a short time. Light infantry units also train to dig in and prepare field fortifications, which can alter vegetation, drainage and wildlife patterns. Engineer units practice obstacle construction, demolition, and road building and maintenance. Logistics units set up small towns with trucks, tracked vehicles, light industrial units, etc. Helicopters practice hovering and landing, which can cause soil erosion, create vegetative bare spots from rotor downwash and disturb bird species. Units must also learn to operate in all types of climates and all types of terrain – mountains, desert, jungle, beaches, etc. All these activities develop military skills, but can often lead to natural resources destruction. Also, significant from the military point of view, they leave a signature which can be useful information for the enemy.

The environmental impacts of military training include vegetation destruction, soil compaction and erosion, increased suspended solids in streams and lakes, increased levels of volatile organic compounds and particulates in the air, higher noise levels and occasional loss of wildlife. Maintenance facilities can impact natural environment by discharging hazardous wastes such as oils, acids, heavy metals, paints and solvents during diagnosis, repair and testing. The housing areas will create the same type of environmental impacts as a small city: municipal solid waste, impacts on drinking water quality, construction waste, etc. The extent of damage to the ecosystem will be determined by the robustness or fragility of the natural resources themselves, type and length of training, weather conditions, recovery potential of the area and the land and wildlife protection and maintenance procedures in effect[2].

Thus military lands are not totally renewable resources. The training capacity of the lands can be undermined if adequate measures are not taken to protect the ecosystems. In many developed countries, the defense forces have formally undertaken the responsibility for maintaining the ecosystems in the military lands. Some of the measures being taken by them are briefly described, taking the example of the US Department of Defense and the US Army. In these organizations, the environmental policies, strategies and action plans have evolved systematically over the past two decades in response to the environmental concerns. The evolutionary process is well documented, and available in public domain. The development and working of these policies can help understand the process of integrating environmental concerns into defense planning in India.

Environmental Policy in the US Department of Defense

The US Department of Defense is the third largest federal land owner in USA and manages about 25 million acres of national lands, which include unique natural and cultural resources. Prior to 1960s, defense lands were exempt from most natural resource planning and environmental protection requirements. As relatively undisturbed natural habitat became increasingly scarce, the priority for its protection also rose, leading to the Sikes Act in 1960. This Act authorized and encouraged, but did not require, the armed services to conserve fish and wildlife, and allow recreation wherever it did not interfere with the military mission. Gradually, the environmental protection laws became increasingly stringent in USA. The Endangered Species Act of 1973 required all federal agencies (including the military) to protect habitat as well as threatened and endangered species and established procedures to consult with the Fish and Wildlife Service, if a proposed activity was likely to harm the listed species. This is a powerful act and has stopped or reshaped many army activities. These acts were mostly seen as reactions to failure, rather than preventing it. By late 1980s the need to be more proactive in environmental conservation and protection was felt. In 1989, the US Department of Defense issued a policy directive for a more proactive approach on environmental management in defense lands. It directed the mission commanders to keep informed of the conditions of natural resources and develop integrated natural resource management plans (INRMPs). More specifically, it asked the Heads of military services to ensure that effective natural resources management is an identifiable function and is specifically accountable in performance evaluations at each command level[3].

All of these culminated, in 1990, in the establishment of the Army Environmental Policy Institute to provide long range strategic planning for future environmental requirements. In 1991, the 'US Army Environmental Strategy into the 21st Century'[4] was declared. The strategy provides a framework to ensure that environmental considerations are integral to the army mission and an environmental stewardship ethic governs all army activities. The strategy takes its direction from its vision that the army will be a national leader in environmental and natural resource stewardship for present and future generations as an integral part of its mission. The strategy identifies four major activities to realize the vision: compliance with all environmental laws, restoration of contaminated sites, prevention of pollution, and conservation of natural resources. In 1994, the army formally took over the responsibility for maintaining the natural resources in the training areas from the engineering community. In 1996, the Department of Defense directed that INRMPs should incorporate the principles of ecosystems management. Guidelines for preparing the INRMPs were issued by the army in 1997. The Presidential Order of April 2000 titled 'Greening the government through leadership in environmental management'[5] directed all federal agencies to develop, document and adopt formal environmental management systems (EMS) to ensure that policies, strategies and procedures established to support federal environment leadership programmes can be effectively monitored and evaluated. The order mandates that an EMS consistent with the Environmental Protection Agency (EPA) requirements must be in place by 2005 in all federal agencies, including the defense agencies. In January 2003, the Department of Defense issued a new directive on EMS implementation criteria[6] and metrics to guide and measure progress in implementing the environmental plans.

As a consequence of the above measures, the US Army has in place three broad environmental programmes[7]. These relate to analysis of environmental impacts associated with new army proposals, management of natural resources through integrated planning, and management of training areas. The first of these is to comply with the requirements of the National Environmental Policy Act. The second gradually integrated INRMPs into science-based ecosystems management and with EMS adopted by the industry to conform to international standards (ISO 14001)[8]. The training area management programme has four broad objectives: integrate environmental planning into all operations, protect natural and cultural resources, ensure that operations comply with international standards, and prevent future pollution. Accordingly, the programme has been designed to include four components: land condition trend analysis (LCTA), land rehabilitation and maintenance (LRAM), training requirements integration (TRI), and environmental awareness (EA). LCTA is used to make decisions regarding training intensity and rehabilitation. LRAM mitigates the adverse effects of training and testing through land maintenance and repair. TRI ensures access to training lands by integrating training activities with the ecological conditions. The EA programme develops material for distribution about environmental stewardship. Several army installations have found it convenient to adopt the watershed management approach for implementing the above four components in a unified framework.

In addition to the generally applicable framework for addressing environmental issues described above, the US Army also has in place a number of specific programmes based on media or resources being deployed. These include[9]:

- Acquisition pollution programme to integrate technology requirements across army commodity areas to maximize environmental benefits over the life cycle of weapons systems. The areas addressed by this programme include elimination of ozone-depleting chemicals, hazardous materials and toxic substances.
- Installation pollution prevention programme to conserve and recover resources and reuse or recycle materials that would normally enter the solid or liquid waste stream by implementing integrated management approaches in all mission areas.
- Air quality management programme to identify sources of emissions, determine the type and amount of pollutants and control emissions into the atmosphere to conform to the environmental laws.
- Environmental noise reduction programme to protect present and future installation missions, personnel, family members and civilian employees by reducing environmental noise impacts where feasible.
- Lead hazard management programme to reduce release of lead, lead dust or lead-based dust paint into the environment from deteriorating paint surfaces and ensure proper disposal of wastes contaminated with lead-based paint.
- Drinking water management programme to preserve rights to and conserve all water resources and protect them from contamination in order to provide drinking water that satisfies the most stringent regulations.
- Waste-water management programme to control or eliminate all sources of pollutants to surface or groundwater using conventional or alternate innovative treatment systems and processes.

- Hazardous waste management programme to substitute nontoxic/ non hazardous materials for toxic/hazardous materials and ensure compliance with all regulations pertaining to generation, treatment, storage and disposal of hazardous wastes.
- Solid waste management programme to minimize generation and disposal of solid wastes by actively encouraging source reduction, reuse, recycling and composting programmes.
- Underground storage tank programme to ensure that hazardous waste is not stored in underground tanks; all underground tanks have double-walled constructions with interstitial space, and remove all abandoned underground storage tanks.
- Installation restoration programme to clean up previously contaminated lands.
- Cultural resources management programme to ensure that the army manages the cultural resources (historic places, native American lands) under its control in compliance with public laws.
- Fish and wildlife and endangered species management programme to ensure that the army meets environ- mental compliance and stewardship responsibilities by carrying out mission responsibilities in harmony with the Endangered Species Acts.
- Notice of violation control and management programme to address promptly non compliance notices from federal environmental agencies or the public and take corrective action.
- Reporting system to deploy a database management system with report code symbols to identify all army environmental programme requirements and track the progress in implementation.

The gradual evolution of the environmental policies and implementation procedures on a system-wide scale by the US Army, as described above, has elevated it to the status of a national leader in environmental management in USA.

Environmental policy in Australian Defense Forces

The Australian Defense Forces (ADF) too has realized that environmental degradation can deny access to lands, reduce the realism of training and limit operational flexibility. Accordingly, the ADF Environmental Vision has been developed which reads, 'Defense will be a leader in sustainable environmental management to support the Australian Defense Forces' capability to defend Australia and its national interests'. The environmental strategy for the ADF[9] is also based on the adoption of the ecosystems approach for integrated natural resources management within the framework of the industrial standard EMS, ISO 14001.

(II) Environment Management System and standards

Increasingly stringent environmental regulations, international obligations and agreements, and private sector recognition that potential economic benefits may result from improved environmental performance, have contributed to the development of systematic new tools for dealing with complex environmental problems. These tools are the EMS. The International Standards Organization (ISO) published the first world-wide standard for EMS, the ISO 14001, in 1996. Both the US Department of Defense and ADF have adopted the industry standard EMS, ISO 14001 as the guiding framework for implementing their environmental policy. The Ministry of Environment and Forests in India is also advocating the adoption of EMS by industry for effective monitoring of compliance and regulation requirements.

The ISO 14001 defines an EMS as the part of an overall management system that includes organizational structure, planning activities, responsibilities, practices, procedures, processes and resources for developing, implementing, achieving, reviewing and maintaining an environmental policy. It is a voluntary EMS standard that provides a framework to move from reactive and fragmented responses to environmental issues common to compliance- based environmental programmes. At the same time the framework provides for a proactive approach that facilitates early identification of impacts, liabilities and opportunities. The ISO 14001 does not specify absolute requirements for environmental performance but demands a commitment to continuous improvements in environmental performance, efficiency of operations and regulatory compliance, through repeatable and consistent control of its operations. Performance levels are established by environmental law or organizational management.

Thus, instead of focusing on specific environmental activities, the ISO 14001 EMS focuses on 'management' activities. The emphasis is on the management structure and how management achieves its stated goals. While the framework of an EMS remains constant over time, each individual EMS should be tailored to reflect those aspects and environmental impacts that have been deemed significant by the organization, based on its corporate culture and strategic goals. An EMS is therefore, not a stagnant system, but provides the mechanisms necessary to continually evolve to meet the ever-changing needs of an organization. The management framework of ISO 14001 EMS is based on a cyclical process of commitment and policy, planning and implementation, evaluation and review, as shown in Figure 1. The essential elements of ISO 14001 are its auditable requirements:

- **Commitment and policy**: An organization should define its environmental policy that ensures its commitment to its EMS and communicates the policy to all employees and public.
- **Planning:** An organization should formulate a management plan to fulfill its environmental policy by identifying its environmental aspects and impacts and legal requirements, and establishing objectives and targets for the significant impacts.

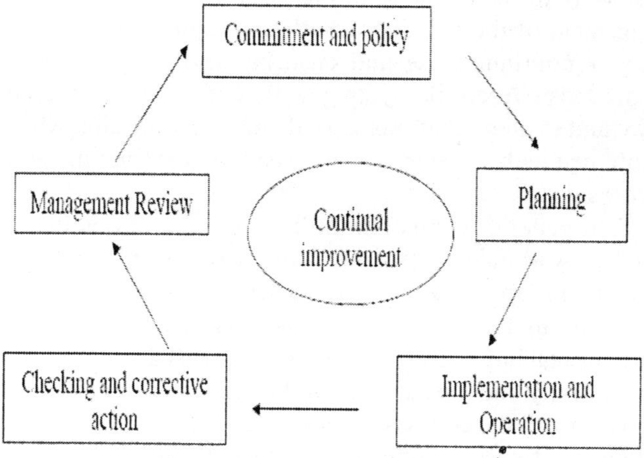

Figure 1. The ISO 14001 EMS process

- **Implementation**: An organization should develop the capabilities and support mechanisms necessary to achieve the environmental policy, objectives and targets, by defining roles and responsibilities, allocating resources, communicating and documenting the procedures, and providing training to its employees.
- **Measurement and Evaluation**: An organization should measure, monitor and evaluate its environmental performance by establishing procedures to control operations and for responding to emergencies.

- **Review and improvement**: An organization should review and continually improve its environmental management systems, with the objective of improving its overall environmental performance.

Formal certification to ISO 14001 can be achieved either by third party audit or self-declaration of conformance with the standard. It is possible to implement ISO 14001 without seeking formal certification.

(III) Towards an Environmental Policy for Defense Forces in India

The defense forces in India have installations and carry out operations across many regions in India. Because of rising pressures on land caused by the growing population and from other economic sectors like agriculture, industry, tourism, etc., newer suitable training areas are unlikely to be available in the future. Sustainable environmental management of the existing installations and sites is therefore critical for their continued use and ensuring India's long-term defense preparedness. Accordingly, they will need to adopt an environmental policy and strategy that promotes the military mission while ensuring compliance with the environmental laws and promoting environmental stewardship.

The general layout of the military installations in India, the nature and intensity of training, the townships, industrial and maintenance units within them, and field operations would be broadly similar to the corresponding features of the US or other modern defense forces. The environmental impacts generated by them will depend on the intensity of the activity and the capacity of the ecosystems, of which they are a part, to absorb the stresses. The stresses are on the soil, water, marine, biodiversity, forest and other eco- system resources. The ecosystems in India are in many ways more sensitive (than is the case in USA) because of different physiography, climate and other natural resource regimes, and significantly higher population densities. The more sensitive ecosystems in India have been identified in the Western Ghats, ravines, the Sunderban deltas with its wetlands and mangroves, the Northeast, the Shivaliks, flood plains, forests, Eastern Himalayan region, red and black soil, and coastal areas. In all these areas, the defense forces have a strong presence. Some of these locations were chosen because of the natural and location advantage they provide for military training. In each location, the environmental issues and priorities are likely to be different. From the perspective of the military mission, therefore, it is important to manage these installations in a sustainable fashion. Equally significant is the need to sustain the ecosystems for future generations as some of them are unique and possess biodiversity and other resources of immense value to mankind.

The environmental policy and strategy to be adopted will need to vary with location, type of installation and nature of ecosystem. A management systems framework needs to be established from a corporate or headquarters perspective to guide its implementation and monitor performance at individual installation level. Drawing on the lessons of the environmental policy and its implementation in the US Army, and the developments in EMS in industry, the Indian defense forces also need to integrate environmental policy and programmes into the military mission while ensuring compliance with national environmental laws and promoting environmental stewardship.

It is suggested that the international standard EMS, ISO 14001 be adopted by the defense forces to design and implement its environmental policy. The documentation and auditable requirements of this EMS (policy statement, environmental management plan, implementation procedures and mechanisms, performance monitoring indicators, review procedures) provide a unified framework within which the environmental management and performance of each installation can be independently articulated, reported and evaluated with respect to its goals. The management plans need to be based on the best science available for the ecosystems and resources in question. This facilitates environmental management on a 'corporate-wide' scale, and also provides for continuous Learning, adaptation and improvements in performance at the installation level.

The challenge is for the defense strategists and the military leadership to articulate the relevance and need for an environmental policy for defense to sustain the military mission, and promote its role in environmental stewardship, obtain the resources, and create organizational structures and systems to implement the policy. The ISO 14001 EMS provides the basic framework to meet the challenge.

Acknowledgements

The article is based on a presentation made at the 29th Defense Management Programme to Senior Officers of India's Defense Forces at the College of Defense Management, Secunderabad. I thank Rear Admiral V. Balachandran, Commandant of the College for suggesting the study.

References

- National conservation strategy and policy statement on environment and development. Ministry of Environment and Forests, GOI, 1992.
- Army Ecosystems Management Policy Study. Army Environmental Policy Institute, 1997, p. 152.
- Environmental Conservation Programme. Department of Defense Instruction 4715.3, 3 May 1996.
- US Army Environmental Strategy into the 21st Century. Army Environmental Policy Institute, 1991.
- Greening the Government through Leadership in Environmental Managreement, Presidential Documents: Executive order 13148, USA, 2000.
- Environmental Management Systems Pilot Study, Final Report. Department of Defense, ISO 14001, 2000.
- The Defense Transformation for the 21st Century Act, Bill Submitted to the US Congress, Department of Defense, 2003.
- Sinclair and Tschirhart, R.,Army environmental policy and ISO 14001. *Federal Facilities Environ. J.*, 2001, 53–65.
- Understanding Environmental Management Systems in Defense. Australian Defense Forces (http://defweb.cbr.def.gov.au/ems)

26. Environmental Issues and Glass Industries in Firozabad

U.S. Pandey
Niharika Chaturvedi

The effect of the glass industry on heavy metal characterization was assessed in the sludge/sediments of 5 discharge pits (sites) for effluents of functional glass factories and/or finishing workshops in the study area. A comprehensive profile of Zn, Mn, Co, Cd, Pb, Cr, Ni, Cu and As levels in industrial sludge was obtained. Correlation coefficients for metal pairs indicate varying levels (positive/negative) of correlation in the samples. Statistically significant difference (P = 0.001) was obtained for each metal on comparing mean metal content among sludge samples. Phytoremedial potential of Typha latifolia, a locally thriving plant species was also valuated. Shoot and root samples from T. latifolia L. were taken for heavy metal analysis and leaf samples for igment analysis to gauge its tolerance of heavy metals. Total chlorophyll and free proline content indicate that the species has adapted itself to multi metal stress in the rhizosphere. BCF > 1 (Bioconcentration factor) and TF < 1 (Translocation factor) were obtained in T. latifolia samples in the case of Mn, Cr and As at all 5 sites; n at 4 sites; Co, Cd and Ni at 2 sites each, and as such it can be advocated as a potential phytostabilising pieces for these metals. The species demonstrated limited capabilities in remediating Pb and Cu.

Key words: Glass industry, Glass Beads, Fuel efficiency, Energy efficiency, Technology transfer, Pollution.

Introduction

This city was built by *Firuz Shah Tughluq*, a *Tughlaq* king. From the earliest it was famous for glass and *bangle* works, and its related small scale industry is famous throughout the world. Due to underdeveloped industry, literacy rates are very low. The city suffers from child labour, low literacy, and pollution. Firozabad City, where people breathe not air but glass, where every second vehicle on the road is found fully loaded with glittering bangles of vivid colours. In this city, if you will stay for two days, you will find it very difficult to imagine that, if it had not been for glass, then what this city would have done. Firozabad is located in north central India, in western Uttar Pradesh state, 40 km away from *Agra* and around 200 km away from *Delhi*, at the northern edge of the *Deccan Plateau*, at 27°092 N, 78°242 E. The height above sea level is 164 meters (540 ft). The district is well connected by rail and bus routes to major cities. The nearest airport is Agra. The boundaries of district touch *Etah* district in north and *Mainpuri* and *Etawah* districts in the east. The *Yamuna* river makes its southern boundary.

The area of the district is about 0.8% of total area of Uttar Pradesh, and the population is 1.1% of the state's total population. Approximately 73.6% of the population lives in rural area. More than half of the workforce is engaged in manufacturing activities. The city is also an important market center. It has severe winter and summer seasons. The district is mostly flat and its slope is from north west to south. Firozabad was known as Chandwar Nagar in ancient times. The name Firozabad was given in the regime of *Akbar* by *Firoz Shah Mansab Dar* in 1566. It is said that *Raja Todarmal* was passing through this town, on a pilgrimage to *Gaya*. He was looted by robbers. At his request, *Akbar*, sent his Mansab Dar Firoz Shah here. He landed near or about Datauji, Rasoolpur, Mohammadpur Gajmalpur, Sukhmalpur Nizamabad, Prempur Raipura. The tomb of Firoz Shah and ruins of Katra Pathan are the evidences of this fact.

Climate: The lowest recorded temperature in Agra was "1 °C; the highest was 48 °C.

Demographics: *As of 2001* India *census*, Firozabad had a population of 278,801. Males constitute 53% of the population and females 47%. Firozabad has an average literacy rate of 59%, lower than the national average of 59.5%: male literacy is 64%, and female literacy is 53%. In Firozabad, 16% of the population is under 6 years of age.

Industry

Evolution of Firozabad Glass Industry

During ancient periods, invaders brought many glass articles to India. These glass articles when rejected were collected and melted in locally made furnace called as "Bhainsa Bhatti." This was the start of the glass industry in Firozabad. In this furnace, wood was used as a fuel. These old traditional furnaces are still in use in *Sasani* near *Aligarh* and at *Purdal Nagar*. During that time only small bottles and *bangles* were made. At this time only one bangle at a time can be made. In this bangle there was no joint. These bangles were called "Kadechhal Ki Chudi". Since then Firozabad is the home of the glass industry, white and coloured glass pieces being manufactured for the purpose of assembling *jhad* and *fanus* (chandeliers) which were in demand by royal courts and nobles for decorating their assembling and drawing rooms. Later on phials for *Itra*, scents, and other cosmetic products were made. Slowly and steadily Indian marriage items like bangles, *kangans, kada*, etc. were produced in bulk for the general public. Today it is nicknamed *suhag nagri* because it fulfills almost all the demand of bangles, kadas, kangans and other items of suhagins(married women).

Since 1989, Firozabad has produced artistic glassware in different colours and shades used in chandeliers and other items. About four hundred glass industries are registered in Firozabad, making different types of glass products. They use natural gas instead of coal. Half of the production of these units is exported. Now-a-days, blowing/modeling is done by the *pot regenerative tank* furnish process in the factories.

Products

All sorts of glass articles, including jars, candle stands, glasses, flower vases, and electric wares such as decorative lights, bulbs and every other sort of glass articles are prepared in this city. These articles are painted with vivid colours and innovative patterns. Etching is done on some products to enhance their beauty. Decoration work is also done on some products by fixing other material on them. Firozabad glass industry broadly consists following types of products: -

- **Glass Art wares** - Glass artware like toys, candle-stands, crosses, Christmas trees, different fruits, mini-trees, images of birds, animals, personalities, Gods and Goddesses are made here for Indian as well as foreign markets.

- **Glass domestic-wares** - Firozabad glass industry for the last two centuries has been involved in manufacturing drinking glasses, beer glasses, wine glasses, dinner sets consisting of bowls, plates, lemon-sets, pudding sets, etc. in attractive cuttings, charming designs.
- **Glass Chandeliers** - Since the Mughal period, Firozabad has been producing different shapes and shades of magnificent hanging chandeliers. Many innovative designs have been added to suit different tastes.
- **Glass Marriage wares** -Bangles, Kangans and karas are made here since ancient period as traditional Indian woman wardrobe accessory. Bangles made here are distributed throughout India. Bangles don't get that much opportunity in the international market, since it is a traditional Indian accessory. But still due to globalization the market for bangles is expanding internationally.
- **Glass hardwares** - All types of glass hardwares are manufactured here for house decoration and other purposes. Glass scientific and lab wares since last 10 years beakers, flasks, test tubes, containers and other scientific wares are being manufactured here, which are needed in various laboratories.
- **Glass Automobile wares-** The glass accessories for automobiles like light bulbs, mine battery bulbs and other light and sight equipments are produced here for two wheelers and four wheelers.
- **Glass Street and domestic light wares-** For the lighting of urban and rural areas and other utility purposes many glass items are being produced including miniature bulbs and high power voltage bulbs. It is estimated that 6% demand of bulbs is being fulfilled from Firozabad
- **Glass Beads** -Firozabad is known as the world capital for beads. Glass beads produced here are exported to all over the world for use in all kinds of items including clothing, accessories, and shoes.

Socio-Economic context

Firozabad has developed as a natural cluster of a diverse range of small-scale glass manufacturers involved in labour intensive production and decoration of bangles, beads, handmade tumblers, tableware and decorative ware. There are also a few medium scale units using semi-automatic equipment to manufacture industrial glass products such as automobile headlight lenses, glass bulbs, shells and consumer products such as tumblers, bowls and ashtrays.

Over the life of the industry and its various sectors there appears to have been little or no change in technology, working practices, product range or market techniques either through internal evolution or induced by outside influences. In terms of the technology of the processes and the working conditions, it is a backwater seemingly cut off from modern methods and marketing practices. This is surprising since Firozabad is located not far away from the national capital (250 Km) and not far away from the tourist Centre of Agra (45 Km). It is located on the national highway (Delhi-Calcutta) and is in the prosperous western part of Uttar Pradesh that is advanced in agricultural and industrial methods. Moreover, a majority of the factory owners belonged to a community that has shown tremendous entrepreneurial capabilities in other parts of the country and in other sectors of industry. The work force is mainly on contract and few companies have permanent employees. Working practices evolved over years, a structured hierarchy within the work force and fears about job losses may all be factors that militate against change and acceptance of new technology.

Economic conditions: Cost of batch, cost of glass. Exports in value and as a percentage of total sales. Most of the outputs from CDGI can be measured; raw material quality, pollution levels, fuel efficiency, glass quality, production efficiency. It is therefore surprising that the Centre has not developed a set of benchmarks and an understanding of the best industry practice in that area. Such data can be used by industry to set performance targets and to measure the progress. The base line data produced in the Report "Base Line Survey cum Diagnostic Study of Firozabad Glass Industry" in 1995 set out the conditions, as they existed in the glass industry in Firozabad. To chart the progress of CDGI and its impact on the glass industry is an essential part of its duties and although this work has not been done, it is not too late to start. The specifications will form an essential part of the move towards achieving an ISO 9000 certificate by the Centre and by the glass industry and its suppliers.

Energy efficiency: The ability to improve fuel efficiency depends upon the type of fuel, good furnace design (including the use of insulation), burner efficiency, temperature control metering and control of the gas usage, furnace pressure control and heat recovery systems. The use of natural gas by some of the plants has resulted in a cleaner fuel with greater control over the combustion conditions when compared with coal. The introduction of natural gas would have taken place even if CDGI was not in existence and it does not seem that its contribution has been very significant. Companies have merely exchanged coal for natural gas without changing furnace design or by introducing control systems that would allow a reduction in the fuel requirement. Only in one plant was an insulated crown seen in which CDGI was instrumental in its application. This should have reduced the fuel consumption but no data was available.

Fuel Efficiencies: The fuel efficiency improvement when converting from coal to gas is disappointing small and suggests that no other changes were made in furnace design or operation. It may be useful to study the energy consumption when melting in single pot furnaces and comparing this with the energy consumption for a series of multi pot configurations. The seeming logic that more pots will give greater fuel efficiency may not be valid. The 12-pot furnace has a lot of empty space in its centre that has to be heated, whereas a smaller group would fit more tightly into the space. A single pot with good insulation may be in terms of investment, furnace and pot life and flexibility of types of glass, quality of glass and fuel efficiency a better option than the industry has thought. This type of study using data from other industry sources would be beneficial to the industry in Firozabad.

It was said that to purchase a meter to measure the gas consumption was too costly for each company. It is suggested that CDGI could purchase one meter and by arranging to fit by-pass lines in each plant, install the meter, carry out measurements during trials and then remove the meter and move to another plant. The fuel consumption in each plant should be broken down into the component parts of the energy required for the process, e.g. batch feed on and melting, glass working, re-heat, annealing, secondary working such as the coating of loams, re-heating and drawing in bangle manufacture. This will identify the most energy consuming parts of the process and help to target changes in working practices or design of equipment. This is an example of CDGI being pro-active and demonstrating their service to the industry.

The work of TERI in one bangle manufacturing plant illustrates what industry needs; an up-dated design using a modern, closed burner with gas filters, pressure control, metering, emergency shut off valve and a recuperator. The experimental day tank at the Centre is a significant development that may be taken up by industry (one plant is planning to install a unit). The day tank is said to have a lesser fuel consumption per tonne of glass melted when compared with a multi-pot furnace, a longer life and a greater utilization of the melted glass. A success with the day tank will boost the standing of CDGI in the industry. The day tank in the Centre has fusion cast glass contact blocks. Whilst these are suitable for a continuous furnace, the temperature cycling in the day tank may be injurious. Sillimanite may be a better choice.

Bangle manufacturers use a high soda glass with Na2O content in the glass being about 20%, which is about 7% more than other typical glass products. The high soda makes the glass easier to melt and fuel efficiency should be lower than for other glasses using the same melting chamber. Indirectly the development of a pot with a longer life will improve fuel efficiency, as less fuel will be required, (because heat transfer is improved when using open pots compared with closed pots), fewer pots will break prematurely and the new pots have a longer life. Again, a study by CDGI of gas consumption in a plant over a long period will establish these facts.

In summary, there is a need to apply energy management practices for the different kinds of furnaces, fuels and secondary processes. The energy management practices would be based upon a thorough energy audit of each type of plant, trials and equipment developments to improve the energy consumption and a best practice established for others to copy. The best plant practice will include heat recovery systems, pollution control, control of the batch preparation and use automated process control equipment to monitor and control temperatures, flows and pressures. The aim will be to demonstrate to other owners that by controlling the process and investing in new equipment they can achieve good quality glass, improve the environment and reduce the cost of operations.

Technology transfer: The transfer of technology to the industry has been slow, mainly because of the slow start up of CDGI, the failure of the open pot furnace and the consequential low esteem by the industry towards the abilities of its staff. All the owners that were met acknowledged the need to improve the assistance from the Centre. The fact that insufficient of them have taken up the services may be because they want a free service or as expressed by one owner that he was too busy to attend. This emphas izes the need for CDGI to visit the plants.

There is also the problem of the team leaders of the work force (are these the equivalent of Union representatives?) not accepting changes in working practices that will undermine their authority and reduce labour numbers. In some cases, owners seemed willing to accept changes but foresaw problems in their implementation. It is probable that the problem at New Bright Glass in which an open pot furnace designed by CDGI was rejected because the working conditions were too hot was in fact a resistance to accept new working practices by the team leaders. If further technology transfer is to be achieved, which involves changes in working practice or conditions then the team leaders as well as the owners will need to be persuaded. This may be achieved by involving all these interested parties at an early stage in the development process.

Until cooperation is achieved between all parties, CDGI, owners and workers, then many of the projects will fail. For example, glass quality has been cited as the drawback to exporting goods and to improve it experiments will be necessary with the batch composition, refining agents, etc. However, if the batch mixers are unwilling to change then experimentation is not possible. In the case of pot furnace technology and day tank technology, new initiatives have been taken up by CDGI and so far the response seems to be positive, though it is too early to state categorically whether the transfer of the new technology in these areas is successful or not.

Pot manufacturing

The use of closed pots was necessary because the combustion conditions using coal firing affected the glass quality and colour. The closed pot relies on heat conduction that is less effective than conduction and radiation that applies with the open pot. The introduction of natural gas, a cleaner flame, allows the use of open pots and thus a possibility to save fuel. This, coupled with the development of a pot with a longer life (using better quality of clays and grog), has significant opportunities for the industry to improve its operations. Trials to date have shown that the pot life increases from some 15 days to 60 days and that pot breakage in the initial warming up stage and transfer are reduced. One Company, Delux Glass, reported that on average they were replacing one broken pot per day. As with the day tank this development will enhance the standing of CDGI. The next step is to encourage the start up of a full-scale manufacturing unit that will require a large area for manufacturing and curing with controlled temperature and humidity. The capital costs and working capital required may be beyond the capability of the present pot manufacturers and CDGI can help through a development fund.

Decoration

The setting up of the Decorating department and the use of, for Firozabad, novel techniques of decorating such as the sand etching, has been taken up by the industry. Trial work should continue on design and technique that can be passed on to industry through contact either with owners or through training programmes. This department, which can experiment with designs and shapes, may assist the development of the bead industry for tourism and export. The use of imported beads for the assembly of chandeliers in Firozabad is an opportunity for the Centre to investigate the manufacture of these beads as it has the glass quality that is required. A demonstration of the production of the beads can be shown to the industry. Similarly, the tourist industry may be interested in glass articles that reflect Indian tourist attractions such as tumblers showing the Taj Mahal. The practice of silvering beads by manually sucking a silver nitrate mixture into the column of beads must cease and CDGI has an obligation to find a solution to this problem.

Objectives

The overall objective of the Study intervention is to analyze the energy and environment performance of the Firozabad glass industry cluster, which in the long run is expected to lead to reduced health implications and a better working environment for employees. This objective is to be achieved through intervention in specific small-scale industry segments.

Description of the activity

The activities undertaken in the small-scale glass industry are part of the study in the energy and environment sphere, which focuses on some of the energy- intensive industry sectors. The SDC's India Programme in this area aims to provide the disadvantaged sectors of the population with the opportunity to benefit from affordable, environment-friendly technologies that save energy, reduce pollution, and promote local and global sustainability. The TERI initiatives may be grouped into the following three distinct phases:

Identification phase (1994–95): During this period, a preliminary diagnostic study of the various units in the glass industry cluster was carried out by TERI, to identify the areas where technological intervention would have the maximum impact in terms of reducing energy intensity and improving environmental performance. Two areas identified for technological intervention were pot furnaces and muffle furnaces, due to their inefficient operation and absence of any ready technological solutions for conversion from coal to natural gas. It was decided that full-scale demonstration plants should be set up to convince the small entrepreneurs of the benefits of technology intervention, thereby motivating them to adopt cleaner technological solutions. Towards the end of the diagnostic study, a stakeholder workshop was organised by TERI—the first of its kind for the Firozabad glass industry cluster—to share concepts and ideas with various stakeholders and experts and to identify the future course of action. The workshop helped to define the overall framework under which the activities would be undertaken. It was also decided to implement the identified technologies with industry associations as key partners and to integrate the concerns of the local industry into the project at every stage.

Action research phase (1996–2001): The major activities during this period related to (a) development of technologies with active involvement of industry associations, entrepreneurs, operators, local masons, and international experts; (b) demonstration of the technologies in the field; and (c) commissioning/fine tuning to show the benefits of the new technologies to local entrepreneurs. A demonstration gas-fired pot furnace was commissioned in 2000 while a gas-fired demo muffle furnace was commissioned in 2001. Since the activities in this phase involved implementation of the technologies in the field, local anchoring at Firozabad was considered essential. A local consultant was employed by the TERI project team, who worked very closely with the project partners and helped in the execution of the project.

Gas fired demonstration muffle furnace Gas-fired demonstration pot furnace

Dissemination phase (2002–ongoing): This phase will focus on assisting the local industry in reducing their energy consumption and improving environmental performance both through adoption of improved designs and through incremental improvements in their existing operating practices. These objectives will be achieved through workshops/training programmes/meetings at the local level and also by providing technical assistance to entrepreneurs willing to adopt the energy-efficient designs. In the case of the gas-fired muffle furnace, technical assistance has already been provided during the past year to some of the entrepreneurs in setting up similar units and helping them operate the furnaces efficiently. Under this phase, it is also planned to develop and test strategies leading towards better socio-economic conditions. The activities in this area will be undertaken with the active involvement of local NGOs.

Overview

Three instruments—regulatory pressure, development and demonstration of cleaner technological options, and the unique collaborative arrangement amongst the participants—were critical to the success of the project.

Regulatory instruments

This case study is a perfect example showcasing the ability of the higher judiciary to break through implementation delays and force action on the environmental front. The Supreme Court of India stepped in when the regulatory bodies were slow in enforcing existing provisions under various legislations related to the environment. In this particular case, it was the Public Interest Litigation for the protection of the famous Taj Mahal that prompted the apex court to pronounce its landmark judgement, which enabled the small-scale industries in Firozabad and other areas of the Taj Trapezium Zone to gain access to cleaner-burning natural gas. The Supreme Court in India has over the last few years passed various other noteworthy judgements on matters concerning pollution from transport and industrial activities, ecological restoration, and protection of India's coastal and forest resources. In the TTZ judgement, although the small-scale industry of Firozabad has benefited from the adoption of cleaner technologies based on natural gas, the case raises some important questions regarding the role of the judiciary vis-à-vis the responsibilities of regulatory bodies in environmental matters. This case study highlights the constructive role played by the judiciary in a situation where the local executing agencies were straying from the desired plan of action. However, a participatory approach involving local institutions, regulatory bodies, and small-scale industries can be more effective and productive than a forced judicial intervention that might raise socio-economic concerns at the local level. This demands optimal functioning of the local agencies and willingness on the part of the SSIs to mitigate pollution through various voluntary measures.

Critical Instruments

- **Technologies** -The intervention in the Firozabad glass industry cluster centred around development and demonstration of cleaner technological options for the small-scale units involved in manufacturing glass bangles and other decorative items. For these industries, the three factors that are of paramount importance are (a) appropriate technological choice; (b) participative technology development; and (c) sustainability issues.

The technological package should be made adaptable to local conditions and must address the specific concerns of the stakeholders (for example, concerns about the environment, energy, and quality) so that the sustainability issues are integrated in the overall framework. In the specific case of gas-fired muffle furnaces, a bottom-up participatory technology development approach was adopted wherein the local furnace builders and entrepreneurs provided valuable input. The credibility that has now been built in the industry cluster through the success achieved on the technological front, allows the project to expand its scope of work into areas that go beyond the traditional limits of technological intervention. It now paves the way for the project to make the intervention more holistic by integrating social dimensions like work environment, occupational health, education, and other such interrelated issues.

TERI has set up demonstration plants in two categories of small-scale industries in the Firozabad glass industry cluster. The results of the demo installations were:

- The demonstration gas-fired pot furnace yielded energy savings of about 55 percent as compared to traditional coal-fired pot furnaces (the base case)
- The demonstration gas-fired muffle furnace yielded energy savings of about 30 percent as compared to traditional coal-fired muffle furnaces (the base case)

The shift to cleaner technological solutions has significantly reduced the emissions of particulate matter and carbon monoxide at the local level, with a markedly reduced pollution load within the factories. As a result, the workers are no longer exposed to carbon monoxide and particulate matter, which were inherent in the old coal-based furnaces. Since energy consumption has decreased by 30 to 55 percent, correspondingly fewer pollutants are generated at the source. From a global perspective, improvement in energy efficiency, specifically in energy intensive small-scale industries, can be regarded as one of the most cost-effective means of reducing the carbon dioxide emissions and thereby mitigating the adverse effects of global warming. It is estimated that adoption of the demonstrated technologies within the Firozabad glass industry cluster has the potential to save around 0.35 million tonnes of carbon dioxide per annum.

The success of the Firozabad example demonstrates a few important lessons, which can be emulated in other small-scale industries in India and other developing countries in the Asian region. These relate to (a) adoption of the cluster approach in small-scale enterprises; (b) the need to undertake initial detailed diagnostic studies; and (c) close involvement of all stakeholders during implementation.

- **Cluster-based approach:** One effective strategy for intervention in small-scale enterprise is to use a cluster-based approach wherein better operating practices can be demonstrated to a group of units located in close proximity. "Industrial clustering" is the existence of a group of factories or units geographically very near to each other and involved in the production of goods and services that have a great deal of commonality. For example, in the case of the Firozabad glass industry cluster, there are a large number of units located within a small town, all involved in manufacturing similar products with a similar level of technology, operating practices, and even trade practices. The benefits of reduced energy consumption and a cleaner working environment were demonstrated to the muffle furnace entrepreneurs of Firozabad. The rate of conversion after the demonstration was fairly fast and, as mentioned earlier, nearly 30 furnaces have already switched over to the use of natural gas. Therefore, in a cluster-based approach, the potential to develop and implement common solutions for adoption of cleaner technologies is very high.

Pre-Studies

Benavides (1992) argues that most small-scale industries do not have a significant impact on the environment. On the contrary, it can be argued that these industries do, in fact, make a significant contribution to environmental contamination at the local level, ie., in the neighbourhoods of small-scale industries. Such contamination can have a serious impact upon the state of health of the people living in these neighbourhoods (Barten, 1992). In these situations, distinctions made between general health hazards and the more specific hazards of industry are rather artificial.

The numbers of small-scale industries in developing countries continue to mushroom, as the shortage of foreign currency forces governments to promote self-help enterprises (Nriagu, 1992:1-37). Invariably, there are few restrictions or controls on what these operations discharge into the environment and quite often the homes and the surrounding play areas are contaminated with toxic metals. The fact that many people live where they work predisposes them to frequent (and potentially high levels of) exposure to metallic wastes and other hazards that result from such small-scale and cottage industries, obscuring the distinction between occupational and environmental exposure. The workers themselves rarely distinguish between occupational and nonoccupational illness, although they may be aware that much of their physical well-being stems from bad working conditions (Shukla, 1991:597-603).

A range of hazardous activities is carried out in small-scale industries such as wood work shops, metal workshops, and garages, set up close to places where women are cooking and selling food. Therefore, besides the hazards of their work, people are exposed to many other hazardous activities carried out in their home environment. The lack of land allocation also causes an insecure situation for informal sector workers since they can be chased away from their premises any time. This state of affairs does not stimulate the operators to invest in their enterprises and thereby improve the occupational and environmental conditions. Any investment is seen as a waste of money. The lack of permanent premises also makes it difficult to obtain credit, and leads to insecurity and low incomes. To what extent these problems affect the mental health of informal sector workers and their productivity is virtually unknown (ILO, 1993). The effect on the health of workers? families is also not considered. It has been suggested that, even when workers and operators in informal sector workshops were aware of health and safety issues, these were not their priority concern. (Ibid). It is the struggle to provide the basic needs of life that causes a daily burden for workers and their families. The exposure of workers and their families to occupational and environmental risks is therefore not chosen voluntarily, but seen as the only possibility to maintain a basic living.

Impacts

- **Need analysis:** The example of the Firozabad glass industry illustrates the importance of undertaking detailed diagnostic studies initially so that the needs of the industry can be clearly understood. Due to regulatory pressure, the small and medium-sized enterprises of Firozabad were looking for assistance in implementing natural gas–based technologies. The SDC/TERI project team has tried to provide this assistance directly by helping the industry to develop technological solutions and assisting them in complying with the Supreme Court directives.
- **Stakeholder involvement:** One of the lessons that can be learned from this case study is the importance of involving various stakeholders

during every stage of project implementation. It is imperative to maintain a good working relationship with all stakeholders including government departments, pollution control boards, and local industry associations. Active involvement of the entrepreneurs, operators, and industry associations in Firozabad has helped in developing solutions that are acceptable to the entrepreneurs. During the pilot phase of the project, participatory design sessions were organized with the peer groups to solicit industry feedback on various designs. Since small-scale industries invariably demand site-specific solutions, ready-made solutions are often not the answer. Hence, the involvement of the industry at all levels—from the entrepreneur down to the operator—is crucial in reaching a satisfactory outcome.

The muffle furnaces used for heat treatment of bangles are not registered and there are no official records that give details regarding their number, location, ownership, etc. Therefore, under the project, a detailed cluster mapping exercise was carried out to identify and document the number and location of these muffle furnaces. It was discovered that there are nearly 800 such furnaces operating in different parts of the city. Within 18 months of setting up the first gas-based furnace in Firozabad, nearly 30 furnaces have already switched over to the gas-based design. It is expected that many more furnaces will be converted to the cleaner option, once commercial issues such as the supply infrastructure for natural gas and the land ownership issue are resolved. Supply of natural gas directly to the muffle furnace owners at their existing sites will not be easy since most of the units are located in congested residential areas where laying of gas pipelines is difficult. In the case of the pot furnace, since the initial cost of the demonstrated technology is nearly three times that of the conventional technology, the rate of diffusion is expected to be relatively slow. Therefore, it is important to interact with the entrepreneurs as well as the financial institutions to make them aware of the long-term benefits and attractive payback of the improved design and to motivate them to adopt/finance such innovative designs. There are nearly 60 such pot furnaces that can benefit by adopting the improved design, either as full-scale replication units or through partial modifications/incremental improvements to reduce energy consumption.

Polluted air from various sources, including the teeming industrial units in its vicinity was believed to cause the damage. But, beyond the fear of a medieval glory being hidden behind fumes, the looming reality is that the numerous polluting industrial units operating in the entire Taj Trapezium Zone – an area of 10 400 square kilometres around the Taj Mahal – were affecting the health of millions of inhabitants of the area.

One of the reasons for this is the pollution caused by the small-scale glass industry cluster in the city of Firozabad, which in was the nerve centre of this trade India. T E R I technological innovation that substitutes coal with natural gas for these units provides some respite to the local industry.

Figure-1 A polluting glass baking unit in Firozabad

A City in the Line of Fire

Of the total glass produced by small-scale units in India, 70% comes from Firozabad. The cluster of glass furnaces based in this little town on the outskirts of Agra meets the country's requirement of glass bracelets, showpiece items, tableware, headlight covers, laboratory ware, vacuum flasks, and the like. However, the going got disrupted because of the landmark Supreme Court judgment in December 1996. In response to a public-interest litigation seeking to protect the Taj Mahal from industrial pollution, the apex court directed all polluting units in the Taj Trapezium Zone to stop the use of coal and switch over to natural gas, which the GAIL (India) Ltd was directed to supply.

An Overdue Intervention

While the bigger glass melting units in Firozabad employ tank furnaces, the smaller ones use pot furnaces. Also, a large number of muffle furnaces are being used for baking bangles. The technology in all these cases, however, had remained the same over the past many decades, and very little effort had been made to improve the working conditions and the technology. The traditional designs of coal-fired pot and muffle furnaces are very poor in terms of energy efficiency and environmental performance. As per the Supreme Court judgement, it was mandatory for polluting furnaces to shift to the use of natural gas within a definite time frame. But there was no off-the-shelf technology available for the pot and muffle furnaces to facilitate the shift. To these industrial units looking for a way out of their predicament, TERI's intervention to provide them with energy-efficient and natural-gas-based technological solutions was m uch-needed indeed (Figures-2 and 3).

Figure-2 A conventional coal-fired pot furnace

Figure-3 The new natural-gas-fired pot furnace

Working within the comfort zone of the traditional glassmakers, TERI researchers developed a model for environmental performance—natural-gas-based pot and muffle furnaces that are energy-efficient, which also make good economic sense. By agreement with the glass industry association and the district authority of Firozabad, TERI set up two demonstration plants, one each for pot and muffle furnaces. Aiming at helping these units switch over to the new technology, the pilot programme was a joint initiative involving TERI and the Swiss Agency for Development and Cooperation.

The Innovation

- **Pot furnace:** A state-of-the-art natural-gas-fired pot furnace was set-up at Firozabad by TERI with support from British Glass. Keeping the traditional design and mode of operation unchanged, the new system minimized energy wastage by using pre-heated air for combustion, and utilizing the heat of the outgoing flue gases. Also, the installation of efficient combustion systems and use of better refractory and insulation materials reduced functional and structural losses. The refurbished furnace was centrally fired by a single burner mounted on the furnace crown for better heat distribution. Improved structural design and use of better quality material for the furnace crown extended its life to three years.

Similarly, better quality refractory materials like silimanite and zirmul were used for the floor in place of conventionally used refractories.

To check the heat from escaping through the stack, a heat recovery system in the form of a metallic recuperator was installed to pre-heat incoming combustion air to 600 °C.

- **Muffle furnace:** A natural-gas-fired muffle furnace was designed and demonstrated by adopting a participatory technology development approach (Figures below).

Figure -4 Natural-gas-based demonstration mufle furnance

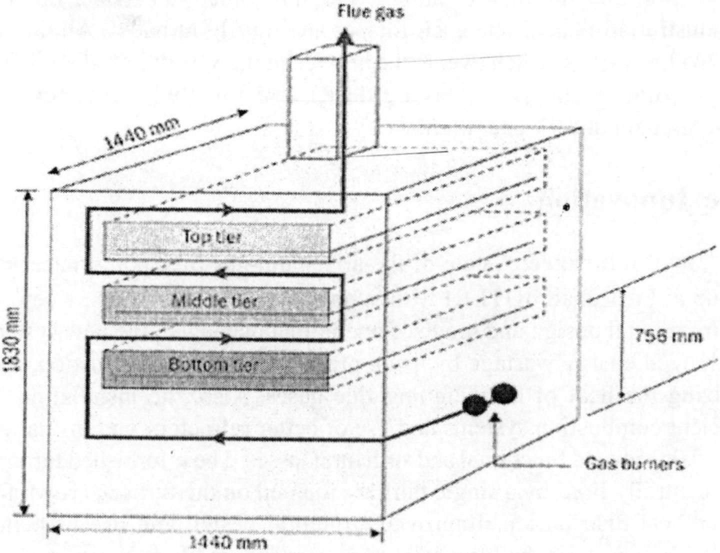

Figure- 5 Schematic diagram of the TERI model of gas-fired muffle furnace

The gas-fired design is very similar to the existing coal-fired units, both in terms of construction and operation. The gas-fired furnace has the provision to use better-quality muffles made of silicon carbide that can enhance the muffle life to more than three years. The design was optimized to achieve complete combustion of gas and better distribution of heat. The technological innovation has made it possible to increase production by 10%. The small-scale bangle manufacturers in the town of Firozabad, about 50 km away from Agra, have been facing serious survival issues due to stringent pollution control norms, disallowing them from the use of coal and wood in heating their furnaces.

While proposition of getting a connection to the CNG pipeline laid out by GAIL in the town appeared to be too expensive for these units earlier, GAIL has placed a complete stop on issuing new gas connections now, citing acute shortage of gas supply from ONGC.

Moreover, ever since the bangle metalising plants in the town succeeded in nominally raising their tariffs citing a rise in raw material prices a few weeks back, the glass bangle baking units too have been asking the bangle exporters and contractors for a raise in their service charges but the exporters are not yet ready to allow for it resulting in a stalemate which has forced the units to go on frequent strikes, the latest being called this week.These units were classified as small-scale units but were not easily extended the facility of SSI loans by banks, which termed them as cottage units which restricted the upgradation of these units from coal-fired technology to CNG, due to financial considerations.

Pollution by small Glass Industries

The glass industries located at Firozabad account for roughly 70 percent of the total glass production in the small-scale sector. These small-scale units in Firozabad suffer from deficiencies like technological obsolescence, poor energy performance, high levels of pollution, and poor management practices. The poor energy and environmental performance is directly related to the lack of technical capacity in these enterprises to identify, access, adapt, and adopt better technologies and operating practices. The glass industry in Firozabad is presently undergoing a very dynamic phase wherein the units are switching over from coal to natural gas as the primary source of energy. The conversion to natural gas has been prompted by the mandate from the Supreme Court of India that requires industrial units in Firozabad to change over to a cleaner fuel. This judgement covers industries located within the Taj Trapezium Zone (TTZ)—an area of 10,400 sq. km. around the Taj Mahal.

The ruling by the Supreme Court was delivered on 30 December 1996 in response to a Public Interest Litigation seeking to protect the Taj Mahal from environmental pollution.

The apex court banned the use of coal/coke in the entire TTZ and directed specified units located in the TTZ, which were using coal/coke, to switch over to natural gas (to be supplied by GAIL3), to relocate outside the TTZ (to alternative plots to be allocated by the Uttar Pradesh State Industrial Corporation), or to shut down. The Taj Trapezium Zone is home to over 40 protected monuments, three of which are listed as World Heritage Sites (WHS), namely the Taj Mahal, the Agra Fort, and Fatehpur Sikri. Two more monuments are proposed to be listed as WHS very shortly. Due to the importance of the area, conditions within the TTZ have received the attention of policy makers at both the local and state levels. Various initiatives have been undertaken to preserve the rich cultural heritage of this region as well as to improve the overall living conditions of the people. In the past, however, most of these initiatives were ad hoc and lacked a coordinated approach to the problems prevailing in the area, which are largely interrelated. The region gained attention after the 1996 Supreme Court judgement because all industry located in the region was required to adopt natural gas technology. As Firozabad

- These costs pertain only to the hardware costs for replication of the respective furnaces and do not reflect the amount incurred towards implementation of the good practice.
- U.S.$1 = Rs 48 (approximately)
- GAIL (Gas Authority of India Limited) is a Public Limited Company that installed

India's first major incountry gas pipeline for supplying natural gas to industrial enterprises in North India. Lies within the TTZ, the small-scale glass factories of this city were directly affected by the Supreme Court orders. Since the factories lacked the expertise to develop cleaner technologies, TERI, under a programme initiated by the Swiss Agency for Development and Cooperation (SDC), decided to assist in developing technologies that are energy efficient and environmentally friendly.

Pollution and health risks

On environmental issues the most important initiative taken seems to be the conversion of coal fired furnaces to gas fired ones in pursuance of the Supreme Court orders banning the use of coal in the furnaces. In 1997, 64 units were offered gas connections out of which 61 have taken the connection. In the second phase, 204 units were short -listed for connections out of which so far 74 units have responded and completed all the formalities. The complaint of the small units has been that they are unable to meet the conditions of the Gas Authority of India Ltd. (GAIL) to provide a bank guarantee of Rs 6 lakh and a deposit of Rs 2 lakh. They also complained that the pricing policy of GAIL was unfair since the consumers in Baroda pay much less that the consumers in Firozabad. They felt that a price equalization scheme on the lines of the scheme prevailing for coal and petrol should be introduced for natural gas. The U P Pollution Control Board appointed the CDGI in 1996 to monitor ambient air quality in three places in Firozabad. The data collected up to March 1999 did not indicate any significant change in the environment since 61 units converted to gas in 1999. However, it is too early to come to any conclusions as the number of units converted into gas is few and the period of observation (less that one year) is too short. The work of CDGI stopped in March 1999 since the U P Pollution control Board did not provide further funds. The funding was resumed and the work started once again on 10 October 1999.

GAIL needs to be pursued to provide gas connection to the remaining industries and reduce the financial rates for the deposit bank guarantee for the small units. Pollution of the atmosphere is either through working practices in the plant, such as raw material handling, batch mixing, secondary heating of glass, or through emissions from the chimney. Workers in the Firozabad glass industry are exposed to many potential hazards. The extent of pollutants and their effect on the workers can only be measured by a comprehensive study involving a sampling of factory air and by medical examination, including blood samples. Some of the areas of hazard noted during the visit are:

- The use of lead, arsenic, barium, cadmium, fluorides in the batch.
- Quartz particles.
- The fumes of fired decorated ware may contain lead or cadmium.
- The silvering of beads.
- The frosting of glass using hydrofluoric acid.

Chimney emissions as experienced in the areas around Firozabad will be mainly oxides of nitrogen, sulphur and carbon, particulate matter, fluorides, boron compounds and lead compounds. The chimney emissions from coal-fired furnaces are the major problem that should be much reduced because of the conversion of furnaces to gas. The work of CDGI in conjunction with the U P Pollution Control Board will be a useful guide to the extent of the improvement. No information seems to be available on legislation relating to the limits of pollution or dangerous substances. Information was obtained from CDGI on emission standards issued by the Pollution Control Board, which quantifies particulate matter and sulphur dioxide but it is not clear if this is legislation or guidelines. It would be in the interests of the small-scale glass industry if CDGI was to obtain benchmark data from other countries in which legislation has been passed and use this to compare progress in the local industry.

Waste gas analysis of chimney emissions is routinely carried out and although some of the information gathered is more useful to the plant operations, it is a basis for gathering information on pollution levels in general. The work should be extended to include all possible harmful substances, including those generated within the plant.

Beneficiaries: The refurbished natural-gas-based furnaces are excellent environment performers in that there was a substantial reduction in energy consumption as well as emissions of pollutants. The glass units could benefit immensely as the gas-fired furnaces demonstrate energy savings up to 50% and 30% in the case of pot and muffle furnaces, respectively. TERI's initiative has motivated 45 muffle furnace units to switch over to the TERI model despite paucity of piped gas supply in the area. In the case of pot furnace, besides the demonstration plant, another unit based on the TERI model has been set up while two more are under construction.

Applications/Benifits: Though both the conventional pot furnaces (that are in use in Firozabad) and the TERI-designed pot furnaces run on natural gas, the latter has an improved design. An incremental investment of 2.4 million rupees for TERI's pot furnace offers payback in 16 months. This is due to the reduction in gas requirement by nearly 35% as compared to the conventional gas-based furnaces. The capital cost required for setting up a gas-based muffle furnace is in the range of 15 000–30 000 rupees, depending on the construction material used. The investment is at par with the existing coal-fired systems.

While the new technologies definitely make economic sense, they also boast of an envious combination of being environmentally sound and, most important, much less harmful to the workforce that would now sustainably continue to manufacture glassware for the nation.

Polluted air from various sources, including the teeming industrial units in its vicinity was believed to cause the damage. But, beyond the fear of a medieval glory being hidden behind fumes, the looming reality is that the numerous polluting industrial units operating in the entire **Taj Trapezium Zone** – an area of 10 400 square kilometres around the Taj

Mahal – were affecting the health of millions of inhabitants of the area. One of the reasons for this is the pollution caused by the small-scale glass industry cluster in the city of Firozabad, which was the nerve centre of this trade in India. T E R I technological innovation that substitutes coal with natural gas for these units provides some respite to the local industry.

Working within the comfort zone of the traditional glassmakers, TERI researchers developed a model for environmental performance—natural-gas-based pot and muffle furnaces that are energy-efficient, which also make good economic sense. By agreement with the glass industry association and the district authority of Firozabad, TERI set up two demonstrations plants, one each for pot and muffles furnaces. Aiming at helping these units switch over to the new technology, the pilot programme was a joint initiative involving TERI and the Swiss Agency for Development and Cooperation.

Conclusion

The development, demonstration and fine-tuning of the technologies required a time period of 3 to 5 years. In all the cases, energy savings of the order of 30– 60 % were achieved under field conditions. In case of brick and foundry sector — for which emission norms exist — the emissions were found to be much below the most stringent emission norms. In the case of brick kilns and glass furnaces substantial reduction in heat exposure to workers was achieved. In all the cases, air pollution in the working environment was reduced, resulting in better and safe working conditions for the workers. Another important result in all the cases was reduction in wastage, improvement in product quality, reduction in processing time — all leading to better productivity and better selling price for the product. The following are admissible:

- *Natural gas fired pot furnace for glass industry.*
- *Natural gas fired muffle furnace for glass industry.*
- *Biomass gasifier based silk reeling oven.*
- *Divided blast cupola with pollution control system for iron foundries*
- *Biomass gasifier based silk dying system.*
- *Development and demonstration of appropriate clean technological solutions in collaboration with industry thereby utilising the available traditional know-how;*
- *Formation of a technology provider network for making available the developed technological solutions to the end users;*

- *Creating awareness among the owners about the environmental concerns and available solutions;*
- *Training workers in the new technologies and creating awareness among them regarding the health risks involved with existing technologies;*
- *Facilitating credit for SIS firms for the adoption of clean technologies*
- *Mobilising community pressure on the polluting industries to adopt clean technologies;*
- *Formulation of realistic and practicable emission regulations and improving the implementation by the strengthening of regulatory agencies (e.g. pollution control boards)*
- *Energy savings of the order of 50-60% compared to traditional clamps.*
- *Increase in production volumes*
- *Better product quality and increase in the percentage of saleable fired bricks (from around 70-80% in clamps to about 90% in VSBK).*

Suggested Approach

The experience from the work carried by Researcher, and other institutions elsewhere, shows that an approach based only on conventional command and control regulation will not work for environmental management in small scale industries. A comprehensive approach for environment management can consist of:

- Development and demonstration of appropriate clean technological solutions in collaboration with industry thereby utilising the available traditional know-how;
- Formation of a technology provider network for making available the developed technological solutions to the end users;
- Creating awareness among the owners about the environmental concerns and available solutions;
- Training workers in the new technologies and creating awareness among them regarding the health risks involved with existing technologies;
- Facilitating credit for firms for the adoption of clean technologies;
- Mobilising community pressure on the polluting industries to adopt clean technologies;

- Formulation of realistic and practicable emission regulations and improving the implementation by the strengthening of regulatory agencies (e.g. pollution control boards)

References

- William A Pizer & Jhih-Shyang Shih & Richard D Morgenstern, 1997. "Are We Overstating the Economic Costs of Environmental Protection?," Working Papers 97-12, Center for Economic Studies, U.S. Census Bureau.
- Chambers, Robert G. & Chung, Yangho & Fare, Rolf, 1996. "Benefit and Distance Functions," Journal of Economic Theory, Elsevier, vol. 70(2), pages 407-419, August.
- Wayne B. Gray & Ronald J. Shadbegian, 1993. "Environmental Regulation and Manufacturing Productivity at the Plant Level," NBER Working Papers 4321, National Bureau of Economic Research, Inc.
- Wayne B Gray & Ronald J Shadbegian, 1993. "Environmental Regulation And Manufacturing Productivity At The Plant Level," Working Papers 93-6, Center for Economic Studies, U.S. Census Bureau.
- Fare, Rolf, et al, 1993. "Derivation of Shadow Prices for Undesirable Outputs: A Distance Function Approach," The Review of Economics and Statistics, MIT Press, vol. 75(2), pages 374-80, May.
- Fare, Rolf, et al, 1989. "Multilateral Productivity Comparisons When Some Outputs Are Undesirable: A Nonparametric Approach," The Review of Economics and Statistics, MIT Press, vol. 71(1), pages 90-98, February.
- Boyd, Gale A. & McClelland, John D., 1999. "The Impact of Environmental Constraints on Productivity Improvement in Integrated Paper Plants," Journal of Environmental Economics and Management, Elsevier, vol. 38(2), pages 121-142, September.
- Palmer, Karen & Oates, Wallace E & Portney, Paul R, 1995. "Tightening Environmental Standards: the Benefit-Cost or the No-Cost Paradigm?," Journal of Economic Perspectives, American Economic Association, vol. 9(4), pages 119-32, Fall.

27. Water Pollution: Sources, Effects and Measures

Vipin Agrawal
Amita Agarwal

Water is most precious natural resource of the nature. No life will be existed in the earth without water. People depend upon water because of its need for domestic uses, irrigation, industrial uses, sanitation purpose, shipping and drinking purpose. As the development takes place, whole environment will be polluted due to increase concentration of different pollutant. There is different type pollution like soil pollution, air pollution, water pollution etc. The polluted water affects all living beings including vegetation. This paper highlight the effects of water pollution three ways- effects of polluted water on people, effects of polluted water on soil and crops and effect of pollution of water on aquatic life. Paper highlights the ways to check the water pollution and its adverse effects.

Key Words: Environmental pollution, Degradable Pollutants, Non-Degradable Pollutants Water Pollution, Sewage.

Introduction

Environmental pollution is a worldwide problem whole world has been facing this problem during the time of development. It may be defined as follows-

"Pollution is an undesirable change in the physical, chemical or biological characteristics of our air; land and water that may or will harmfully affect life or that of desirable species, our industrial progress, living conditions and cultural assets or deteriorates our raw material resources". This pollution is the resultant adverse effects of pollutants.

Any substance or factor affecting Environment so that human activities are adversely affected is called "Pollutants". Pollutants may be divided into two groups:

I. Degradable Pollutants: The decomposed of organic matter such as plants and crop residues, animal wastes etc. helps in the enhancement of soil fertility but the evolved injurious gases during decomposition may pollute the environment, such type of substances are called Degradable Pollutants.

II. Non-Degradable Pollutants: Chemical and industrial wastes, which do not get decomposed completely or undergo incomplete decomposition, are called Non-Degradable Pollutants". The e.g. of such pollutants are plastic substances, BHC, DDT, Aluminium, Arsenic, Silicate substances etc.

The natural and human activities are mainly responsible for environmental pollution. From the point of view of agriculture and environment interaction, the most important environmental pollutions are- Air Pollution, Water Pollution, Soil Pollution, Sound Pollution, and Radioactive Pollution.

Water Pollution

People depend upon water because of its need for domestic uses, irrigation, industrial uses, sanitation purpose, shipping and drinking purpose. The major use of water is in agriculture in our country and its use is estimated up to 73% by 2000 A.D. The water pollution may be defined as "The addition of any substance to water or any change in its physical and chemical properties which interferes with its proper use is known as water pollution". Water always contains impurities in the dissolved or suspended form. In case of dissolved phase most common are dissolved gases and minerals where as suspended matter and even microbes contribute to suspended impurities. These impurities may be derived from atmosphere, soil and its contact with other sources during storage or supply. According to health department, the quality parameters for drinking water are given below:

Table-1
Quality Parameters for Drinking Water

Colours, odour, taste	**should not be unacceptable**
Turbidity	**not more than 10 PPM on silica scale**
Temperature	4°C to 10°C
Total solids	500 to 1000 PPM
pH-7 to 8.5	
Hardness	upto 100 PPM
Chloride	upto 250 PPM
Sulphates	upto 250 PPM
Lead	upto 0.1 PPM
Fluoride	upto 1.5 PPM
Copper	upto 1 to 3 PPM
Iron and Manganese	upto 0.3 PPM
Magnesium	upto 125 PPM
Arsenic	upto 0.05 PPM
Selenium	upto 100 PPM
Zinc	upto 1.5 PPM
Dissolved Oxygen (DO)	minimum 6 PPM
Biological Oxygen Demand (BCD)	maximum 2 PPM

The quality of irrigation water is evaluated on the basis of the following parameters:

i. On the basis of electric conductivity
ii. On the basis of sodium absorption ratio
iii. On the basis of Residual Sodium Carbonate:

Table-2
Quality Parameters for of Irrigation Water
(On the basis of electric conductivity)

Electrical conductivity at 25oC (mmhos/cm)	**Class**	**Remark**
<0.25	C_1	Low salinity water
0.25-0.75	C_2	Medium salinity water
0.75-2.25	C_3	High salinity water
>2.25	C_4	Very high salinity water

Table-3

Quality Parameters for of Irrigation Water

(On the basis of Sodium Absorption Ratio)

Limits of SAR	Class	Remark
<1.25	S_1	Low sodium water
10-18	S_2	Medium sodium water
18-26	S_3	High sodium water
>26	S_4	Very high sodium water

Note : Sodium Absorption Ratio

$$SAR = \frac{Na^+}{\sqrt{\frac{Ca + Mg}{2}}}$$

Table-4

Quality Parameters for of Irrigation Water

(On the basis of Residual Sodium Carbonate)

Limits of RSC (me/L)	Class	Remark
<1.25	**1st**	**Safe**
1.25-2.5	2nd	Marginal
>2.5	3rd	Unsuitable

Note: Residual Sodium Carbonate [RSC (me/L)] $\doteq$ $(CO_3^{--} + HCO_3^-) - (Ca^{++} + Mg^{++})$

Water Pollutants

Generally, two types of pollutants are found in the water:

- **Physical pollutants:** Such as dust particles, plant and crop residues, microbes and insects etc.
- **Chemical pollutants:** Such as Chloride, Sulphates, Carbonates, Bicarbonates, Nitrate, Insecticides, Weedicides, Radioactive elements and other chemical products.

Sources of Water Pollution

The various types of water pollutants are derived from the following sources of pollution:

(1) Natural Sources: The soil erosion and decomposition of minerals and bio-mass are responsible for water pollution and natural source of water pollution. When underground water comes in contact with the minerals it gets contaminated. The water of lakes, ponds and reservoirs is diluted in rainy season but due to evaporation etc. it gets concentrated in the hot season thereby its use may be harmful for crops. The presence of Arsenic, Lead, Mercury, Cadmium, Barium, Cobalt, Tin etc. is responsible for water pollution.

(2)Human Sources: Use of water for domestic as well as for industrial purposes the problem of water pollution becomes more significant. The sources of such impurities are given below:

- **Sewage**: Sewage is derived through water from domestic, animals or industrial plants and the most common pollutants included in this category are human excreta, animals dung, plastic cutting, cloth, soap detergents and dust particles.. Due to anaerobic decomposition of sewage and other wastes there is remarkable increase in the demand of oxygen (BOD).
- **Industrial Effluents:** The organic and inorganic pollutants are present in the effluents discharged from tanneries, textile mills, paper and pulp mills, steel industries, breweries, chemical and fertilizer industries etc. The pollutants include metallic wastes, phenols, oils, greases, plastics, toxins, resins, salts, acids etc.
- **House Hold Detergents:** Various types of detergents, insecticides, soaps, paints varnishes etc. are used in houses for cleaning of floor or furniture or clothes. All such pollutants are drained out and accumulated near by water and become the source of pollution.
- **Thermal Pollution:** The water is used in power plants such thermal and nuclear and the discharged water give of polluted water after use. The water of some thermal plants is returned after use at a very high temperature to the rivers/streams/lakes etc. Such type of [pollution is called thermal pollution mainly due to heat which acts as pollutants which affects the aquatic life. The waste water of nuclear power plants also leaves the adverse effects on aquatic life but not due to heat.
- **Oil Pollution:** the marine pollution is chief cause of oil pollution in water. Due to accidents or deliberate discharge of oil polluted wastes creates pollution. Hydrocarbons and benzyrene accumulation may cause cancer due to consumption of fishes collected from polluted water sources.

- **Disposal of Dead Bodies:** The disposal of dead bodies in rivers, lakes etc. is quite common practice in India. The ash or the products of decomposition becomes the source of water pollution in such cases.
- **Mining Processes:** The River passing through mining zones get contaminated with the various types of released minerals being of harmful and toxic nature. The Damodar River passes through coal mining zone and water is contaminated remarkably.
- **Nuclear and Radioactive Substances:** The radioactive substances are spread in atmosphere during nuclear weapon use or bomb blasts. Such dust particles of radioactive substances fall down and create water pollution in the affected area.
- **Agricultural Discharges:** The use of fertilizers and pesticides in agriculture become the main cause of water pollution. These chemicals through rainfall, irrigation, drainage, leeching etc. transported into the rivers, lakes, streams or nearby water source and create problems to natural environment of plants and animals.
- **Acid Rain Pollution:** Mixing of gases through vapours in the atmosphere and formation of acid because the harmful effect on plants and animals in the form of "Acid Rain" on the land.
- **Water Mineral Contact:** The poor quality water containing remarkable quantity of dissolved salts is generally found in arid and semi arid climate. The abundance of salts from mineral sources in such areas is the main cause of release of such dissolved salts in the underground water. Brackish nature of sea water is also the effect of water mineral contact.

Effects of Water Pollution

Water is an important natural resource. Therefore, people depend on water for various purposes even ignoring its quality. The polluted water affects all living beings including vegetation. Some effects of polluted are described below:

I. Effects of polluted water on people

People use water for drinking and other domestic purposes. The impurities of water such as presence of salts, change in colour, bad smell due to decomposition create various problems related to mainly human health some bad effects of polluted water are discussed below:

- **Various Diseases**: due to addition of sewage and wastes in water, oxygen levels are decreased which is indicated by BOD values. The number of microbes (Escherichia Coli) is also increased and consumes the oxygen. Hence, reduction in oxygen level may result to increase the anaerobic condition. On the other hand addition of phosphates and nitrates' etc. through domestic wastes or their products of decomposition water bodies become active and this phenomenon is called "Eutrophication". The luxuriant growth of algae in water is the result of eutrophication and more growth of the same would decrease oxygen level of water. The enhancement in the anaerobic condition may result into water borne diseases as polio, cholera, typhoid, dysentery, diarrhea amoebiasis etc. The use of nitrate rich fertilizers is the cause of an enhancement in the NO_3 level of ground water. Sometimes underground water may contain high level of nitrate. Such type of water is unfit for drinking and may cause diseases. The nitrates are converted to nitrate by microbial flora of intensive in case of nitrate rich water is used for drinking. These nitrites form methaemoglobin combining with hemoglobin of blood and produce" metha emoglobinaemia" disease which leads damage to respiratory and vascular system, blue colour of skin and may become cause of cancer in some cases. The enhancement of phosphate and nitrate levels in water may deplete oxygen due to excessive algae development which may result death of fish and various types of aquatic life.

The use of pesticides, Weedicides and other chemicals also add toxic substances. The most common chemicals are BHC, DDT, Aldrin etc. The chronic d accumulation of such chemicals may create kidney mal functioning, excessive concentration of amino acids in blood and urine, abnormalities of brain tissues.

- **Changes in the Colour and Taste of Water:** The development of blue green algae, decomposition of organic substances and contamination of various harmful gases such ashydrogen sulphide , ammonia etc. in polluted water are responsible for the change in colour and development of smell. The change in colour and bed smell may result the water unsuitable for use.
- **Change in Food Quality:** The use of polluted water for the preparation of food is harmful for health. The food preservation is affected badly by the polluted water. The cost of purification of polluted water becomes higher and even then it remains of questionable quality for domestic and industrial use.
- **Hazardous for Health:** The contamination of water due to mercury, lead, fluride, other toxic substance and harmful gases may cause the health problems. The dysphasia, ataxia, mental

deterioration, convulsion are due to mercury contamination whereas lead pollution causes damage to liver and kidney, mental retardation abnormalities in fertility and pregnancy. The mottling of teeth, pain in bones and joint and bending of legs from knees are due to excess fluoride are in water. The cattle grazing around fluoride sources, the contaminated water is responsible to develop fluorosis.

II. Effects of Polluted Water on Soil and Crops

The quality of irrigation water affects the properties of soil and crop performance. The use of saline water develops the soil salinity and increases the electrical conductivity of the soil. An increased concentration of salts in the soil decreases the availability of nutrients through water absorption by plants which in turn results decrease in productivity. The sodic water irrigation creates drainage problem and leaching of excess salts is reduced. The soil expandable sodium percentage become more than 15 and pH values rise more than 9and anaerobic condition prevail in the soil horizon. The anaerobic conditions reduce the population microbes and result the poor microbial activity. The water containing more concentration of boron is responsible for the higher boron accumulation specifically in heavy textured of soil.

The most common effects of saline/ sodic water on crop is reduction in germination, decreased plant population and ultimately reduction in the yield. The adverse effect of saline/ sodic water on the growth parameters such as plant height, number of tillers and branches and dry matter accumulation and those of yield parameter such as number of grain leanth of year test weight of grains, size of pots are more common. The smaller plant with lesser Leaf Area Index result lesser assimilates for their conversion to seed. The use of boronated water develop chlrosis in the leaves and some time yield reduction become more than 50 %.the contaminated water in the residues of pesticide ,Weedicides and other chemical also contain hazardous concentration of the element, which affects the soil physiochemical property of plant growth.

The presence of higher concentration of Carbon dioxide in atmosphere produces carbonic acid due its solubility in the water. The carbonic acid would develop the acidity in soil solution or underground water. The use of such water results thin and lesser plants root and rootlets. The availability of a number of nutrients is reduced in the acidic medium. The increase in water-table near canal area contributes toward sodium accumulation and deterioration of soil structure which in turn decreases the crop production.

III. Effect of Pollution of Water on Aquatic Life

Aquatic life includes animals and vegetation both. The thermal pollution of water has adverse effect on aquatic life such as early hatching of fish eggs, decrease in the aquatic species diversity due to shifting of algae forms and other organisms towards more heat tolerant forms. Deposition of dust on the surface of water in lake and ponds reduces the photosynthesis of aquatic vegetation.

Ways to Check the Water Pollution and its Adverse Effects

The biodegradable, non- biodegradable or slow degradable pollutants are responsible for the extent of water pollution. The various ways/method suggested checking the water pollution is as follows:

- The most scientific way to check water pollution is to control pollution at source by stabilization of the ecosystem.
- The various kinds of pollutants such as industrial effluents sewage/sludge, thermal pollutants etc. should be recycled to useful use.
- The urban and rural organic waste may be recycled to generate gas and electricity. Similarly proper technology may be developed for polluted water improvement through utilization of domestic and industrial waste water in agriculture, de-toxification of toxics substances by biological methods and aquaculture processes.
- The common pollutants namely radioactive, chemical and biological can be removed by various methods such as electrolysis, absorption, ion-exchange reverse osmosis etc. The use of 'Solar Power' for cleaning the impurities of water has been recommended where a combination of sunlight and titanium dioxide is used to destroy pesticides, solvents, explosives, cyanides and dioxins etc.
- Restricted use of chemicals in agriculture should be recommended to check water pollution.
- The drinking water sources should be well protected so that contamination through drainage water may be checked.
- The disposal of solid wastes, sewage, dead bodies in water sources should be restricted.

- The quality of irrigation water should be evaluated before use. However some practical suggestion should been mentioned below-

i. **Proper Use of Rain-Water-**The rainwater should be utilized for the leaching of salts during rainy season before next crop. The moisture of rainy season should be conserved for the preparation of fields and for the sowing of Rabi season crop. It is an observation that saline water affects adversely the germinations of seeds; however plants show tolerance later on. There fore, early sowing of crops during Rabi season is a practice to utilize the moister of rain water for germinations of seeds.

ii. **Pre Irrigation with Good Quality Water-** The use of good quality water for pre sowing irrigational owes the leaching of salt from the root zones and provide chances of batter utilization of saline water at latter growth stages.

iii. **Mixing and Cyclic Uses of Canal of and Good Quality Water-** If possible canal or good quality water should we used for pre-sowing irrigation. However under un-assured or short supply of good quality water and only availability of poor quality water one should follow the proper schedule of irrigations. The canal / good quality water may be used separately or mixed together with saline water for improving the poor quality of the latter. It has been suggested that irrigation under problematic water quality situations should be scheduled as two canal water irrigation followed by one saline water irrigation or one canal water irrigation followed by one saline water irrigation during crop growth period.

iv. **Scientific Method of Irrigation-** The common and old surface irrigation methods includes check basins border strips and furrows. However drip system of irrigation is getting attentions of farmer in arid and semi-arid tracts as it maintain proper condition for water uptake by plant and avoid leaf injury. This method have some difficulty such as clogging of drippers and more accumulation of salt in surface soil which may be harmful for the germination of next crop without leaching the use of pitchers and earthen cups proved better for saline irrigation purpose on micro-scale only.

v. **Requirement of Manure and Fertilizers-** At least 25-50 % more nitrogen should be applied under saline water irrigation. The foliar sprayof nitrogenis also useful. Under saline water with more chloride in water, the phosphrous dose should be increased 25-50 % more. Application of 10 tonnes /ha FLM or 5 tonnes/ha green manure of Dhaincha along other nutrition has been found useful to minimize the bad effect of poor quality of water. The use of bio-fertilizers has been suggested for pulse crops.

References

- K. Perualammal,J. Muthumadan and K. Alagar Samy- 'Nationalisation of Rivers in India'
- Wayne B Gray & Ronald J Shadbegian, 1993. "Environmental Regulation And Manufacturing Productivity At The Plant Level," Working Papers 93-6, Center for Economic Studies, U.S. Census Bureau.
- J.Analah, 'Population Explosion and Urbanization'.
- Boyd, Gale A. & McClelland, John D., 1999. "The Impact of Environmental Constraints on Productivity Improvement in Integrated Paper Plants," Journal of Environmental Economics and Management, Elsevier, vol. 38(2).
- Prasad,M.G.K. (1990),Pollution control in India- An Overview, Financial Express.
- Bhisam Pal and V. Agarwal'Environment', shree Kirshna Publication, Agra.
- Asthana D.K. and Meera Asthana(2003)Environment- Problems and Solutions, S. Chand and Company Ltd. New Delhi.
- S.A.Sujata, 'Concept and Strategies of Sustainable Agriculture'.

28. Environmental Pollution in India: Effects, Causes and Concerns

Pawan Kumar Dubey
Prerana Bansal

Due to many environmental tragedies of the mid-20th century, many nations, instituted comprehensive regulations, designed to repair the past damage of uncontrolled pollution and prevent future environmental contamination.

Pollutant is a substance or effect which adversely asters the environment by changing the growth rate of species, through interfere with the food chain, or interferes with health, comfort, amenities, or property values of people. Pollutants are residues of the things we make use, and throw away. As per the Environmental Protection Act, 1986 of India "environmental pollutant" means any solid, liquid or gaseous substance present in such concentration as may be, or tend to be injurious to environment and "environmental pollutant" means the presence in the environment of any environmental pollution. There are two basic types of pollutants, biodegradable pollutants and non- degradable pollutants.

Key Words: - Biodegradable pollutants, Non- degradable Compound, Global warming, Photochemical smog.

Introduction

Pollution is an undesirable change in the physical, chemical, or biological characteristics of our air, land and water that will harmfully affect, the human life and the desirable species, or that may waste or deteriorate our raw material resources. Pollution is the deliberate or accidental contamination of the environment with man's waste. Pollution is the introduction by man of waste matter or surplus energy into the environment, which directly causes damage to man and his environment. It refers to the contamination of Earth's environment with materials that interfere with the human health, the quality of life or the natural functioning of ecosystems such as living organisms and their physical surroundings.

Pollutants

Pollutant is a substance or effect which adversely asters the environment by changing the growth rate of species, through interfere with the food chain, or interferes with health, comfort, amenities, or property values of people. Pollutants are residues of the things we make use, and throw away. As per the Environmental Protection Act, 1986 of India "Environmental Pollutant" means any solid, liquid or gaseous substance present in such concentration as may be, or tend to be injurious to environment and "environmental pollutant" means the presence in the environment of any environmental pollution. There are two basic of pollutants.

Biodegradable Pollutants

These are materials, such as sewage, that rapidly decomposes by natural processes. These pollutants be come a problem when added to the environment faster than they can decompose. In other words, this category includes the substance for which there due to many environmental tragedies of the mid-20th century, many nations, instituted comprehensive regulations, designed to repair the past damage of uncontrolled pollution and prevent future environmental contamination. These state laws helped to limit uncontrolled pollution, but progress has been slow and many severe contamination problems remain due to lack of funds for cleanup and enforcement exist natural waste treatment mechanisms. Heat or thermal pollution can be considered in this category since it is dispersible by natural means. However, pollution can be considered in the degradable type of pollution. When these input exceeds the decomposition or dispersal capacity of the environment. Moreover, there are limits to the amount of organic matter that the air. If we are to avoid the overall limits of the biosphere, we have to 'preserve about 4 to 5 acres of biologically productive land and freshwater space per person.

Non-degradable Pollutants

Non-degradable pollutants are the materials that either do not decompose or decompose slowly in the natural environment. Once contamination occurs, it is *through* difficult or impossible to remove these pollutants from the environment. Non-degradable compounds such as dichlorodiphenyltrichloroethane (DDT), dioxins, polychlorinated biphenyls, (POBs), and radioactive materials can reach dangerous levels of accumulation as they are passed up the food chain into the bodies of progressively larger animals. For example, toxic material accumulates even greater amount in larger fish or other carnivores that eat the small fish. This process is known, as bioaccumulation.

These pollutants also combine with other compounds in the environment to produce additional toxins. The only possible abatement for such pollution is expensive removal or extraction from the environmental life-support system. Removal of many such pollutants' from the biosphere would be virtually impossible. The obvious and sensible solution, which is of course not so easy to practice, is to outlaw the dumping of such materials to the environment (or at least control the rate of input so as to avoid toxic build up) or to stop production of such substances entirely after finding more degradable substitutes.

Effects of Environmental Pollution

Humans are at the top of the food chain, they are particularly vulnerable to the effects of non-degradable pollutants. This was clearly illustrated in the 1950s and 1960s when residents living near Minamata Bay, Japan, developed nervous disorders, tremors, and paralysis in a mysterious epidemic. More than 400 people died before authorities discovered that a local industry had released mercury into Minamata Bay. This highly toxic element accumulated in the bodies of local fish and eventually in the bodies of people who consumed the fish. More recently, research has revealed that many chemical pollutants, such as DDT and PCBs, mimic sex hormones and interfere with the human body's reproductive and' developmental functions. These substances are known as endocrine disrupters.

Primary and Secondary Effects

The effects of these pollutants may be immediate or delayed. A primary effect of pollution occurs immediately after contamination occurs, such as the death of marine plants and wildlife after an oil spill at sea. Secondary effects may be delayed or may persist in the environment into the future, perhaps going unnoticed for many years. DOT, a non-degradable compound, is seldom poisons birds immediately, but gradually accumulates in their bodies. Birds with high' concentrations of this pesticide lay thin-shelled eggs that fail to hatch or produce deformed offspring. These secondary effects threatened the survival of species such as the bald eagle and peregrine falcon, and aroused public concern over the hidden effects of non-degradable chemical compounds.

- **Effects on Natural Resources:** Pollution also has a dramatic effect on natural resources. Ecosystems such as forests, wetlands, coral reefs, and rivers perform many important for Earth's environment. They enhance water and air quality, provide habitat for plants and animals, and provide food and medicines. Functions of any one or this entire ecosystem may be impaired or destroyed by pollution. Moreover, because of the complex relationships among the many types of organisms and ecosystems, environmental contamination may have far-reaching, consequences that are not immediately are difficult of the predict For instance, scientists can only speculates on some of the potential of the ozone layer, the protective layer in the atmosphere that shields Earth form the Sun's harmful ultraviolet rays.
- **Cost of Pollution:** Another major effect of pollution is the tremendous cost of pollution cleanup and prevention. The global effort to control emissions of carbon dioxide, a gas produced from the combustion of fossil fuels such as coal or oil, or of other organic materials like wood, is one such example. The cost of maintaining annual national carbon dioxide emissions at 1990 levels is estimated to be 2 percent of the gross domestic product for developed countries. Expenditures to reduce pollution in the United State in 1992 totaled as $109 billion, in which $105.4 billion on reduction, $1.9 billion on regulation and $.17 billion on research and development. Twenty-nine percent of the total cost went toward air pollution control, 36 percent to water pollution control, and 36 percent to solid waste management.
- **Social Implications:** In addition to its effects on the economy, health and nature, pollution also has social implications. Research

has shown that low-income populations and minorities do not receive the same protection from environmental contamination, as higher-Income communities. Toxic waste incinerators, chemical plants, and solid waste dumps are often located in low-income communities because of a lack of organized, informed community involvement in municipal decision-making processes.

- **Causes of Environmental Pollution:** Pollution exists in many forms and affects many different aspects of Earth's environment; Point-source pollution comes from specific, localized, and identifiable sources, such as sewage pipelines or industrial smokestacks. Nonpoint-source of pollution comes from dispersed or uncontained sources such as contaminated water .runoff from urban areas or automobile emissions.

Air Pollution

Human contamination of Earth's atmosphere can take many forms and has existed since humans first began to use fire for agriculture, heating, and cooking. During the Industrial Revolution of the 18th and 19th centuries, however, air pollution became a major problem. Urban air pollution is commonly caused by smog. The dark london smog, that is generally a smoky mixture of carbon monoxide and organic compounds from incomplete combustion (burning) of fuels such as coal. As the smog ages and reacts with oxygen, organic and sulfuric acids condense as droplet, increasing the haze. Smog developed into a major health hazard by the 20th century.

A second type of smog, photochemical smog, began reducing air quality over large cities like Los Angeles in the 1930s. This smog is caused by combustion in car, truck, and airplane engines; which produce nitrogen oxides and hydrocarbons from unburned fuels. Sunlight causes the nitrogen oxides and hydrocarbons to combine and turn oxygen into ozone, a chemical agent that attacks rubber, injures plants, and irritates iungs. The hydrocarbons are oxidized into materials that condense and form a visible, pungent haze.

Eventually most pollutants are washed out of the air by rain, snow, fog, or mist, but only after traveling large distances, sometimes across continents. As pollutants build up in the atmosphere, sulfur and nitrogen oxides are converted into acids that mix with rain. This acid rain falls in lakes and on forests, where it can lead to the death of fish and plants, and damage entire ecosystems. Eventually the contaminated lakes and forests may become lifeless.

Air Pollution and Global Warming

One of the greatest challenges caused by air pollution is global warming, an increase in Earth's temperature due to the build of certain atmosphere gases such as carbon dioxide. With the heavy use of fossil fuels in the 20th century, atmospheric concentrations of carbon dioxide have risen dramatically. Carbon dioxide and other gases, known as greenhouse gases, reduce the escape of heat from the planet without blocking radiation coming from the Sun. Because of this greenhouse effect, average global temperatures are expected to rise 1.4 to 5.8 degrees Celsius (2.5 to 10.4 degree Fahrenheit) by the end of 21st century. Although this recent appears to be a small change, the increase would make the Earth warmer than it has been in the last 125,000 years, possibly changing climate patterns, affecting crop production, disrupting wildlife distributions, and raising the sea level.

Air Pollution and Damage to Stratosphere: Air pollution can also damage the upper atmospheric region known as the stratosphere. Excessive production of chlorine-containing compounds such as chlorofluorocarbons (CFCs) (compounds formerly' used in refrigerators, air conditioners, and in the manufacture of polystyrene products) has depleted the stratospheric ozone layer, creating a whole above Antarctica that lasts for several weeks each year. As a result, exposure to the Sun's harmful rays has damaged aquatic and terrestrial wildlife and threatens human health in high latitude regions of the northern and southern hemisphere.

Water Pollution

Industrial wastes, sewage, and agricultural chemicals such as fertilizers and pesticides are the main causes of water pollution. In developing nations, more than 95 percent of urban sewage is discharged untreated into rivers and bays, creating a major human health hazard. Water runoff, a non point source of pollution, carries fertilizing chemicals such as phosphates and nitrates from agricultural fields and yards into lakes, streams, and rivers. These Combine with the phosphates and nitrates from sewage to speed the growth of algae, a type of plantlike organism.

The water body may then become choked with decaying algae, which severely depletes the oxygen supply. This process, called Eutrophication can cause the death of fish and other aquatic life. Agricultural runoff may be to blame for the growth of a toxic form of algae called pfiesteria piscicidia, which was responsible for killing large amounts of fish in bodies of water from the Delaware Bay to the Gulf of

Mexico in the late 1990. Runoff also carries toxic pesticides and urban and Industrial wasters into lakes and streams.

Soil erosion, the wearing away of top soil by wind and rain, is also contributes to water pollution. The marine fisheries supported by ocean ecosystems are an essential source of protein, particularly for people in developing countries. Yet pollution in coastal bays, estuaries, and wetlands threatens fish stocks already depleted by over fishing.

Soil Pollution

Soil is a mixture of mineral, plant and animal residue that forms during a long process that may take thousands of years. It is necessary for most plant growth and is essential for all agricultural production. Soil pollution is a buildup of toxic chemical compounds, salts, pathogens (disease-causing organisms), or radioactive materials that .can affect plant and animal, life. Un-healthy soil management methods have seriously degraded soil quality, caused soil pollution and enhance erosion. Treating the soil with chemical fertilizers, pesticides, and fungicides interferes with the natural processes occurring within the soil and destroys useful organisms such as bacteria, fungi, and other microorganisms. Improper irrigation practices in areas with poorly drained soil may result in salt deposits that inhibit plant growth and may lead to crop failure.

- **Solid Waste:** Solid, wastes are unwanted, solid materials such as garbage, paper, plastic and other synthetic materials, metals and wood. Billions of tons' of solid waste are thrown out annually. Cities in economically developed countries produce far more solid waste per capita than those in developing countries. Moreover, waste from developed countries typically contains a high percentage of synthetic materials that take longer to decompose than' primarily biodegradable waste materials of developing countries.

Areas, where wastes are buried called landfills are the cheapest and most common disposal method for solid wastes, worldwide. But landfills quickly become overfilled and may contaminate air, soil, and water. Incineration or burning of waste reduces the volume of solid waste but produces dense ashen wastes (some of which of become airborne) that often contain dangerous concentrations of hazard materials such as heavy metals and toxic compounds. Natural biological processes to decompositions of organic wastes are an effective strategy for dealing with organic garbage and produce a material that can be used as a natural fertilizer. Recycling, extracting and reusing certain waste materials, has become an important part of municipal solid waste strategies in developed countries.

Worldwide expanding recycling programmes can help reduce solid waste pollution, but the key to solving severe solid waste problems lies in reducing the amount of waste generated. Waste prevention, or source reduction, such as altering the way the products are designed or manufactured to make them easier to reuse, reduces the high costs associated with environmental pollution.

- **Hazardous Waste:** Hazardous wastes are solid, liquid or gas wastes that may be harmful to people or the environment and tend to be persistent or non-degrade in nature, such wastes include toxic chemicals and flammable or radioactive substances, these waste including industrial wastes from chemical plants or nuclear reactors, agricultural waste such as pesticides and fertilizers, medical and household hazardous wastes such as topic paints and solvents.

Hazardous wastes of particular concern are the radioactive wastes from the nuclear power and weapons industries. Till today there is no safe methods for permanent disposal of old fuel elements from nuclear reactors. Most are kept in storage facilities at the original reactor sites where they were regenerated.

Noise Pollution

Unwanted sound, or noise, such as that produced by airplanes, traffic, motor vehicles, trains or Industrial machinery, is considered a form of pollution. Noise pollution is at its worst in densely populated areas. It can cause hearing loss, stress, high blood pressure, sleep loss, distraction, and lost productivity. Measured in decibels, noise intensity can range from zero, the quietest sound the human ear can detect to over 160 decibels. Conversation takes place at around 40 decibels, a subway train is about 80 decibels, and a rock concert is from 80 to 100 decibels. The intensity of a nearby jet taking off is about 110 decibels. The threshold for pain, tissue damage, and potential hearing loss in human is 120 decibels. Long-lasing, high-intensity sounds are the most damaging to hearing and produce the most stress in humans.

Solutions to noise Pollution includes an adding insulation and sound-proofing to doors, walls, and ceilings,; using production, particularly in industrial working areas; planting vegetation to absorb and screen out noise pollution; and zoning urban areas to maintain a separation between residential areas and zones of excessive noise.

Control of Environmental Pollution

Control of Air Pollution:

- Encourage your family to walk to the neighborhood market.
- Whenever possible take your bicycle.
- Don't let your father drop you to school, take the school bus.
- Encourage your family to form a car pool to office and back.
- Reduce the use of aerosols in the households.
- Look after the trees in your neighborhood.

Control of Water Pollution:

- Utilizing an approach to improve storm water management capacity throughout the system.
- Repair and replacement of leakage and malfunctioning equipment.
- Increasing overall capacity of the sewage collection system (often a very expensive option).
- Industrial specialized treatment system.

Control of Soil Pollution:

- Reducing chemical fertilizer and pesticide use.
- Reusing of materials.
- Recycling and recovery of materials.
- Reforesting.
- Solid waste treatment.

Control of Noise Pollution:

- The noise levels in any area /zone shall not exceed the ambient air quality standards in respect of noise as specified in the schedule.
- The authority shall be responsible for the enforcement of noise pollution control measures and the due compliance of the ambient quality standards in respect of noise.
- The respective State Pollution Board of Pollution Control Committees in consultation with the Central Pollution Control Board shall collect, compile and publish technical and statistical data relating to noise pollution and measures devised for its effective prevention, control and abatement.
- Restrictions on the use of loud speakers/ public address system.
- Consequences of any violation in silence zone / area.

- Complaints to be made to the authority.
- Power to prohibit etc. continuance of music sound or noise.

Conclusion

Because of the many environmental tragedies of the mid-20th century, many nations, instituted comprehensive regulations, designed to repair the past damage of uncontrolled pollution and prevent future environmental contamination. These state laws helped to limit uncontrolled pollution, but progress has been slow and many severe contamination problems remain due to lack of funds for cleanup and enforcement.

International agreements have also played a role in reducing global pollution. The Montreal Protocol on Sub-' stances that deplete the Ozone Layer (1987) set international target dates for reducing the manufacture and emission of the chemicals, such as CFCs, known to deplete the ozone layer. Regulations and legislation have led to considerable progress in cleaning up air and water pollution in developed countries. Vehicles in the 2000s emit fewer nitrogen oxides than those in the 1970s did; power plants now burn low-suffer fuels; industrial stacks have scrubbers to reduce emissions; and lead has been removed from gasoline.

Nongovernmental citizen groups have formed at the local, national, and international level to combat pollution problems worldwide. Many of these organizations provide information and support for people. Green peace is an activist organization that focuses international attention on industries and governments known to contaminate land, sea, or atmosphere with toxic or solid wastes. Friends or the Earth international is a federation of international organizations that fight environmental pollution around the world.

References

- Chand, Atten (1985),"Environmental challenges", UDH Publisher, New Delhi.
- Katyal,Timmy and Sataka, M. (1989), Environmental pollution, Anmol Publications, New Delhi.
- Prasad,M.G.K. (1990),Pollution control in India- An Overview, Financial Express.
- Sharma.P.D.(1990),"Ecology and Environment," Rastogi Publications Meerut.
- Asthana D.K. and Meera Asthana(2003)Environment- Problems and Solutions, S. Chand and Company Ltd. New Delhi.
- Population Explosion and Urbanization-J.Analah.

Section-VI
Sustainable Development and Miscellaneous Issues

29. Foot Print of Sustainable Development and Its Promoter GIS

Anupam Pathak

Sustainable development is development sources and integrates it into a map project. Many that meets the needs of the present without countries, except ours, have an abundance of compromising the ability of future generations to geographic data for analysis. GIS is a key tool used meets their own needs. It contains within it two key in assessment, prioritization, mitigation, planning, concepts, the concept of needs, in particular the science and training. Economic inequality, social essential needs of the world's poor, to which instability and environmental degradation are overriding priority should be given; and the idea of common features of unsustainable development. limitations imposed by the state of technology and Poor people bear the brunt of these problems social organization on the environment's ability to because their livelihoods are precariously meet present and future needs. Sustainable balanced on volatile economic opportunities and development is maintaining a delicate balance environments vulnerable to change. What we need between the human need to improve lifestyles and is a national strategy for sustainable development feeling of well-being on one hand, and preserving (strategic and participatory process of analysis, natural resources and ecosystems, on which we and debate, capacity strengthening, planning and future generations depend. GIS organizes action towards sustainable development), a tool to geographic data so that a person reading a map assist the farmers in overcoming their problems can select data necessary for a specific project or and start to strengthen their capacity for task. GIS can be an important tool for helping sustainable development. Using GIS in sustainable people map out plans for successfully achieving agriculture can lead to great benefits, especially management strategies that are sustainable both at for how they can be seen all the cultures they local and global levels. A good GIS program is represent, the soil which is substances in the soil, able to process geographic data from a variety of availability of water, crop rotation, pests.

Key words: Sustainable development, Environment, Geographic Information System, Mitigation

Introduction

Negative effects of growth, industrialization, expansion of agriculture and unbridled over- exploitation of nature became increasingly apparent by 1960s. Problems related with the environment tool up myriad shapes and variety. At places it was reduced fish stock or disappearing bird life, at other it was photochemical smog which caused respiratory ailments, at still other places it was degeneration of soil and reduced agricultural productivity causing hunger, starvation and misery. Where the root- did causes of the problem lie? What was the most effective way to solve them? Different reasons were proposed and different ways were adopted to handle them. There was by no means a consensus among people facing these problems. People are usually reluctant to act for issues with which they have no concern or experience. A general concept of universal applicability had to with which they have no concern or experience. A general concept of universal applicability had to emerge before efforts to conserve the environment, wildlife and natural resources could be intensified.

Sustainable development has become the "buzzword" of both the academic and the business world. "Sustainability" has been present for the last decades in academic papers, syllabuses of Faculties, boardrooms of local authorities and corporations and offices of public relations officers. Unfortunately, sustainability has become a "fashionable" concept in theory, but it is considered extremely expensive to be put in practice by major corporations, firms and local or national governments. What people tend to neglect and forget is the evolution of the concept of sustainability. Although the history and evolution of a concept might seem unimportant, it could help us predict the future trends and s t flaws that will appear. And it will help us ensure that the 21 century will be "the Sustainability Century" (Elkington, 1997)

The "Warning"

More than 200 years ago, the first questions arose regarding the impact of the evolution of our civilization could have on the environment and resources of our planet. In 1798, Thomas Robert Malthus (1766-1834), demographer, political economist and country pastor in England wrote *An Essay on the Principle of Population*. He predicted that the world's population would eventually starve or, at the least, live at a minimal level of subsistence because food production could not keep pace with the growth of population. He believed that the population was held in check by "misery, vice and moral restraint". Malthus wrote that "population, when unchecked, increased in a geometrical ratio and subsistence for man in an arithmetical ratio". Technological advances since that time have proved him wrong. Through better farming techniques, the invention of new farming equipment, and continuing advances in agricultural science, "production has increased much more rapidly than population, so much so that in real terms, the price of food is much lower today than it was two hundred years ago, or for that matter, even fifty years ago".

The main purpose of the above mentioned Malthusians was to provide a useful reminder to the society and to the local, national and international authorities that if we continue our consumption trends we could find ourselves in trouble.

The concept of **sustainable development** emerged from the post-War environmental movement, which recognized the negative impacts of human growth and development on the environment and communities. Since publishing the first ever national strategy for sustainable development in 1994, the UK Government has played a lead role in promoting sustainable development at home and overseas.

1972: Limits to Growth

Commissioned by the Club of Rome, Limits to Growth attempts to model the consequences of a growing human population in a world of finite resources, concluding that current patterns of growth cannot be sustained indefinitely.

1987: Our Common Future

The term sustainable development came to prominence through the United Nations Brundtland Commission. The commission's 1987 report, **Our Common Future** defined sustainable development as "development which meets the needs of the present without compromising the ability of future generations to meet their own needs".

1992: Rio conference

The concept received further attention at the United Nations **Conference on Environment and Development** in Rio de Janeiro in 1992, the first international attempt to develop strategies for a more sustainable pattern of development. Representatives of 178 national governments, including more than 100 heads of state, and many organizations representing civil society attended the conference. The world had never previously witnessed a larger gathering of national leaders. At the summit, governments around the world committed to sustainable development. The UK government was the first to produce its national strategy in 1994.

1999: A Better Quality of Life

In 1999, the UK government outlined how it proposed to deliver sustainable development in A Better Quality of Life. This set out a vision of simultaneously delivering economic, social and environmental outcomes as measured by a series of headline indicators.

2002: Johannesburg summit

The **World Summit on Sustainable Development** (WSSD) took place in Johannesburg from 26 August to 4 September 2002.

The summit delivered three outcomes: a political declaration, the Johannesburg Plan of Implementation and the establishment of numerous partnership initiatives. Key commitments covered sustainable consumption and production, water and sanitation, and energy.

The outcomes complemented the Millennium Development Goals, reinforce Doha and Monterrey agreements and set challenging global goals and targets on accessing water, sanitation and modern energy services; increasing energy efficiency and use of renewable energy; sustainable fisheries and forests; reducing biodiversity loss on land and in our oceans; chemicals management; and decoupling environmental degradation from economic growth – that is, achieving sustainable patterns of consumption and production.

The UK's international priorities on sustainable development have principally been framed by the Millennium Development Goals, the Doha Development Agenda of the World Trade Organization, the Monterrey Consensus on Financing for Development and the Plan of Implementation of the 2002 World Summit on Sustainable Development (WSSD).

Securing the Future

2005 saw the publication of **Securing the Future**, a revised UK Government strategy for sustainable development. At the same time, a strategic framework was agreed by the UK Government and the devolved administrations in Scotland, Wales and Northern Ireland, providing a consistent approach and focus across the UK.

GIS Promoter of Sustainable Development

GIS for sustainable development examines how GIS applications can improve collaboration in decision making among those involved in promoting sustainable development. GIS (Geographic Information System) is an information system based on the computer, which uses digital representation and analysis of geographic features that are present in surface and ground events that occur on them. The meaning of digital representations is the conversion of analog form (continuous line) in a digital.

Material and Methods

GIS represents the combination of data content and indexing management, with spatial representation and analysis techniques to facilitate understanding of real world entities and interactions between them. An advanced GIS requires two main components, the hardware and software, subordinate to a third, the human. GIS with remote sensing can produce digital maps of cover/land use, hazard and risk maps of floods, droughts, demarcation maps of areas affected by disasters. However, the results of a GIS is not limited to maps, using such a system can be obtained dedicated thematic GIS databases, products derived from GIS layer-info and most importantly, estimation of data.

Result and Discussions

"Every object present on Earth can be geo-referenced" is the cornerstone of any GIS databases. The database is a collection of information about things and combinations of these, Geo reference refers to the location of a level or coverage in space, defined by a referential coordinate system. A GIS is a computer system designed to work with data referenced by coordinates spatial/geographical, which is both a database system with specific capabilities for spatial reference data and a set of operations that work with these data. He is often considered a high-level map.

GIS technology integrates common database operations such as query and statistical analysis with unique visualization and geographic analysis benefits offered by maps. These features distinguish GIS from other systems and make it have value to a wide range of people to explain events, forecast results and planning strategies. Sustainable development is the balance of meeting humankind's present needs while protecting the environment to ensure the fulfillment of future generations' needs. The growing human population and its demands on the earth's resources generate a need for sustainable practices. Implementing these practices often requires collaboration between different organizations. ESRI's commitment to developing interoperable technology sets the stage for cooperation between organizations so that they can make well-informed decisions. GIS software allows users across the globe to share ideas on how to meet their resource needs, plan efficient land use, and protect the environment to guarantee the survival of future generations. GIS is an important tool for helping people map out plans for successfully achieving management strategies that are sustainable both at local and global levels. Geographic Information Systems provide policy makers and planning agencies with visualization tools to manage growth and change. GIS is key tool used in: **Assessment, Prioritization, Mitigation, Planning, Science and Training.**

Mitigating the effects of natural hazards and providing potential risk analysis for communities are common application areas for GIS. Information from a central database can be visualized and analyzed to mitigate areas that are environmentally at risk. The GIS can be used to assess contamination of soil, water, or the atmosphere. With the help of GIS, you can monitor pollution to environment relationships, identify potential risks, prioritize potential pollution scenarios and model the rise of pollution levels over time and predict environmental impact.

Geospatial analysis is essential for planning that encompasses the demands of building a hydroelectric dam, planting urban trees to cool down a city, or establishing guidelines for the protection of habitat. GIS tools are essential for performing geospatial analysis. Environmental planning scenarios include waste management plan for best Research Journal of Agricultural Science, 42 (3), 2010 treatment and disposal site location, most efficient routes for fleet, and analysis of impact on public and land, disaster management response plan for a toxic spill (includes impact areas, infrastructure, what-if scenarios, escape routes, and responder centers), vegetation management plan for a right-of-way corridor that includes a schedule of planting and cutting and generates regulatory compliance reports

Environmental scientists can use GIS to study change, impacts, and relationships in the environment. GIS is an integral component of scientific methodology, allowing scientists to capture, manage, manipulate, and visualize geographic information. It makes it possible to collect large volumes of information about observable events and build and test theories about geographic patterns and processes that show relational patterns. You can access many types of data from many resources, quantify uncertainty, test hypotheses, perform geographically weighted regression models, perform spatial clustering algorithms and testing in order to determine distribution, visualize cartography and scientific information, perform spatial analysis and modeling, collaborate with many other scientific agencies and of course, publish and serve findings.

Conclusion

A GIS may have the following advantages for the sustainable development: Project Planning-GIS benefits are often found in the detailed planning of the project, which has many space components, where problem analysis is necessary before starting the project. The system can generate thematic maps based on one or more base maps, for example, generate maps of land use on soil composition, vegetation and topography. It is also possible to calculate areas, lengths and distances. Decision Making-A GIS is not an automatic decision making but is a tool for query, analysis and layout of the map data used to support decision-making process. Visual analysis-DTM (Digital Terrain Modeling) is an important tool of GIS. Using modeling DTM/3D, land can be better viewed, this leads to a better understanding of the relationships within that field. Using GIS many calculations and modeling can be done much easier, for example lake volume, the volume of soil erosion, etc. Improve organizational integration - Facilitating the communication and sharing information, by creating a shared database, data can be collected only once and can be used several times. As a conclusion, saying "the benefits of GIS" for sustainable development, we can refer to that data are much better organized, are eliminated redundancies in data storage, users have increased productivity and ultimately, analysis, statistics and searches are made much easier and accurate, meaning better decisions. When referring to the drawbacks, the biggest risk is to forget that digital products aren't more accurate than maps derived from, that must necessarily be considered to collect accurate data in the field. Incorrect data can lead to larger errors, resulting in a wrong system. And if there's something not good, the methods and techniques have no fault; everything was started from field work.

References

- Adger Neil, Winkles Alexandra (2007) – Vulnerability, Poverty and Pustaining Well-bing, Published in Handbook of Sustainable Development, Edward Elgar Publishing Limited, Cheltenham.
- Asefa, Sisay (2005) – The Concept of Sustainable Development: An Introduction, Published in the Economics of Sustainable Development, W.E. Upjohn Institute for Employment, W.E. Upjohn Institute for Employment Research, Michigan.
- Bartlett, Albert (2006) – Reflections on Sustainability, Population Growth, and the Environment, Published in the Future of Sustainability, Springer, Dordrecht
- Baumol William, Litgan R., Schramm C. (2007) – Good Capitalism, Bad Capitalism and the Economics of Growth and Prosperity, Yale University Press, New Heven & London.
- Khoshoo T.N., 1990, Man in Nature- Past, present and future.
- Meadows D.H., et al, 1972, The Limits to Growth. Universe Books, N.Y.
- Milbreth LW., 1989, Envisioning a Sustainable Society: Learning Our Way Out. State University of New York Press, Albany, N.Y.
- Miller G. Tyler, 2001, Environmental Science. 8th Edn. Brooks- Cole, Thomas Learning Inc. USA.
- Meera Asthana, D.K. Asthana- Environmental Studies S. Chand & Co. New Delhi.
- ***. http://en.wikipedia.org/wiki/Geographic_information_system
- ***. http://www.ccdmd.qc.ca/en/gis/module1.html
- ***. http://www.gisdevelopment.net/education/papers/index.htm
- ***. http://egsc.usgs.gov/isb/pubs/gis_poster.

30. Green Marketing-as one of the Sustainable Future of Marketing Styles in India

Satendra Bhardwaj
Prakash Pandey
Ruchi Sharma

Green marketing must satisfy two objectives: improved environmental quality and customer satisfaction .Paul Hawken, Amory Lovins, and L. Hunter Lovins write in their book Natural Capitalism: Creating the Next Industrial Revolution that greener, more sustainable products need to dramatically increase the productivity of natural resources, follow biological/ cyclical production models, encourage dematerialization, and reinvest in and contribute to the planet's "natural" capital.

The Chapter will help to know about what is the relevance of Social Marketing in order to protect the environment and to improve the quality of life and are concerned with issues that include conservation of natural resources, reducing environmental pollution, protecting endangered species, and control of land use. Many companies are finding that consumers are willing to pay more for a green product. This chapter will also examine some of the reasons that why organizations adopting green marketing.

Key words: Environment, Consumer, Pollution and Green Marketing.

Introduction

Green marketing means the development, pricing, promotion and distribution of products that do not harm the environment. Environmentalists evaluate products to determine their impact on environment and marketers' commitment to the environment. Environmentally safe products are approved and companies receiving the green signal can use it in advertising and on packaging. According to a recent Environmental Leader and Media Buyer Planner study 33% of marketers find green marketing techniques to be more effective than 'normal' marketing activities. In fact, more than 80% plan to spend more on green advertising this year than last year; manufacturers plan to spend 'significantly higher' budgetary amounts on green efforts. One interesting finding is that smaller businesses are spending more (26% of budget) on green marketing efforts compared to larger businesses (6% marketing budget). The majority of green marketing dollars are currently spent in the online marketplace (74%) with print (49%) and direct advertising (40%) also ranking high.

Green marketing is also known as Environmental Marketing and Ecological marketing. It came into prominence in the late eighties and early nineties. It involves the study of the positive and negative aspects of marketing activities on pollution, energy depletion and non-energy depletion. Green marketing consists of all activities designed to generate and facilitate any exchanges intended to satisfy human needs and wants, such that the satisfaction of these needs and wants occurs, with minimal detrimental impact on the natural environment.

Objectives

1. To introduce the terms and concepts of green marketing.
2. To discuss briefly why going green is important.
3. To examine some of the reasons that why organizations adopting green marketing philosophy.
4. To mention the problems with green marketing.

Importance of Green Marketing

Green Marketing is becoming increasingly important due to the following reasons

1. **Opportunities.** Business firms perceive green marketing to be an opportunity that can be used to achieve their objectives. For

example, Xerox introduced a high quality recycled photocopier paper in order to satisfy the demand for less environmentally harmful products.

2. **Social Responsibility.** Many firms are beginning to realize that as members of the wider community they must behave in an environmentally responsible fashion. Therefore, environmental issues are being integrated into the firm's corporate culture.
3. **Government Pressure.** Government agencies are creating more and more regulations to control hazardous wastes produced by industry. These regulations seek to protect consumers and the society from negative impact of business activities on the environment. In some cases, Governments try to induce firms and individuals to become more responsible towards the environment.
4. **Competitive Advantages.** Firms marketing environment friendly products and services will have competitive advantages over firms marketing non-environment friendly products and services. For example, McDonald's replaced its clamshell packaging and waxed paper to minimize ozone depletion.
5. **Cost Factor.** Firms also use green marketing use to reduce costs. Disposing of harmful by products such as polychlorinated biphenyl (PCB) contaminated oil is becoming increasingly costly and difficult. Therefore, firms that can reduce harmful wastes can gain substantial cost saving. More efficient production processes can not only reduce waste but also the need for raw material. A firm may develop a technology for reducing and recycling waste. For example, firms that clean the oil in large industrial condensers increase the life of those condensers, remove the need for replacing the oil, as well as the need to dispose of the waste oil. This reduces operating costs for the owners of condensers and generates revenues for the firms cleaning the oil.

Additional Benefits are as Follows

Today's consumers are becoming more and more conscious about the environment and are also becoming socially responsible. Therefore, more companies are responsible to consumer's aspiration for environmentally less damaging or neutral products. Many companies want to have advantages of green marketing areas follow:

1. It ensures sustained long-term growth along with profitability.
2. It saves money in the long run, thought initially the cost is more.

3. It helps companies market their products and services keeping the environment aspects in mind. It helps in accessing the new markets and enjoying competitive advantage.
4. Most of the employees also feel proud and responsible to be working for environmentally responsible company

Sustainability of the Environment due to Green Marketing in the Following ways

1. **Eliminate the Concept of Waste.** Waste and pollution arise due usually from inefficiency. Therefore, make things without waste instead of handling the waste.
2. **Reinvest the Concept of a Product.** Products should be either consumable or durables. Consumables can be either eaten or placed in the ground so that they turn into soil without any harm to the environment. Durables could be made, used and returned to the manufacturer within a closed loop system.
3. **Make Prices Reflect the Cost.** Every product should reflect or at least approximate its actual cost-not only the direct cost of production but also the cost of air, water and soil.
4. **Make Environmentalism Profitable.** Consumers are beginning to recognize that competition in the market place should not be between companies harming the environment and those trying to save it.

Problem in Green Marketing

Much organization wants to turn green, as an increasing number of consumers, want to associate themselves with environmental-friendly products, alongside, one also witnesses confusion among the consumers regarding the products. In particular, one often finds distrust regarding the credibility of green products. Therefore, to ensure consumer confidence, marketers of green products need to be much more transparent, and refrain from breaching any law or standards relating to products or business practices.

Conclusion

In terms of defining the future for green marketing, ultimately it will be the consumer who will decide. To succeed, consumers will need to recognize that there is a compromise involved in making a greener purchase, and they will need to have confidence in the environmental benefits involved. Consumers will need to be confident that:

- Environmental issues are real problems, that something needs to be done, and that they are involved in doing so.
- The company's market offering meets these needs and has a proven eco-performance over competition or previous offerings.
- Purchasing the product will make a difference and it is worth doing.

Peat tie replaced the 4 Ps with the 4 Ss in 1995

- Satisfaction of customer needs.
- Safety of products for consumers, workers, society and the environment.
- Social acceptability of products, their production and the other activities of the company.
- Sustainability of the products, their production and the other activities of the company.

Green marketing is gaining significant attention from both CEOs and consumers. Given that a carefully crafted green marketing strategy can earn credibility with customers and provide a platform for revenue growth, it's an area worthy of additional consideration.

References

- G. Fowler, "Green Sales Pitch Isn't Moving Many Products," Wall Street Journal, 6 March 2002.
- K. Alston and J. P. Roberts, "Partners in New Product Development: SC Johnson and the Alliance for Environmental Innovation," Corporate Environmental Strategy 6, no. 2: 111–28.
- J. Ottman, Green Marketing:Opportunity for Innovation (Lincolnwood [Chicago]:NTC Business Books, 1997).
- P. Hawken, A. Lovins, and L. H. Lovins, Natural Capitalism: Creating the Next Industrial Revolution (Boston: Little, Brown, and Company, 1999).

- Business Week, "Alternate Power: A Change in the Wind," 4 July 2005, 36–37.
- T. L. Friedman, "Geo-Greening by Example," New York Times, 27 March 2005; and T. L. Friedman, "The New 'Sputnik' Challenges: They All Run on
- Oil," New York Times, 20 January 2006.
- Gupta, C.B and Nair, N. Rajan "Marketing Management" Sultan Chand and Sons First Edition New Delhi 1996.
- Madan Pankaj, Mittal amit and Verma Hemraj "Marketing Management"Global Vision Publishing House,First Edition, Delhi 2010.

31. Current Challenges and Development in Corporate Sustainability Reporting

H.P. Malonia
Vishal Agrawal

Reporting and external corporate communication play an important role in corporate sustainability. As well as economic reasons, the vision of sustainable development also emphasises the importance of corporate sustainability reporting. On the one hand, companies depend on supply of resources from various stakeholders, so that management is challenged to secure social acceptance by communicating the benefits that the company creates for society and the sustainability effects of its activities. On the other hand, the vision of sustainable development requires participation, which in turn requires the reporting and communication of sustainability - relevant issues and activities. No participation is possible without communication.

Keywords: Sustainability, Sustainability reporting, Sustainability communication, Corporate sustainability.

Introduction

Because of their social embedded ness in their stakeholder environment, companies cannot act in isolation from their social environment. The purposes of a company are determined by various stakeholders who pursue different economic, ecological and social goals and who demand information related to their goals. In consequence, corporate sustainability management is challenged not only to manage the economic, ecological and social effects of corporate activates systematically, but also to provide stakeholders with information about sustainability - relevant issues and how the company is dealing with them. Sustainability management designs processes and structures in order to ensure sustainable corporate and business development and to contribute positively to the sustainable development of society generally. This means that information creation and information flows have to be organized in line with reporting, communication and dialogue requirements with key stakeholders.

Goals and Benefits of Sustainability Reporting

1. Legitimating of corporate activities, products and services which create environment and social impacts
2. Increase in corporate reputation and brand value
3. Gaining a competitive advantage
4. Signalling superior competitiveness, with sustainability reporting activities as a proxy indicator for overall performance
5. Comparison and benchmarking against competitors
6. Increasing transparency and accountability within the company
7. Establishing and supporting employee motivation as well as internal information and control processes.

Historical Development of Sustainability Reporting

Reviewing the historical development of sustainability reporting over recent decades, companies have changed the perspectives and directions of their non-financial reports in response to different societal challenges. Figure -1 illustrates the different stages and forms of reporting, particularly in Europe, with regard to the three-pillar approach to sustainable development.

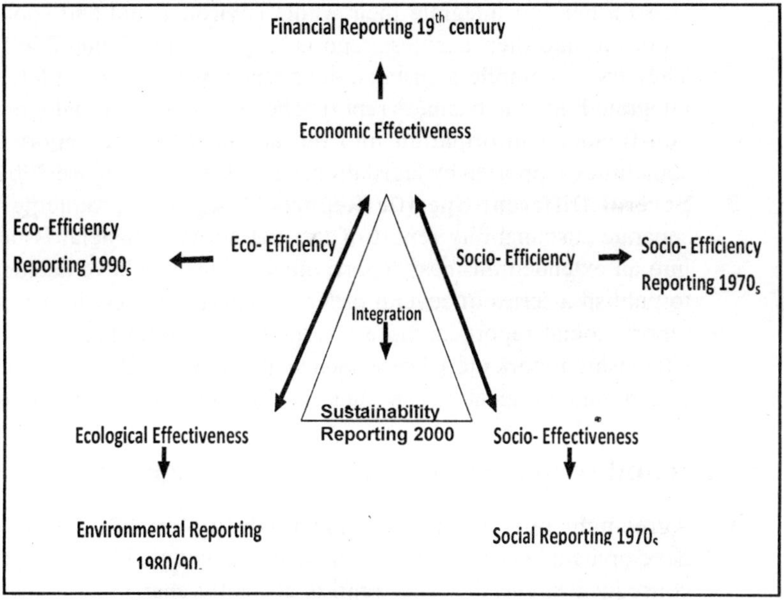

Figure-1
Source: Based on Schaltegger et. al. 2002

Leading corporations are currently attempting to integrate environmental, social and financial accounting information in very different ways, which has led to the extension of existing reports, to a large variety and various combinations of different reporting formats, and to the development of new kinds of reports:

1. **Specific Sustainability Report in Addition to the Financial Report-** some companies have started to publish so-called sustainability reports in addition to their financial report. These reports constitute a single publication which simultaneously provides information about the company's ecological, social and economic sustainability activities and performance, often following the format of "Triple P-Report" (People, Planet and Profits)
2. **Extended Business and Financial Reports-** Because of the increasing financial importance of environmental and social issues, an increasing number of companies are integrating sustainability aspects into their financial report, in either the balance sheet or profit and loss account. Some companies have decided to go a

step further and integrate their whole environmental and social reporting into their business reports (e.g. Renault Group 2002). They try to compile a sustainability report which is completely integrated into the business report. The development to integrate non-financial information into the annual financial report is sometimes supported by legislation, such as in France since 2001.

3. **Several Different Specific Reports:** Instead of producing a separate sustainability report of integrating non-financial issues into an extended business report, other companies have decided to publish a series of several different reports (e.g. environment report, social report, corporate responsibility report, corporate citizenship report, etc.).These each deal with a specific challenge of corporate sustainability and address different stakeholder groups.

Sustainability Reporting -Current Challenges

1. Agreement over the terms "Sustainability" or "Sustainable development" is usually rather difficult and not made explicit. In consequence, sustainability reports currently change their main focus fairly fast and often between different sustainability perspectives. This challenges management to establish an approach to identify what contextual priorities should be chosen in each reporting period and how to define and communicate its understanding of corporate sustainability.
2. It is often difficult to identify and analyse sustainability issues, which requires a change in current and traditional terms and perceptions. Management is furthermore challenged to link strategic analysis and management together with information management, corporate accounting and sustainability reporting.
3. The complexity of corporate sustainability as a set of interrelated goals leads to problems for management in operationalisation, measurement and communication. Sustainability reporting must therefore be backed up with a systematic accounting and information management system which provides a comprehensive basis for all sustainability issues but which nevertheless, in not excessive in size.
4. Developing solutions often requires interdisciplinary teamwork and lateral organisation processes. Experts who are accustomed to communicating in their own scientific or professional "language" have to open up to the approaches, terms and interpretations of other sciences and professions. Successful sustainability reporting

and communication, like sustainability management in general, therefore requires the development of interdisciplinary employee skills with a strong fundamental training in sustainability management, communication and soft skills.

5. On the one side, information about the sustainability of a company is not simple for stakeholders to access directly and they can often do this only with difficulty and its acquisition can involve very high costs in both time and money. This leads to information asymmetry between the company and its stakeholders. Situations of asymmetric information tend to create a climate of low credibility which the company has to overcome with specific communication and management activities such as warrants, verifications, labels certified by credible NGOs, etc.
6. Companies, on the other hand, do not always have sufficient knowledge about the information needs of stakeholders. As a result current sustainability reports do not always meet stakeholder's information needs and often only a small part of the desired readership is actually contacted. Indeed, the latter is not a sustainability related phenomenon but the common fate of communication. So far, only a limited number of systematic and comprehensive surveys have been conducted on the reception of environmental and sustainability reports. In any case, a stronger involvement of stakeholders into the reporting process is needed in order to enable a better target-oriented form of reporting.
7. Currently, most sustainability reports are non-specific, aiming at a diffuse and excessively wide group of potential readers (lack of target group orientation). This creates a risk of information overload and often leads to an additive and separate treatment of ecological, social and economic matters. The term "carpet bombing syndrome" (Sustainability and UNEP 2002) illustrates the fact that some companies have "flooded" their readers with increasingly extensive sustainability reports- recognized by some, but in practice mostly read by only a few. To counteract these companies should strive for a well-balanced and clearly arranged presentation of sustainability information which addresses a clearly defined group of readers. This very often requires direct consideration of specific interlink ages between economic, social and ecological business achievements, including synergy effects and conflicts between objectives. In practice, these interfaces are discussed only rudimentarily (e.g. INEM 2001, Sustainability and UNEP 2002).

8. An improvement in the often-criticised lack of comparability of sustainability reports requires a consensus or a generally accepted standard about what information shall be disclosed and in what format. The Global Reporting Initiative (GRI) is attempting to create such a standard. In addition to information which is related to activities and management systems, significant sustainability performance indicators are of particular importance (but are often not in the centre of attention). The improvement of data quality and the quality of data collection procedures is an additional requirement for the comparability of published sustainability information. The comparability of ecological and social performance information is often limited because the procedures and practices of data collection and information management can vary over time or between companies.

So far, small and medium sized enterprises (SMEs) are particularly reluctant to issue sustainability reports. Since SMEs constitute a large part of the economy and of its social and environmental effects, it is necessary to emphasise the benefits and to keep the costs of sustainability reporting low (e.g. ACCA 2004, European Environment Agency 2003, and GRI 2004)

Sustainability Reporting-Current Developments

1. **Guidelines and Standards:** Various institutions have published guidelines, standards, regulations, sets of criteria, etc, which should help to harmonize corporate sustainability reporting and to provide some guidance for management. A guideline is a non-binding guidance document based on practical experiences. Standards are developed by standardisation organisations and are often the basis for certification procedures.

   Current international examples are the guidelines of the Global Reporting Initiative (GRI 2002) and the World Business Council for Sustainable Development (WBCSD 2002), and the standard ISO 14063 for environmental communications (ISO 2004).
2. **Regulations:** As well as guidelines and standards, the number of European countries which regulate corporate environmental and sustainability reporting is increasing (e.g.IIIEE 2002, KPMG 2005)
3. **Assessment and Verification :** Assessment and verification of sustainability information can help to improve the credibility of sustainability reporting. The assessment and verification of

information which is disclosed in corporate reports can in principle be conducted by independent external experts from accounting firms, independent association, or renowned NGOs.

4. **Ranking and Ratings:** Now-a-days sustainability reports have themselves increasingly become a subject of ranking and reporting competitions. Ranking of sustainability reporting reflect the expectations of some stakeholders and the developments of reporting in practice. Such rankings can also involve recommendations for improvement in future. By this means, rankings aim to improve the quality of sustainability reporting and, to some extent, also contribute to a certain degree of standardisation. In most cases, the criteria for both ratings and rankings of sustainability reports are often generated from reporting guidelines (e.g. CEFIC 1998, GRI 2002) as well as from other standards and research activities on sustainability reporting (e.g. AA1000, SA8000).

5. **Professionalization of Sustainability Communication :** Corporate sustainability reports are usually developed by either employees from the environmental or sustainability department or from the corporate communication unit, or by an external public relations agency. All three groups tend to lack some relevant knowledge for sustainability reporting if they are not trained adequately or if they do not collaborate in an interdisciplinary working group. Successful sustainability reporting requires the sustainability department's knowledge about sustainable development in general and the sustainability issues which are relevant for the company, in particular.

   The challenge of sustainability reporting and communication cannot be met successfully without managers and employees, who have a combined knowledge in sustainable development, corporate sustainability management, and communication in media-conform ways as well as in the organisation of inter-disciplinary team processes.

6. **The Role of Reporting and the Link to Accounting :** If sustainability reporting aims to go beyond diffuse qualitative statements then it must include quantitative measures describing the state of practice, the goals and the progress made. Furthermore, sustainability reporting of quantitative measures requires an accounting system which provides the necessary information.

Sustainability accounting can be defined as a subset of accounting that deals with the activities, methods and systems that are required in order to record, analyse and report; firstly, environmentally and socially induced economic impacts; secondly, a company's ecological and social impacts, production site, etc; and thirdly, and perhaps the most important measurement of the interactions and links between the social, environmental and economic issues which constitute the three dimensions of sustainability

Conclusion

In the light of the increasing and currently still underestimated relevance of sustainability reporting for the reputation and social acceptance of a company, it can be expected that an increasing number of companies will be addressing this topic. Sustainability reporting is more than the publication of a print report, and should be embedded in a comprehensive sustainability communications approach and in the company's general communication concept if it is to become more effective. Guidelines and standards as well as auditing and verification processes may provide assistance for management, but will not be sufficient so long as the link between corporate strategy, information management and the reporting activities are not designed in a systematic manner. In order to create corporate credibility, the sustainability reporting activities themselves have to be credible. This requires that the underlying corporate activities are not just for show but are systematically designed for effect and for the improvement of corporate sustainability.

References

- Austrian Institute for Sustainable Development (OIN) (2003) Guidelines: Reporting about Sustainability. 7 Steps to a Successful Sustainability Report Vienna. OIN
- Bebbington J and Thomson I (1996) Business Conceptions of Sustainability and the Implications for Accountancy (Research Report No. 48) London, The Chartered Association of Certified Accountants, Certified Accountants Educational Trust
- Clausen J, Loew T and Westermann U (2005) Sustainability Reporting in Germany Summary of the Results and Trends of the 2005 Ranking, Berlin
- CSR Europe (2000) Communicating Corporate Social Responsibility, Transparency, Reporting, Accountability Brussels.
- FORGE Group (2002) Guidance on Corporate Social Responsibility Management and Reporting for the Financial Services Sector, London.

- Gray R (1992) Accounting and Environmentalism. An Exploration of the Challenge of Gently Accounting for Accountability, Transparency and Sustainability Accounting, Organizations and Society Vol. 17, NO. 5, 399-425
- Koll A (2004) A decade of Sustainability Reporting: Developments and Significance, International Journal for Environmental and Sustainable Development Vol. 3, No. 1, 51-64
- KPMG (2005) KPMG International Survey of Corporate Sustainability Reporting 2005 De Meern, KPMG

32. Sustainability Accounting and Reporting:

An Overview of A Emerging Subset of Accounting and Reporting Practices

M.K. Agrawal
Pradeep Kumar Singhal

Companies are key contributors to economic, environmental and social wellbeing. Corporate activities pervade the present and are likely to be critical in the future, so that corporate sustainability is necessary for long-term sustainable development of the economy and society. In this context, sustainability accounting and reporting which serve the collection, analysis and communication of corporate sustainability information become crucial triggers for management towards corporate sustainability. If corporate sustainability is seen as being the result of management attempts to address sustainability challenges, then it makes sense to discuss and define sustainability accounting and reporting on the basis of the challenges embedded in the sustainability triangle and addressed by cornerstone publications.

Key words: Corporate sustainability, Sustainability Accounting, Sustainability reporting, Eco-efficiency, Socio-efficiency.

Corporate Sustainability

From a pragmatic point of view, corporate sustainability can be viewed as the result of management attempts to tackle challenges posed by the need for corporations to move towards the goal of sustainability. However, it remains unclear when a company can be considered to have reached the state of being sustainable. Sustainable development of a corporation requires the initiation and establishment of organisational development and organisational learning processes. If this view is taken to its extreme, corporate sustainability cannot reflect a given state to which management may strive, but will always have to be a moving target for organisational development. Nevertheless, for reasons of clarity it is helpful for a company which is striving towards corporate sustainability to distinguish between the target state of corporate sustainability and the process of sustainable development. The term corporate sustainable development is therefore used here to mean the processes which are implemented in order to reduce negative impacts and to increase the positive effects of corporations towards attaining a sustainable economy, environment and society, whilst corporate sustainability represents the desired outcome of such processes. In corporate practice, the focus is usually on the processes rather than on the end state, representing in essence an incremental process of continual development towards sustainability.

The distinction between corporate sustainability and corporate sustainable development is to some extent also reflected in environmental standard ISO 14031's distinction between operational performance indicators (OPIs, which map performance and outputs) and management performance indicators (MPIs, which map the route that management is taking to improve its future OPIs).

Given the broad and ambitious goal of sustainable development in general, corporate sustainability is a challenging concept which is in need of operationalisation. In this context, information about sustainability impacts and sustainability performance can help managers to incorporate deliberative, sustainable thinking into, their decision-making, planning implementation and control activities. This is the sharp end of the debate about corporate sustainability. As a consequence, sustainability accounting and reporting which serve the collection, analysis and communication of corporate sustainability information-become crucial tools for management in moving towards corporate sustainability.

Sustainability Accounting and Reporting

With the growing communication efforts being made by companies which place importance on sustainability, it is not surprising that sustainability accounting and reporting have achieved respectable and for many, astonishingly fast - management relevance. Furthermore, this development is characterized by a broad variety of different perspectives to address relevant company sustainability issues; it will be disturbing for deep green and very ambitious actors that new approaches towards measuring, analyzing and communication sustainability issues are mainly being developed on the basis of the history and growing body of literature on EMA and reporting. However, this development can also be interpreted as an evolutionary process founded in the environmental origins of sustainability accounting and reporting.

Under this view, the term sustainability accounting is used to describe new information management and accounting methods that aim to create and provide high quality information to support a corporation in its movement towards sustainability. *Sustainability reporting,* by contrast, describes new formalized means of communication which provide information about corporate sustainability.

The *linkage of sustainability accounting and reporting* is crucial for two reasons. Firstly, accounting information which is not communicated cannot exert any influence and is thus unable to contribute towards the company's sustainable development. Secondly, reporting is needed in order to substantiate information about the actual status of and progress towards, corporate sustainability; otherwise the information tends to be considered to be rather superficial.

Historical Development of Sustainability Accounting and Reporting

The concept of *Sustainability accounting* has emerged over a period of years from both philosophical accounting discussions and developments in accounting. First, it needs to be recognised that accounting has long been presented in a conventional way for use by both management and external parties. *Financial reporting* is based on accounting information which is gathered within organisations and then prepared for presentation to external parties through disclosure in external reports.

A second type of accounting, *cost accounting,* was initially closely related to financial accounting in that it provided information about inventory values for inclusion in the annual financial reports (Wells 1978). Cost accounting was then adapted from its initial financial accounting purpose in order to assist with management control, to emphasise performance reporting based on financial representations of the expected and actual performance of both organisations, and of parts of organisations such as divisions or departments, and their comparison to provide the basis for management action based on the differences reported.

Since this early adaptation of financial accounting for management control, *management accounting* has developed separately to focus on generating information for management planning, controls and decision - making

Corporate (environmental and sustainability) performance measurement and management which as an integrative approach tries to link strategic management, management accounting, and reporting, in order to organize the flow of information between its justification, creation and communication. In this view, the term 'reporting' is not limited only to external reporting as it is in financial reporting but rather encompasses the whole information communication process, internally as well as with external stakeholders.

The term *sustainability reporting* is usually used to refer to the publication of external reports, as either printed brochures or electronic versions on the internet. However, one main effect of sustainability reporting is the involvement of management and employees in setting sustainability goals for the corporation, collecting data, and creating and communicating sustainability information. The design of external sustainability reporting should therefore consider its interplay with internal communication and reporting processes.

Changes to conventional accounting have taken the form of *environmental accounting and reporting* as the foundation for external environmental reporting, with a major emphasis on environmental impacts and extended performance expressed in physical and qualitative terms and *triple bottom line reporting* which introduces separate economic, social and environmental statements for organisations (Gray and Milne 2002). Environmental Management Accounting (EMA) and environmental reporting constitute in any case an important part of sustainability accounting and reporting.

The Zenith of accounting and reporting at present in *sustainability accounting and reporting* with its conceptual emphasis on accounting for ecosystems and for communities, and consideration of eco-justice, as will as more conventional issues of effectiveness and efficiency (Gray and Milne).

Basic Structure of Sustainability Accounting and Reporting

If corporate sustainability is viewed as the result of management's attempts to address sustainability challenges, it makes sense to discuss sustainability accounting and reporting on the basis of the challenges embedded in the sustainability triangle (see Figure-1). The vision of corporate sustainability today is a broad approach relating to the contextual integration of economic, environmental and social characteristics. It comes as a surprise to realise that the best known aspect of accounting for corporate sustainability is the heuristic, multi-criteria *triple bottom line perspective* which aims to integrate the economic, social and environmental aspects of business management. This differs from the preceding political and macro perspective in which the orientation towards future and present needs, as formulated in the Brundtland report, has dominated for much longer (UNWCED 1987).

Sustainability Triangle

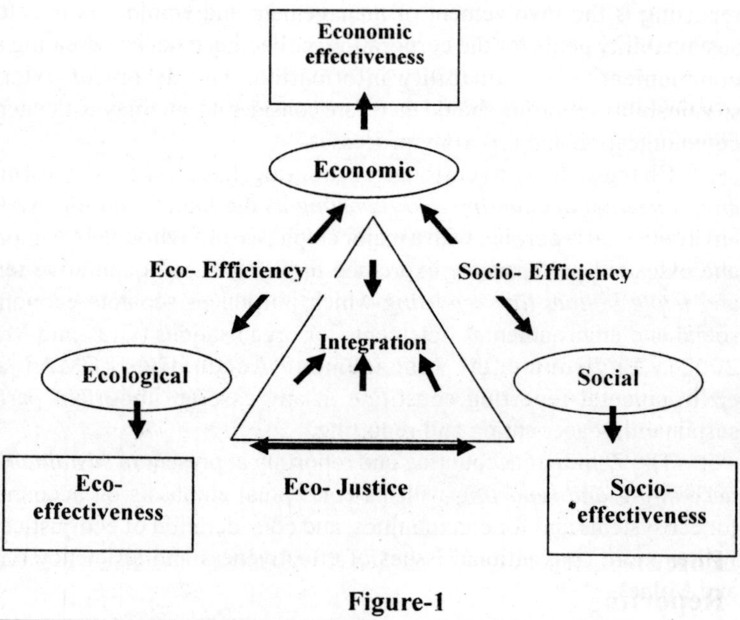

Figure-1

Source: *Schaltegger et al. 2003b, Schaltegger and Burritt 2005, Schaltegger et al. 2003a).*

Challenges Deriving from the Sustainability Triangle

The Sustainability triangle visualises the three perspectives of sustainability not just by plotting ecological, social and economic goals in a triangle but by also addressing the interrelationships between these three dimensions. The challenges to corporate sustainability relate to the economic, ecological and social consideration in the triangle and their interrelationships.

The *ecological challenge* is to increase the ecological effectiveness, or effectiveness of business activities. Eco-effectiveness describes how well environmental impacts have been reduced. *Ecological effectiveness* measures the absolute environmental performance (e.g. tonnes of CO_2 emissions reduced in the last period) and is a general description of the extent to which the targeted objective of minimizing environmental impacts has actually been achieved.

The *social challenge* of corporate sustainability is to increase the company's social effectiveness, or socio- effectiveness. The social challenge related to corporate sustainability is to ensure the existence and success of the enterprise whilst at the same time taking account of the diversity of social, cultural and individual social demands.

The economic challenge aims to improve eco-efficiency and socio-efficiency. The economic sustainability challenge is concerned with undertaking effective environmental management and social management as economically as possible.

Eco-efficiency is defined as the ratio of an economic (monetary) measure to a physical (ecological) measure. It can be defined as the ratio of value added to environmental impact added per *unit; socio-efficiency* (also known as 'societal efficiency') can be defined as the ration of value added to social impact added.

The *integration challenges* are the contextual integration challenge which is about bringing together the first three challenges, and the *methodological integration challenge,* which focuses on integrating environmental and social management into conventional economically-orientated business management.

Potential Relevance of Sustainability Accounting and Reporting

The table-1 gives an overview of the potential relevance of sustainability accounting and reporting

Table-1

Business environment	Societal expectation	Relevance of sustainability accounting	relevance of sustainability reporting
trust me	None	Internal efficiency improvements	Internal communicating to achieve efficiency improvements
Tell me	Communicate	Information creation for highly visible and formally required issues	Sustainability as an important internal and external communication element
Show me	Communicate and illustrate	Information creation for an over-arching picture of sustainability performance	Essential communication element as part of a set of "voluntary" communication activates
Prove to me	Measure, account for communicate and illustrate	Basis of sustainability performance management Basis to create transparency Basis for verification	Additional element in a systematic set of trust building activities (such as stakeholder dialogues and involvement)

In a world in which society trusts business managers without having any specific sustainability expectations, management will focus on environmental and social information which has been identified as being of internal organisational and direct economic relevance. Not Only does the role of sustainability reporting depend on societal expectations, but stakeholder reactions also exert a substantial influence on what management considers is sufficiently important to be accounted for. The importance of social and environmental measurement, sustainability accounting, the quantity of information required, and the quality requirements of information created, all increase with changes in societal expectations.

It should be mentioned that corporations do not have merely a passive role in identifying their societal environment and adapting to it through their accounting, reporting and management systems. Company managers can also influence their business and societal environment and contribute to a change in the way their management is approached, it is possible for trust in the business world to result from creating transparency, involving and communicating with stakeholders in a trustworthy manner, and accounting for and revealing sustainability performance improvements on the basis of best practice measures. By voluntarily taking the actions associated with a "prove to me" world, without having been forced into this, management can contribute towards achieving the needed attitudinal and behavioural business climate Management can build up relationships such that it can operate its business in an environment of stakeholder trust. Central to this will be an understanding of the dynamics of institutionalising high-trust relations, in particular the understanding of embedding accounting and reporting in the intrinsic satisfactions that stakeholders gain from their social involvement with organisations (Fox 1975:72).

References

- Dyllick T and Hockerts K (2002), Beyond the Business Case for Corporate Sustainability Business Strategy and the Environment Vol. 11 No. 4, 130-141
- Linking Sustainability Management to Business Strategy Business Strategy and the Environment, Vol. 11, No. 5, 269-284
- Forum for the Future (2005) What does Sustainability Accounting Look like in practice? [online] Available from internal URL <http://www.forumforthefuture.org.uk/Sustainability Accounting_ page 1352.aspx>
- Gray R and Bebbington J (2000), Environmental Accounting, Managerialism and Sustainability: Is the Planet Safe in the Hands of Business and Accounting? in Freeman M and Jaggi.
- Epslein M (1996), Measuring Corporate Environmental Performance, Best Practices for Costing and Managing an Effective Environmental Strategy Chicago, Irwin
- Gray R and Milne M (2002), Sustainability Reporting Who's Kidding Whom? Chartered Accountants Journal Of New Zealand Vol. 81, No. 6, 66-70
- Maunders K and Burritt RL (1991), Accounting and Ecological Crisis Accounting, Auditing & Accountability Journal Vo. 4, No. 3, 9-26
- Milne MJ (1996), On Sustainability, the Environment and Management Accounting, Management Accounting Research, Vol. 7, No. 1, 135-161.

33. Tourist Information System (TIS) For the Sustainable Development and Planning of Pune District

Virendra Nagarale
Dattatraya Harpale

Geographical information is necessary for the sustainable planning and management of the resources in the area. The area selected for proposed research is Pune District. Such study would be helpful to Government for planning and making strategic for District. Moreover, such study may provide potentiality to increase at glance current information of whole district. Besides this planners, administrators, layman and those who are interested in research in the field of economic and social planning, strategic study and social and economic development may get the benefit. In this present study try to provide an integrated plan for tourist places in the Pune district with some special case studies.

With the help of such type of T.I.S. (Tourist Information System) one can conclude or identify accounts of the needs and potentials of rural areas and economically disadvantage. To give attention towards man power development and training through such study can be done by this project. It brings socio-cultural benefits to the local community in terms of employment opportunities, income generators and environmental awareness. Preparation of new tourist maps, updating the database and the publicity of tourist places are some of the latest advancement in the study of the tourism development have been happening because of GIS.

Keywords: Tourist Information System (TIS), Digital Elevation Model (DEM), Short Path Trace Analysis, Geographical Information system (GIS).

Introduction

Tourism is considered as one of the world's largest industry. India is a developing nation. The Government and their agencies as well as private sector units and individuals are taking various measures to promote tourism. Promotion of tourism can contribute immensely to our economy. Many years tourism was neglected at various levels but now a day's concentrated effort are being made to improve the position and standard of tourism and for also the social benefit of the people. Pune district is highly rich by its scenery, spiritual base and its culture. Tourism industry will change the future of the district and will improve the social, cultural, economic status of the district; this was the main motive behind selection of this study area.

From the last two decades of 20th century, computer revolution has boosted the tourism development through media advertising publicity. In the 21st century GIS has added in this list. In the present study GIS will be used as a tool for the sustainable development and planning of Pune district for known as well as unknown tourist places. Governments of underdeveloped and developing nations have interest in the development of tourism as a solution to grow their economics statue to regional level.

Study Area

The study area, Pune district, usually termed as the cultural capital of Maharashtra. It is situated at 559 m ASL and lies between 17° 54' to 19° 24' N Latitude and 73° 33'to 75° 10'E Longitude. The total geographical area of the district is 15643sq.km. The district headquarters or the district place is Pune. The district consists of 14 tahasils. The area of district is surrounded by Thane district in the north and northwest, Raigarh district in the west, Satara district in south, Solapur district in the south and southeast and Ahmadnagar district to the east. (Fig. 1)

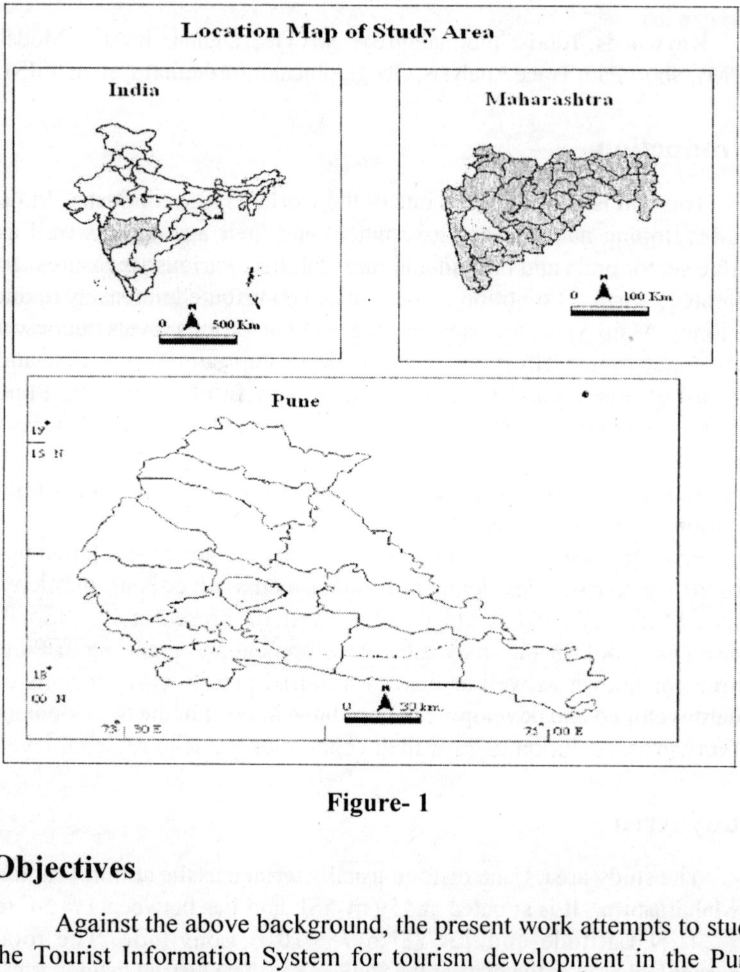

Figure- 1

Objectives

Against the above background, the present work attempts to study the Tourist Information System for tourism development in the Pune district. The broad objectives of the proposed study are:

1. To collect the information about tourism facilities available in the study region.
2. To find out potential tourist sites and their suitability in the study area with the help of GIS technique.
3. To prepare a new tourist map of study area with adequate and up to date Tourist Information System (TIS) with the help of GIS technique.

Methodology and Database

In order to understand for the sustainable development and planning of Pune district and Tourist Information System in Pune district, the methodology adopted for the present study is divided into three phases are namely pre-field work phase, field work phase and post field work phase.

In the first phase i.e. pre-field work phase literature review i.e. previous work carried out by other researchers are obtained from various journals, internet, visit to the MTDC resort to know about annual tourist

In the first phase i.e. pre-field work phase literature review i.e. previous work carried out by other researchers are obtained from various journals, internet, visit to the MTDC resort to know about annual tourist flow, collection of survey of India toposheet (SOI) having scale 1:250000 (47/E,47/F,47/I,47/K,47/M,47/N,47/O), Atlas, Gazetteers, District Census Handbook, Tourist maps, etc. use for collection of information, District Resource map of Pune district published by Geological Society of India, Government published map of Pune district P.W.D. map, Digital Elevation Model (DEM) and other were completed with help of S.O.I. toposheets.

In the second phase i.e. fieldwork phase extensive field surveys were undertaken, to existed tourist places and newly found tourist places. The questionnaire filled in this phase. During this field surveys tourist facilities regarding destination photographs, GPS reading altitude and the related information is noted which were also useful to site suitability study (Fig. 2).

In the third phase i.e. Laboratory worked is carried out. Government published map of (PWD) Pune district is scanning, then digitization to generate thematic layer i.e. point layer- tourist places, tahasil headquarter, line layer- roads, railways, rivers and polygon layer- dams, reservoirs etc. and map were georeference same time. Integrating all above information and maps with help of GIS software, Auto CAD Map- 2000i, ILWIS- 3.2 a, Surfer 7, various thematic layers were generated. Finally these thematic layers are analyzed to demarcate tourism potential zones and site suitability.

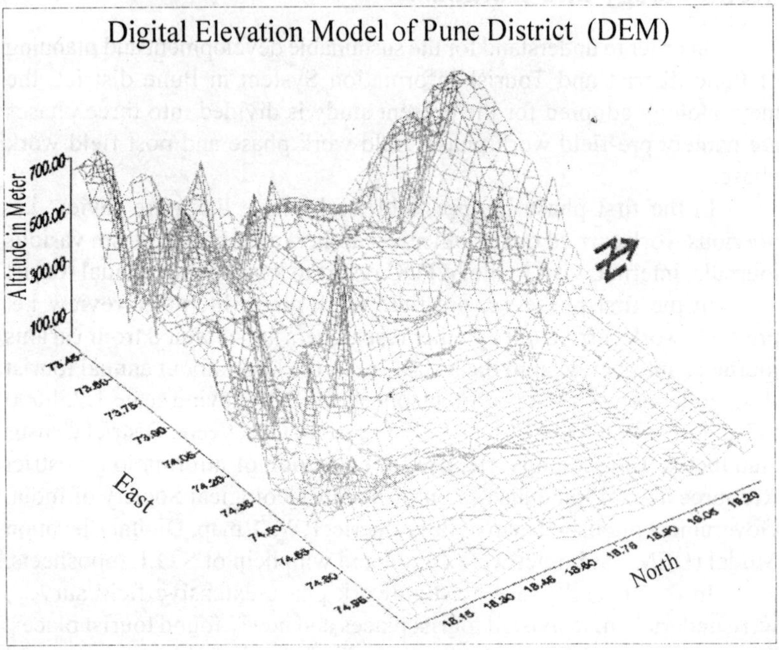

Figure- 2

Potential Tourist Centers in Pune District

The term potential means something existing but not yet fully exploited. There are various criteria to selecting new potential suitable sites i.e. on the basis of phsiography, climate, and purpose of visit, on the basis of natural and cultural resources. The district has vast and rich tourism resource potential of different cultural background in all its 14 tahsils. We classified potential suitable sites on the basis of natural and cultural resources, i.e. Pune, Haveli, Junnar and Baramati. Potential tourist centers are selected in this manner they represent the entire Pune and same problem found in similar type of tourist centers in throughout the district i.e. Malshj, Pabal, Ramdara, Keled, Bhuleshwr, Ketkawale (Fig. 3). The study shows that the tourist centers known to outsider's tourists are limited to Lonavala, Khandala, Bhimashankar and Pune city only. This prevents full exploitation of the tourist potential of the other tourist centers.

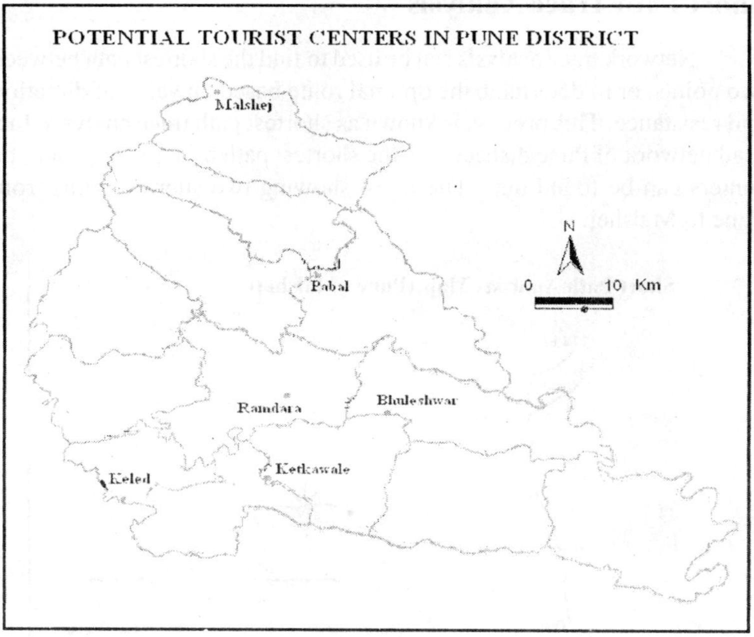

Figure - 3

Potential suitable sites are very rich in natural and cultural resources, these tourism resources are not fully utilized because inadequate tourist facility, lack of information regarding this centers. New potential suitable sites can provide more scope for a generation of employment opportunities due to increases tourist facilities in terms of different socio economic and marketing indicators also bring money and reducing the regional disparities in Pune district. New potential suitable sites will be reducing the pressure on existed tourist centers and also reducing the migration of local peoples towards the Pune city.

GIS Analysis of Tourist Centers in Pune District

Analysis is an integral part of any GIS system. It can solve many diverse situations and problems. Analysis is the process of analyzing the system and receiving the desired information. This can help in various ways.

Short Path Trace Analysis

Network trace analysis can be used to find the shortest path between two points, or to determine the optimal route based on value of direction and resistance. This process is known as shortest path trace analysis. In a road network of Pune district, e.g. the shortest path between any tourist's centers can be found out. The fig. 4 showing two shortest paths from Pune to Malshej.

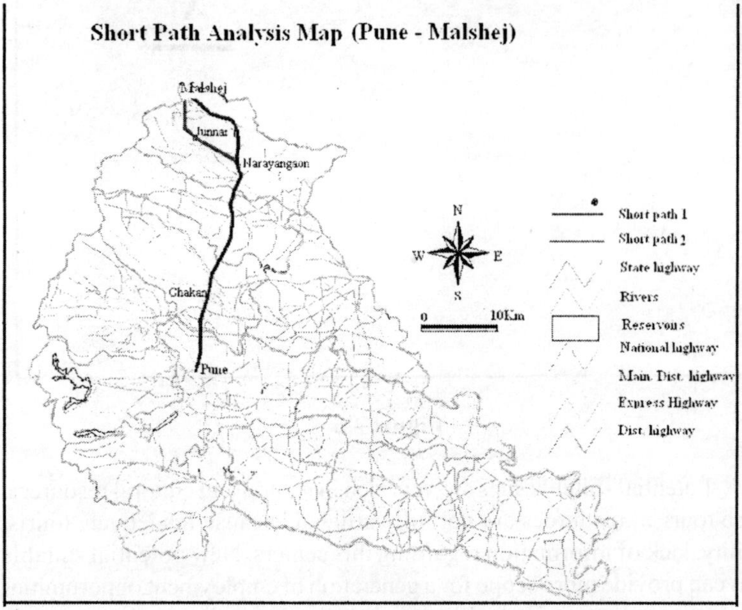

Figure- 4

Tourist Information System

Using this tourist information system one can may be layman or tourist get correct information about particular location. Explain it with e.g. **Karla.**

Karla: An ancient complex of Buddhist caves situated on Sahyadri range during back to 2nd B. C. and 7th A.D. The chaityagriha located in this caves is supposed to be the largest chaityagriha in India (Fig. 5). The chaityhall measure 38 m in length, 14 m in breadth and 15 m in height. The wooden arches used in the construction of this chaityhall have remained intact without suffering any ill effect of weather. 15 pillars on each side separate the nave from the aisles, each pillar has a tall base, an octagonal shaft, and richly ornamental capital on which kneel to elephants, each bearing two figures generally a man and women. Just outside the main cave is the Sri Ekviradevi temple. The people are Koli (spider) community mostly visited Karla caves as well as the temple of Sri Ekviradevi.

Tourist Information System (TIS) Map of Karla

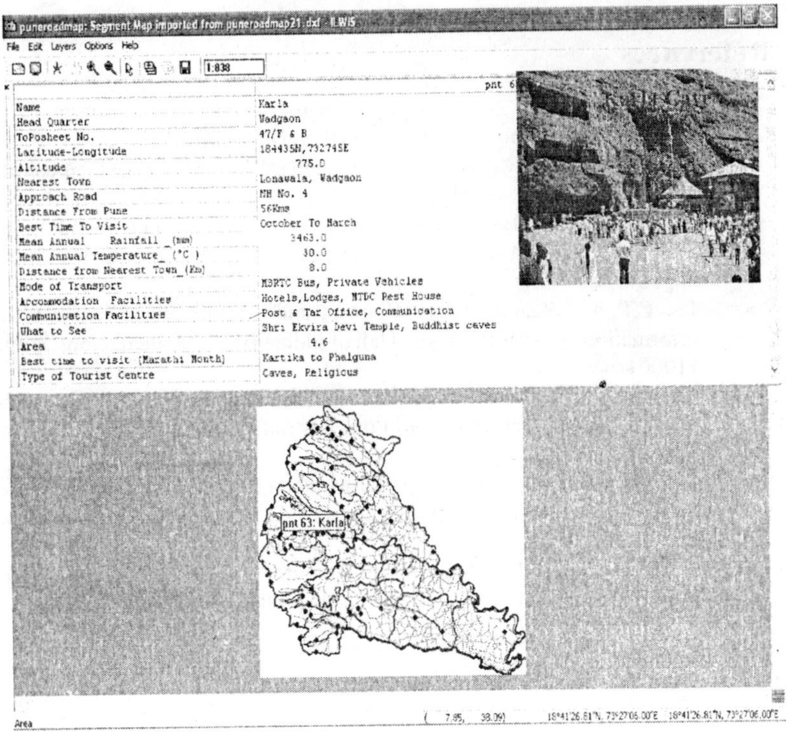

Figure- 5

Conclusion

Present study tries to provide an integrated plan for tourist places in the Pune district with some special case studies. Various tourist zones according to tourist category can be demarcate or generated with the help of GIS. If someone wants to know the religious places along the Mumbai-Pune Express way map can be displayed. Also encouraged and facilitated youth geographers to travel and foster national integration. As far as potential places are concerned, after field work it is observed that few people only visited these tourist places who know about the places, but other have no any information about the places which should be published or advertised or marketing and the information should be reached to the tourists and they can visit the places. Also take an account of domestic needs of the Pune district by giving information such as cheap accommodation (particularly pilgrim tourist plaçes), easy way to transportation

References

- Akmanchi Anand, (2002). Pune fire emergency management information system, Project report submitted to IIRS, Dehradum.
- General Geography (Pune District) Standard Three (1993) Maharashtra State Bureau of Textbook Production and Curriculum Research, Pune.
- Harpale D.V. (2009) "Identification of New Tourist Centers and Their Site Suitability A Case Study of Pune District Maharastra State unpublished thesis
- Lo, C.P. and Yeung (2004). Concept and Techniques of Geographic Information System, Prentice –Hall of India Private Limited, New Delhi, 110001
- Nagarale, V.R. (2007). 'Site Suitability for Tourism Development with the help of GIS, A case study of Pune District Maharashtra.' UGC Minor Research Project.
- Pani Padmini, (2004). Geospatial Modeling for Identifying Aesthetic Spot for Tourism in a hilly Region, Project report submitted to IIRS, Dehradum.
- Reddy Anji (2001). 'Textbook of Remote Sensing and Geographical Information System (Second edition) B S Publications, Hyderabad.

34. Population in the context of Environment and Sustainable Development

Arun Kumar Singh

Discussion and debate on the interrelationships among population, environment and development has been initiated way back in 2nd half of eighteenth century with the writings of Thomas Robert Malthus and his contemporary philosophers and thinkers. But they were concerned mainly with the balance between population and natural resources conceptualized as means of subsistence and livelihood (food and water). Discourse on environmental issues and its relationship with population in the context of development could take place with the founding of the UNO (1945) and became explicit at the first session of the Population Commission in 1947. The Rio Conference (1992) was a milestone in the evolution of an international consensus on the relationships among population, environment and development, based on the concept of sustainable development. Cairo Conference (1994) is a landmark as it talked about equity, equality and empowerment along with right-based approach to development. Accordingly our population policy has also changed in approach. Present paper based on secondary data sources (reports, research papers and published materials) tries to examine how population issues form the base for environment and sustainable development.

Key words: Population growth, Sustainable development, Poverty, Education, Population Policy.

Introduction

Last century has witnessed an unprecedented population growth, economic development and environmental change. In the last hundred years (1900-2000) world population grew from 1.6 billion to 6.1 billion persons (United Nations, 2001) and India's population increased from 236 million to 1028 million. In both the cases there was around four times increase in population. During the same period world real gross domestic product (GDP) increased 20 to 40 times (DeLong, 1998 quoted from UN Publication 2001)), allowing the world not only to sustain a fourfold population increase but also maintain the quality of life. But population growth and economic growth occurred unevenly throughout the world and not all regions have benefitted equally because of differential level of knowledge base, skill and technological development.

Table-1 Demographic Features and Economic Growth Rate

S.No.	Country	Per capita GDP Growth Rate (1990-2007)	Average annual Population Growth Rate (1990-2007)	TFR
1	China	8.9%	0.9%	1.7
2	India	4.5	1.8	2.8
3	United States	2.0	1.0	2.1
4	Indonesia	2.3	1.4	2.2
5	Brazil	1.2	1.5	2.3
6	Pakistan	1.6	2.2	3.5
7	Bangladesh	3.1	2.0	2.9
8	Russian Fed.	1.2	-0.2	1.3
9	Nigeria	1.2	2.6	5.4
10	Japan	0.9	0.3	1.3

(**Source:** after Shiva Kumar et al-2010)

Population and Environmental Issues in the context of Development

In the second half of 20th century it was realized that global population growth had been reaching unprecedented high levels, a situation which could be termed as grave. UN Conference on the Human Environment popularly known as **Stockholm Conference (1972)** made some specific recommendations which constituted the basis for activities of the United Nations on environmental issues during the 1970s and

1980s.The Declaration affirmed that **"the natural growth of population continuously presents problems for the preservation of the environment, and adequate policies and measures should be adopted to face these problems."** The Declaration called for policies that were without prejudice to basic human rights and were deemed appropriate by the Governments. **Bucharest Conference (1974)** could be termed as the first UN conference on world population and tried to examine the interrelationships of population, natural resources and environment. The debate reflected profound divergences in the perception of population environment interactions, and contained most of the core elements of the future debate on the issue. The preparatory process for the Mexico City Conference included an Expert Group Meeting in Geneva (1983) on Population, Resources, Environment and Development. Core of the deliberations were poverty, the food versus feed controversy, food self-sufficiency and particularly the role of population growth. The discussion on resources and the environment covered the resource base, environmental degradation and non-renewable resources. Attention was centred on the various mechanisms that could expand resource availability as well as those activities that had caused a degradation of the environment. The **United Nations Conference on Environment and Development, held at Rio (1992)** was a milestone in the evolution of an international consensus on the relationships between population and environment, based on the concept of **"sustainable development"** derived from Bruntland Report and entitled as *Our Common Future* (1987).The Report defines sustainable development as development that **"meets the needs of the present without compromising the ability of future generations to meet their own needs".** It further adds that poverty, environmental degradation and population growth were inextricably related and none of those problems could be successfully addressed in isolation. Hereafter sustainable development was taken as a constructive alternative to the dilemma of economic growth versus environmental protection. It also identified population policies as an integral element of sustainable development. Forging a balance among population, sustained economic growth and sustainable development was the central theme of the **International Conference on Population and Development held at Cairo in 1994**. The Expert Group Meeting found that although rapid population growth, persistence of poverty and environmental degradation were inextricably interrelated, the magnitude of their reciprocal impacts in different social and ecological settings had not been sufficiently documented. Thus, in order to promote sustainable development, the Expert Group Meeting emphasized the urgent need to

strengthen data collection and research efforts in that domain and to test the efficiency of proposed policies and strategies in concrete settings. The Programme of Action recommended several actions to help achieve integration of population, development and environment. First, it called for development strategies that realistically reflected the short-, medium- and long-term implications of, and consequences for, population dynamics and patterns of production and consumption. Second, it emphasized measures aimed at the eradication of poverty, with special attention to the areas within or on the edge of fragile ecosystems. Third, it stressed policies to address the ecological implications of inevitable future increases in population numbers as well as changes in population concentration and distribution, particularly in ecologically vulnerable areas and urban agglomerations.

Issues involved in Environment and Sustainable Development

Sustainable development aims at improving human well-being, particularly by alleviating poverty, increasing gender equality, and improving health, human resources, and stewardship of the natural environment. If we do not put the human population at the core of the sustainable development agenda, our efforts to improve human well-being and preserve the quality of the environment will fail. The then Indian Prime Minister Mrs. Indira Gandhi emphasized in Stockholm Conference (1972), that the removal of poverty is an integral part and essential ingredient of environmental strategy. Even mandate of 1971 general election was the result of the slogan (Garibi Hatao given by Mrs.Gandhi). Poverty and degraded environment are closely inter-related, especially in an area where people depend for their livelihoods primarily on the indigenous natural resource base of their immediate environment. Restoring natural systems and improving natural resource management practices at the grassroots level are central to a strategy to eliminate poverty and strategy of sustainable development. **Removal of poverty is therefore a prerequisite for the protection of the environment.** While conventional economic development leads to the elimination of several traditional occupations, the process of sustainable development, guided by the need to protect and conserve the environment, leads to the creation of new jobs and of opportunities for the reorientation of traditional skills to new occupations. Women, while continuing to perform their traditional domestic roles' are increasingly involved in earning livelihoods. In many poor households they are often the principal or the sole breadwinners. A

major thrust at the policy level is necessary to ensure equity and justice for them. Literacy and a basic education are essential for enabling the poor to access the benefits offered by development initiatives and market opportunities. **Basic education is therefore a precondition for sustainable development.**

The integration of agriculture with land and water management, and with ecosystem conservation is essential for both environmental sustainability and agricultural production. An environmental perspective must guide the evaluation of all development projects, recognizing the role of natural resources in local livelihoods. This recognition must be informed by a comprehensive understanding of the perceptions and opinions of local people about their stakes in the resource base.

To ensure the sustainability of the natural resource base, the recognition of stakeholders and their roles in its protection and management is essential. Water governance arrangements should protect ecosystems and preserve or restore the ecological integrity of all natural water bodies and their catchments. This will maintain the wide range of ecological services that healthy ecosystems provide and the livelihoods that depend upon them. Biomass is a major source of fuel and energy in rural area so appropriate mechanisms needs to be evolved to make the consumption of biomass sustainable, through both resource management and the promotion of efficient and minimally polluting technologies. The traditional approaches to natural resource management such as sacred groves and ponds, water harvesting and management systems, etc., should be revived by creating institutional mechanisms which recapture the ecological wisdom and the spirit of community management inherent in those systems.

The strong relationship between health and the state of the environment in developing countries is becoming increasingly evident. This calls for greater emphasis on preventive and social medicine, and on research in both occupational health and epidemiology. Most of the developing countries are repositories of a rich tradition of natural resource-based health care. This is under threat, on the one hand from modern mainstream medicine, and on the other from the degradation of the natural resource base. **Traditional medicine in combination with modern medicine must be promoted while ensuring conservation of the resource base and effective protection of IPRs of traditional knowledge.** Developing countries should also strive to strengthen the capacity of their health care systems to deliver basic health services and to reduce environment-related health risks by sharing of health awareness and medical expertise globally.

Schumacher(1974)considers **education as the greatest resource as it helps in transmitting the ideas of values and ethos. Basic education which promotes functional literacy, livelihood skills, understanding of the immediate environment and values of responsible citizenship is another precondition for sustainable development.** Such education must be available to every child as a fundamental right, without discrimination on the basis of economic class, geographical location or cultural identity. An understanding must be promoted among key decision makers of the potential of education to promote sustainability, reduce poverty, train people for sustainable livelihoods and catalyze necessary public support for sustainable development initiatives. The empowerment of women and girls must be supported by actions to improve their access to basic and higher education, training and capacity building. Collaborations and partnerships between and among scientists, government and all stakeholders, on scientific research and development and its widespread application need to be improved.

India's Population Policy: Quest for Sustainable Development

India has the distinction of being the first developing country to adopt a well defined population policy and to launch the official programme of addressing population related issues facing the countries way back in 1952 but the focus was on population control through reducing birth rate. Population policy was a part of social and economic development policy till a national policy was launched in April 1976. The strategy adopted were **'Coercion'** and the number of sterilizations were more than the total number done in the previous five years and more than the number done in any other country in the world until that time. If the effectiveness is to be measured in terms of number then it could be held as a success (Gwatkin, 1979) but from a rights based perspective it was a failure culminating into debacle of the regime and it is believed that the mandate was the first in the world in which fertility behavior was a deciding factor (Cassen, 1978).

Last decade of 20th century witnessed changes in the family welfare policy and programme. The components of sustainable development like health care, primary education and provision of certain basic amenities such as drinking water and roads became the responsibility of the grass root level functionary (panchayats) and there was provision of reservation of one third of the seats in panchayats for women and weaker section of the society under 73rd CAA. Cairo Conference (1994) is a landmark event where it was explicitly agreed that the advancement and protection of women's human rights should be central to government efforts to address population and development issues. ***Thus at the grass root level equity, equality and empowerment became reality.*** Population policy of any country can no longer confine itself to the reduction of birth rate alone. It must be a comprehensive strategy for overall improvement of social, economic and demographic fabric of a nation. The official realization, that population is not merely about numbers but about the health and quality of life of people in general and women in particular. So there is a need for a better and more widespread understanding that the number of children desired by any couple depends on a large and complex interrelated number of socio-economic and cultural factors, and that any policy action seeking to control population must seriously take all these variables into account. Mrs.Gandhi(1980) rightly remarked " ***if there is any chance of children dying more, then people will not be willing to risk that. So, they must be ensured better health, education, and of course, even employment later on for sustainability of the society. It is part of national planning as a whole.(Chhabra 1981)"*** An important part of empowering women in matters pertaining to population is to explicitly recognize and respect their rights over their bodies and their reproductive behaviour. This recognition must permeate society in general, and religious, judicial and law-enforcement institutions in particular, through continual campaigning and dialogue. Thus the strategy for decreasing the rate of population growth, limiting our number in the young and reproductive age groups, enhancing our employment potential, diversifying economic base, improving level of literacy and education and balancing sex ratio should constitute the elements of any population policy. Unfortunately since independence we have been emphasizing the role of positive measures (contraception, sterlisation and abortion) at the name of population policy to keep the number under control without paying much attention to the dynamic interaction between the demographic structure and social and economic development and implication of such interaction for limiting the rate of population growth. In this context the new National Population Policy 2000 (NPP 2K) is a comprehensive document addressing the population problems by taking into account the related issues like child survival, maternal health, women empowerment and employment and contraception. The policy has well spelt immediate goal (to provide integrated service delivery for basic reproductive health and child care to all), midterm goal (to bring total fertility rate to the replacement level) and long term goal (to achieve population stabilization by 2045) exactly on the line of Cairo recommendation. These goals intended to be achieved by 2010 include the following:

- meet the needs for basic reproductive and child health services, supplies, and infrastructure;
- make schooling up to age 14 free and compulsory and reduce dropout rates for primary and secondary school to below 20% for both boys and girls;
- reduce the infant mortality rate to below 30 infant deaths per 1,000 live births;
- reduce the maternal mortality ratio to below 100 maternal deaths per 100,000 live births;
- promote delayed marriage for girls until at least age 18 and preferably after age 20;
- increase the proportion of deliveries that take place at medical institutions to 80% and the proportion of births attended by medical personnel to 100%;
- achieve universal access to family planning information, counseling, and services and offer a wide range of contraceptive choices;
- achieve universal registration of births, deaths, marriages, and pregnancies;
- contain the spread of HIV and better integrate the management of reproductive tract infections and sexually transmissible infections;
- integrate traditional Indian medicine into outreach and the provision of reproductive and child health services;
- create a people-centered approach to all social programs relating to family welfare; and
- Converge social sector programmes with family welfare programmes so as to make it a people's movement.

Initiatives for strengthening governance to achieve the goal at local level

Effective management of available resources (physical or cultural) in an indigenous environment at a sustainable basis requires participation by all stakeholders. At the local level, strengthening democratic institutions generally leads to better and more sustained management of natural resources. To enhance effectiveness of people's participation in local governance, committees comprising both elected and executive members of local bodies and representatives of community groups, must be formed. Appropriate capacity building would enable them to undertake local development activities according to community priorities, monitor project implementation and manage community assets. As mentioned earlier such conditions have already been created in India through the 73rd and 74th amendments of its Constitution and its effective implementation of the provisions needs to be ensured across geography. To achieve the objectives as listed out a multi pronged strategy is suggested. The operational strategies and strategic themes as outlined in the NPP2K emphasise decentralized programme implementation based on an assessment of community needs and local contexts. The operational strategies outlined in the policy envision a decentralized, integrated approach to health service delivery implemented by local institutions like the Panchayati Raj Institutions (PRI) and village self help groups. The policy also lays emphasis on information, education and communication services i.e using public media and other tools to increase knowledge and awareness of the need for family planning and the options available to clients For the first time in the history of India's population policy, the national rural health mission (NRHM) has incorporated a framework after wider consultation of with diverse stakeholders including civil society groups. In addition a community monitoring group has been formed to monitor activities at village, block and district levels.

Conclusion

Discussion initiated through this paper clearly indicates that the issues like population, environment and sustainable development cannot be seen in isolation and it is the population who has to play an instrumental role in maintaining the balance between the environmental issue and developmental initiatives. This is why population issues need greater attention of the planners and policy makers if society intends to achieve the objective of sustainable development. Such themes have rightly been addressed in the NPP2K.

References

- Agarwala,S.N.(1985):India's Population Problems, Tata Mcgraw-Hill Publishing Company Limited, New Delhi.
- Cassen, R.H. (1978). India: Population, Economy, Society. New York: Holmes and Meier Publishers, Inc.

- Chhabra,Rami (1981): "Population Policy in India:Two Comments" Population and Development Review Vol 7No 1(March 1981).
- Chandna,R.C.(2008):Geography of Population: Concepts, Determinants and Pattern, Kalyani Publishers,Ludhiana.
- Gwatkin, Davidson R. (1979). Political Will and Family Planning: The Implications of India's Emergency Experience. Population and Development Review, 5, (1), 29-59.
- Schumacher,E.F(1974):Small is Beautiful, Sphere Books Limited, London.
- Shiva Kumar et al (2010): Handbook of Population and Development, Oxford University Press, New Delhi.
- United Nations (2001): World Population Monitoring, U.N. Publications, New York.
- World Commission on Environment and Development (1987): Our Common Future. Oxford, United Kingdom, and New York: Oxford University Press.

35. Creation of Database and Mapping of Natural Resources for Planning Using Remote Sensing and GIS : A Study of Ropar Block, District-Rupnagar (Punjab)

Inder Jeet Singh
Simran Kaur

Remote Sensing (RS) data and Geographical Information System (GIS) play an important role in the field of natural resources. One of the greatest advantages of using RS data for natural resource monitoring is its ability to generate information in spatial and temporal domain, which is very crucial for successful prediction and validation. However, the use of RS technology involves large amount of spatial data management and requires an efficient system to handle such data. The GIS technology provides suitable alternatives for efficient management of large and complex databases. The present paper is an attempt to create a precise and adequate structured database of the available natural resources in the study area and map these using Remote sensing and GIS technologies to find out the gap areas for planning. Area undertaken for present study is Rupnagar block formerly known as Ropar is a part of the kandi belt of the Himalayas and the alluvial plain of the river Sutlej. It lies on the north eastern corner of Punjab State between 76°25' to 76°45' E longitude and 30°50' and 31°15' N latitude. It is bounded by Himachal Pradesh in the North-East, Union Territory Chandigarh and Haryana in the South *and Gharshankar and Ludhiana in the West. The study is based on databases consists of integrated spatial/ non-spatial data elements comprising of the map inputs from remote sensing as well as conventional sources. After analyzing the data it is found that most of the area under foothill zone has coarse eroded sandy soils mixed with pebbles. The river Satluj forms the main drainage system in the area and roughly flows in northeast-west direction. The central part of the block experiences low ground water level of 5-10 meters below ground level because of excessive extraction of the underground water for the cultivation of high water demanding crops like rice and wheat. In the study area, considerable area (88.85 km^2) of the block is under forest which is 22.76% of the total geographical area of the block. But this figure is much less than the required area of 33% as per national forest policy for maintaining proper ecological balance.*

Key Words: Remote Sensing and GIS, Spatial, Non Spatial, drainage.

Introduction

If one looks at the problem of planning in a holistic 'spatial view', the top-down approach has to take into account the way economy of the nation is spatially organized in the initial stage and the design spatial organization of economy which would fulfill long term objectives of national planning.State in India is constitutionally committed to removal of regional as well as interpersonal disparities in development. For removal or minimization of inter-regional in-equalities, regional planning at different spatial scales is used as an instrument.

Remote Sensing (RS) data and Geographical Information System (GIS) play an important role in the field of natural resources. One of the greatest advantages of using RS data for natural resource monitoring is its ability to generate information in spatial and temporal domain, which is very crucial for successful prediction and validation. However, the use of RS technology involves large amount of spatial data management and requires an efficient system to handle such data. The GIS technology provides suitable alternatives for efficient management of large and complex databases.

The natural resources are continuously shrinking and deteriorating due to their excessive and unscientific utilization. Hence, there is an urgent need for the conservation and management for all kinds of natural resources. The development of modern data- gathering techniques and computer- assisted cartography give new impetus to spatial data storage and exchange. The natural resources development without endangering the environment is a crucial issue (Lillesand et. al., 2004). Such a task requires reliable and up to date information on the special distribution of natural resources, and their complex interrelationships. In addition the information about climatic conditions of the area and socio-economic status of people is also essential for integrated all round development.

Research objectives

The main objectives of the study are to create a precise and adequate structured database of the available natural resources in the study area and map these using Remote Sensing and GIS technologies to find out the gap areas for planning.

Methodology

Database of the paper consists of integrated spatial/ non-spatial data elements comprising of the map inputs from remote sensing as well as conventional sources. The database is designed and organized in a manner such that it minimizes the data redundancy and maximizes the multiplicity of usage of each data element. Information data have been collected from both traditional and modern sources. Traditional includes both primary and secondary sources of information about the landform, drainage, slope, relief have been collected from topo-sheet at 1:50000 scales, available from survey of India. Survey of India topo-sheets- 53 A/8, 53 B/10, and 53A/12 at 1:50000 scales were used to extract the information about road/ rail network, drainage system, settlement system, geomorphology of the block. For data storage, manipulation, analyzing and mapping Arc GIS, 9.2 and Erdas Imagine 8.4 have been pressed into service. Data regarding the biodiversity in the region has been taken from Punjab State Council for Science and Technology. Punjab Forest data has been taken from forest survey of India data 2007.Working plan for strip & block forests of Rupnagar forest division have been consulted. Layers of study area have been extracted after geo-referencing thematic map of soil types of Punjab have been used from National Bureau of Soil Survey and Land-use Planning (NBSS&LUP) Nagpur. Groundwater data of wells showing the depth of water level for the month of June and October for 35 years from agriculture department and under ground water depth have been extracted after geo-referencing the maps of the study area from Ground Water Information Booklet Ropar District, Punjab, Central Ground Water Board, Ministry of Water Resources, Government of India, North Western Region, Chandigarh, 2007.

Study Area

Block Rupnagar formerly known as Ropar is a part of the kandi belt of the Himalayas and the alluvial plain of the river Sutlej. It lies on the north eastern corner of Punjab State between 76°25'to 76°45'E longitude and 30°50' and 31°15' N latitude. It is bounded by Himachal Pradesh in the North-East, Union Territory Chandigarh and Haryana in the South and Gharshankar and Ludhiana in the West. There are 199 villages in Rupnagar Block of Rupnagar district.. The area can be divided into following main physiographic units namely dissected Shiwalik foothills, inter-fluvial alluvial and flood plain (Fig. 2). The river Satluj forms the main drainage system in the area and roughly flows in northeast-west direction (Fig 3 and Fig. 4). The Rupnagar block falls under semi-arid (sub-moist) and less hot zone of Punjab.

Results and Discussions

To *Ginsburg* (1957, p.206)"all the free given material phenomenon of nature within the zone of men's activities are resources. In fact, the resources are of paramount necessity to human wants and make a dynamic concept, not fixed and not a neutral stuff. Its expansion and utilization is based on the attitude, wants and abilities of man. Ginsberg has divided resources into Physical and Biotic resources.

I) Physical Resources

These include all the non-living parts of human environment i.e. rocks, minerals, land forms, the non- living parts of the soil, water, air, etc. In the study area, soils are the main physical resources. Agriculture being the main base of economy of Rupnagar Block, quality and availability of soils play the most important role in economic development of the study area. Besides, sand and stones found on the river beds, used in construction activities, have some commercial significance.

The soil is a very Important natural resource, because agricultural production is barically dependent on furtility of soil. In India, the national Bureau of Soil Survey and Land Use Planning, Nagpur conducts soil surveys and map soils at different spatial scales. soil layers of Rupnagar Block have been extracted from the soils of Punjab map, prepared by

Figure- 2

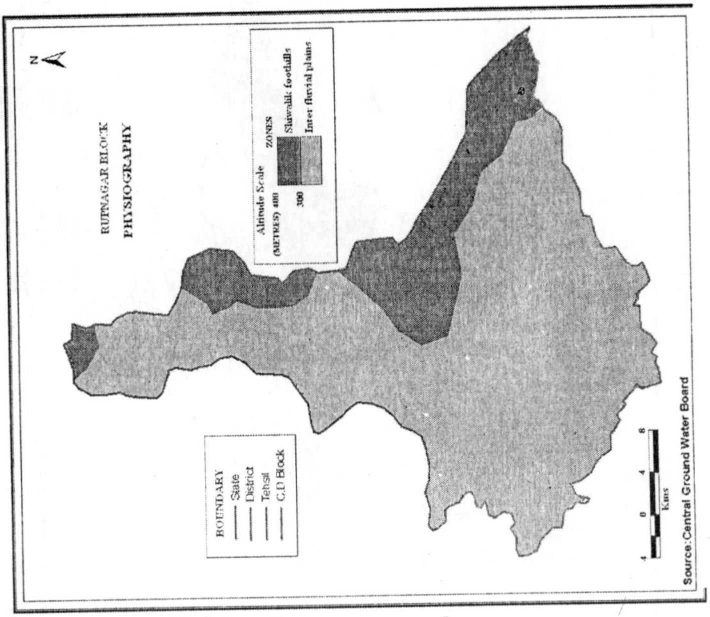

Figure- 1

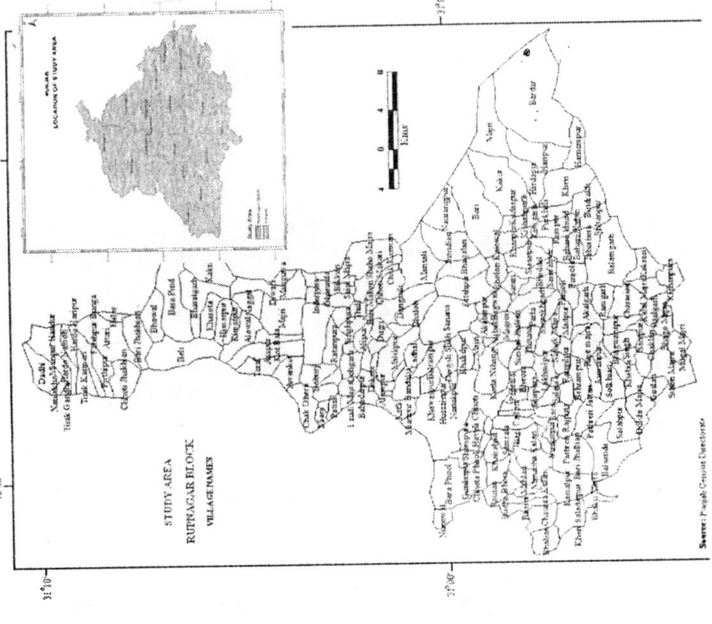

Figure- 4

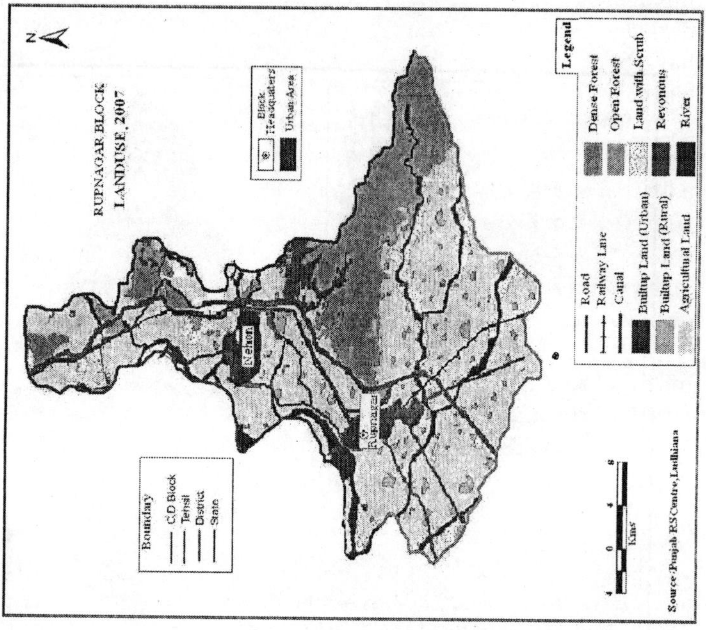

Figure- 3

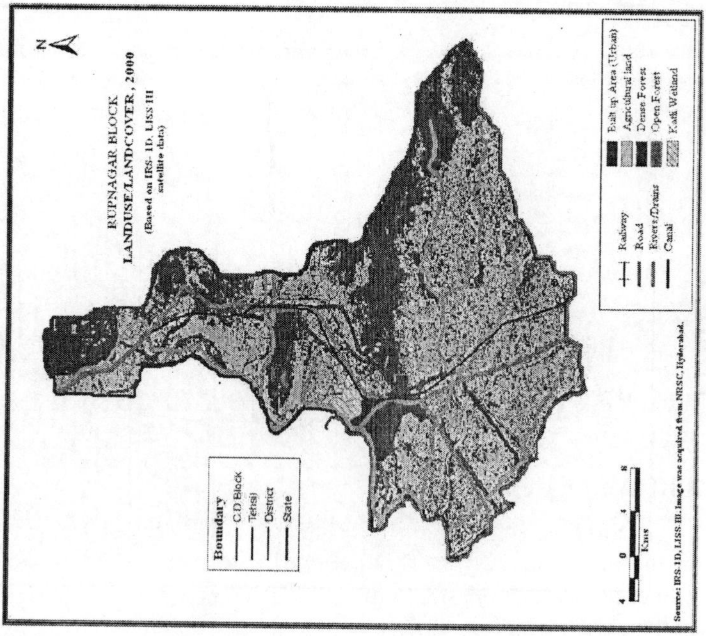

this agency in 1994 and revised in 2004 (Fig. 5). In the following, textural, drainage, erosion and flooding characteristics of soils in Rupnagar Block have been discussed, in brief. The soil Rupnagar block are broadly classified under four major types.

Texture - Soil texture is one of the important, fundamental and permanent characteristic because it largely influences the structure, porosity, adhesion and consistency of the soils. The texture is important because it largely determines the water retention and transmission properties of the soils. The texture of the soils in Rupnagar block ranges from coarse sand in the shiwalik foothill zone to loamy as we move towards the plains and it gets finer as we move towards the natural water bodies.

Soil Drainage - Surface drainage is the disposal of excess rainwater over ground surface through an open drainage system with an adequate outlet. Surface drainage is helpful where

1. Soils are deep with low infiltrations rate
2. Intensity of rainfall is high
3. Drain is level to nearly level, and
4. The water table is high.

Soil Erosion - Soil erosion under geomorphic cycles is a comparatively slow and continuous phenomenon. Although the process of soil erosion is imperceptibly slow and it can scarcely be detected, the loss over periods of time is indeed very great. So, on this basis the soils are sub divided into three categories ranging from slight to moderate to severe erosion, slight erosion can be seen near the natural drainage system and moderate erosion is found in the inter fluvial plains, whereas dissected foothill region experience severe soil erosion.

Flooding (% area flooded)

On the basis of flooding Rupnagar soils have been classified under two sub divisions that are Slight Flooding (<25%) and Moderate Flooding (25-50%). In Rupnagar block, moderate flooding is experienced in the areas adjacent to river Satluj. Loamy to fine loamy type of soil. The areas along the river Satluj have moderate flooded soil, coarse Loamy well drained soil.

I.b) Water Resources

Surface water - The River Satluj forms the main drainage system in the area and roughly flows in northeast-west direction. There are numerous *choes* (seasonal rivulets) which join the river Satluj or die out in the piedmont areas, indicative of high permeability in piedmont zone.

The subsidiary drainage pattern is sub-dendritic, sub-parallel and is mostly seasonal. Most of the streams in the hilly tract have a narrow gorge and a fairly long length of flow. For most part of the year the streams do not carry any water, however, when the discharge is full, it erodes the soil and degrade the cultivated area.

The Nangal Hydel Canal, Anandpur Sahib Hydel Canal and Sirhind Canal passes through the Rupnagar district. There is one main canal in the study area. i.e. Bhakra Main Canal which runs from north to south and cuts the block vertically from centre in two parts (fig. 6). However, the area irrigated by canals makes about 9 percent of the total irrigated area in the Block. Remaining 91percent is irrigated by tube-wells using the ground water (Govt. of Punjab, 2007). Undulating terrain contributes significantly in this context. The drainage pattern including, the river Satluj and Bhakra main Canal. In addition, there are 32 tanks in the block mainly concentrated in the South-Central part of the Block and there are only four tanks which fall under dissected foothill zone which were overlapped on the digitized layer of drainage (Fig. 6).

Ground Water -The ground water distribution in the study area differs widely. Depending upon underlying rock formations, structural fabric, geometry, surface expression etc. Hence, to understand the movement and occurrence of ground water below the surface, the geology, geomorphology, structural set up and recharging conditions need to be studied and understood.

The Central Ground Water Board has calculated the block-wise groundwater resource potential of Rupnagar district, this excludes hilly areas. The net groundwater availability of district is 440.92 million cubic meters, while the gross groundwater draft is 414.09 million cubic meters only. This leaves a balance of 19.29 million cubic meters for the future. The stage of groundwater development for the district as a whole is 94 percent. However, in 2 of the 5 blocks (Chamkaur Sahib and Morinda) stage of groundwater development has exceeded the available recharge (>100%), thus categorized as over- exploited-blocks (table 1).

Rupnagar Block, among all the five blocks in the district, has the highest net availability of groundwater resource for the present as well as future irrigation development. The parameter of stage of ground water development indicates that the availability for future irrigation development is more in southern part of the district. No doubt, further development of ground water resources is feasible in the district but a check is required on groundwater development in the over-exploited blocks.

Table 1
Rupnagar District: Block-wise Distribution of Groundwater Resources, 2004

(Figures in Million Cubic Meters)

Block Name	Net Annual Availability	Gross Draft For Irrigation	Gross Ground Water Draft For All Uses	Allocation for domestic and industrial requirements (next 25years)	Net Availability for future Irrigation development	Stage Availability for future Irrigation development (%)	Category
Anandpur Sahib	54.58	16.67	26.58	11.69	26.21	49	Safe
Chamkaur sahib	138.73	161.48	171.62	12.06	-34.81	124	Over-Exploited
Morinda	58.21	106.44	114.89	9.39	-57.62	197	Over-Exploited
Nurpurbedi	45.68	32.99	40.26	8.51	4.18	88	Safe
Rupnagar	143.72	51.07	60.75	11.33	81.33	42	Safe
Total	**440.92**	**368.65**	**414.09**	**52.98**	**19.29**	**94**	Safe

Source: (Govt. of India (2007), Ground Water Information Booklet Ropar District, Punjab, Ministry of Water Resources, New Delhi)

The ground water in the district is saline in nature with medium to high salinity- (Govt. of India, 2007, p.). It is however, suitable for irrigation in all types of soil.In the entire district, Iron-oxide, an essential nutrient for plants and animals, is below the permissible limit (1.0 mg/ liter) except with an exception in Rupnagar block. Majority of the samples in Rupnagar block show slightly higher values than the permissible limit of 1.0 mg/ liter.As calculated from the average depth of 35years of the underground water monitoring wells it has been observed that there is a slight difference in the pre and post monsoon water table depth, trend of 1 to 4 metre change in the depth have been noticed.

Status of Groundwater Development

The groundwater development for domestic and irrigation is mainly through tube wells, hand pumps and cavity wells. Large quantities of water are abstracted through shallow and deep tube wells. Depending upon the water level, submersible or centrifugal pumps can be used to harness the water. In rural as wells as in urban areas drinking water supply is based on groundwater. Tube wells tapping shallow and deep aquifers are constructed by Tube wells corporation and public health department. Out of the total 889 villages in the district, 821 villages have been connected in the piped water supply. Out of the net irrigated area in Rupnagar Block, only 9% of land is irrigated by canals and rest by groundwater. Thus 91% of the total irrigated area is irrigated by groundwater by means of shallow and deep tube wells. The total number of shallow private tube wells in the district is 21,873. Majority of the private tube wells used for irrigation purpose tap the shallow aquifer. During the past few years, density of shallow tube wells has increased tremendously because of the cultivation of water demanding crops. Rice, maize and wheat are the main crops grown in the district.

Within the Rupnagar Block, the depth of ground water varies from 2-5 meter below ground level in the southern and northern part of the block to 5-10 meter below ground level in the central part of the block because of excessive extraction of the underground water in the pre monsoon season, and the variation equalizes in the post monsoon season with the depth of underground water as 2-5 meters below ground level in the entire block because of groundwater recharge because of monsoon rains. (Ground Water Information Booklet Ropar District, CGWB, Punjab, 2007) shown in **Fig. 7 and 8** respectively.

(II) Biotic resources

These resources are those which affect the economic activities of man and are also affected by human activities. In favorable conditions these resources get enriched but in adverse conditions these resources get degraded for example, forests and wildlife. These are the mobile resources. These can be recreated. Forests once destroyed can be replanted. Animals can be bred. The number can be increased to any desired limit. The occurrence depends upon physical and climatic conditions. The biotic resources have been broadly classified under Forests and Wildlife

Forests -Forests are the lifeline of all development activities and sustenance of human life. Increasing forest cover through proper management, protection and regeneration are essential components of conservation of soil and water resources and above all for regulating climatic aberrations. In Punjab, forests are important for continuing its agricultural activities

Figure-5

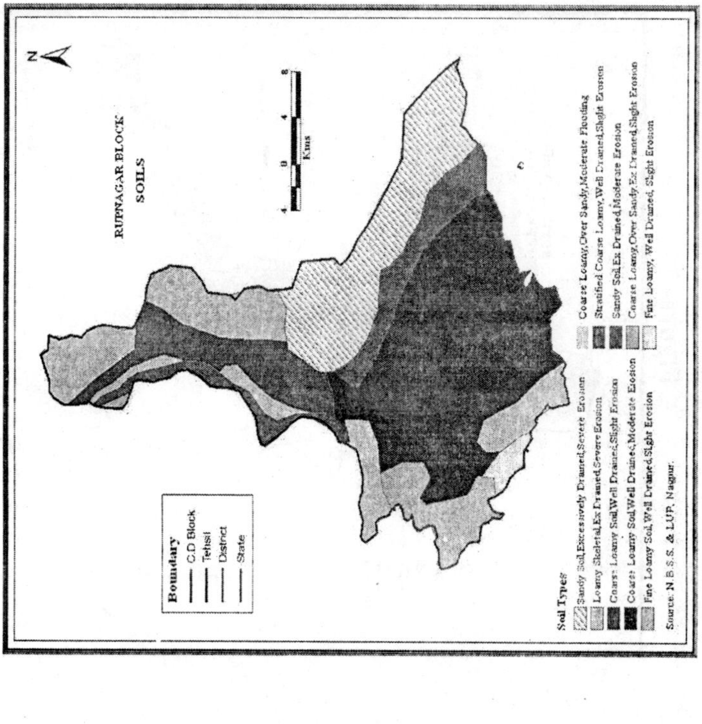

Figure- 6

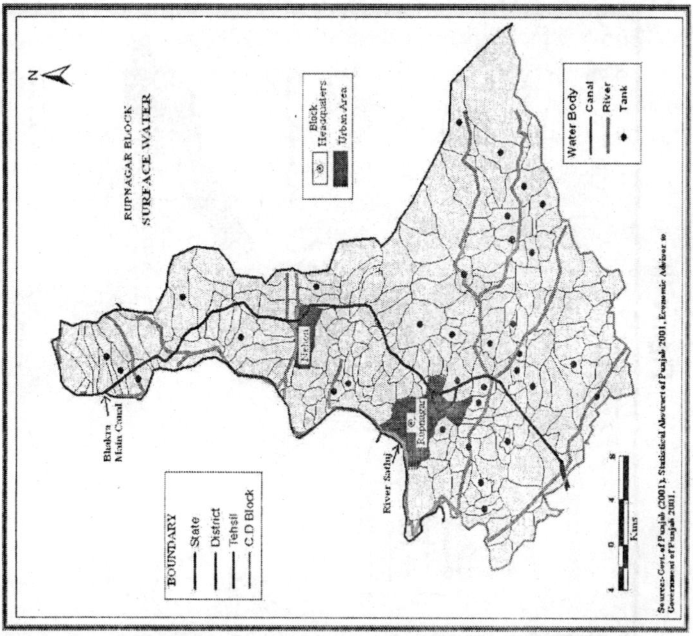

Figure- 7

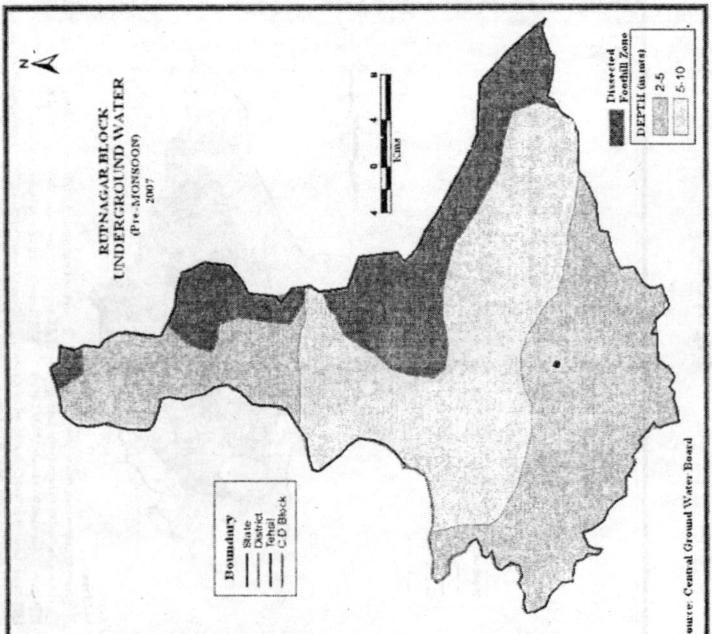

Figure- 8

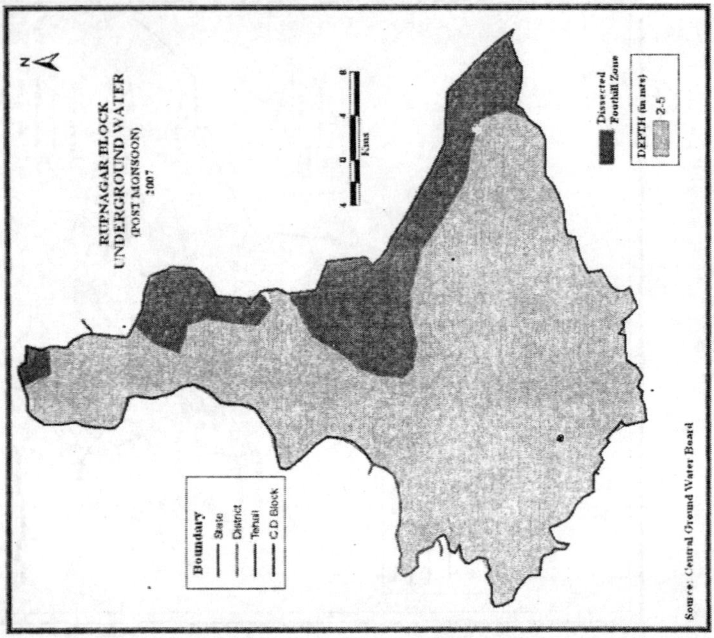

Table 2
Forest Cover: Block in comparison to District/ State, 2001.
(in Square kms)

Recorded forest area Rupnagar block	**Punjab**	**Rupnagar**	**district**
Reserved Forest (RF)	44	-	-
Protected Forest	1137	56	89
Un-classed Forest	1903	460	
Total	**3084**	**516**	**89**

Source: Report on inventory of trees outside forest in Punjab, Punjab Forest Department (Govt. of Punjab) Chandigarh.

According to Forest SOI report (2007-08) based on the digital interpretation of IRS 1C/1D LISS –III & PAN Data, the area covered by Reserved Forests in Punjab is 44 km^2 whereas it is nil in the district of Rupnagar district. Area under Protected Forest in Punjab is 1137 km^2 although Rupnagar district falls in foothill zone of Shiwalik still, very small portion of it comes under this district which is 56 km^2 and Un-classed Forest occupies 1903 km^2 of the state of Punjab out of the total 3084km^2 which is 6.12% of the state's geographic area and it occupies 460 km^2 of the district out of the total 516km^2 which is 24.42% of the district's geographic area and 16.73% of the State's Forest area. The ratio of area under forest in the district is almost four times the than that of the state being a fact that major part of the district falls under the Shiwalik foothill zone. Whereas protected and un-classed forest in Rupnagar block occupies an area of 88.85 km^2 which is 22.76 percent of the total geographic area of the block. See table 2.

In the forests map Rupnagar Block two types of data have been used 1.Revenue records from Govt. of Punjab (2001&2007), Statistical Abstract of Punjab,(2001&2007), Economic Advisor to Govt. of Punjab (2001&2007). 2. Satellite data based on the digital interpretation of IRS 1C/1D LISS –III & PAN data, PRSC, Ludhiana (Fig 3 and 10).

The area under forest as shown in the village directory includes forested areas classed or administered as forest under any legal enactment dealing with forest whether state owned or private. This figure may be slightly different from one furnished by the forest department owing to the reason that contains land, though not wooded, are taken as forest by forest department.

Area under Forest- Most of the villages under and near foothill zone have a sizable forest area of 10 hectares and above with few exceptions. For example Dhaloh, Majri, parts of Bara Pind have no forest area because of sandy soils and non availability of seasonal choes and river.

There are very few villages with very little area under forest i.e., less than ten hectares, Allowal, Tarf, *Cherian*, Allowal, Alampur, Bholukalewal to name a few. 137 villages of the total have no forest area, and 130 out of these fall in Inter-fluvial plains (Fig. 9).

The forests area comprises of both protected and un-classed falling in Ropar Forest Division. The total forest area under Rupnagar block is 8885.14 ha as surveyed and corrected from preliminary working plan. Majority of the block forest areas are evacued lands purchased by the Forest Department during the year 1961 and 1971. The forests are mostly situated in the Shiwalik foothills and a few in the plains roughly in the proportion of 3:1.The smallest forest is Surari (0.4 ha) and is the biggest Narangpur (400 ha). The forests are very important from point of view of preservation of natural environment. Over the years, the forests have been afforested, reforested and treated with different kind of soil conservation measures with varying degree of success. There are a very few forest areas which have not been given a set of treatments in the past. The special object of management of these forests shall be:

- To improve the permanent vegetative cover in the Shiwaiks in the interest of soil and water conservation.
- To in increase the cover of bhabar grass for meeting the demand of small scale cottage village industries.
- To meet the fuel, fodder and small timber, and other requirements of the local inhabitants.
- To plant suitable indigenous species to raise the ecological status of the area.

Satellite based data which have been obtained by extracting the layer of Rupnagar Block out of the Rupnagar district forest cover Map for the year 2007, based on the digital interpretation of IRS 1C/1D LISS –III & PAN Data, (PRSC, Ludhiana) and digitizing after geo-referencing in GIS environment, (Fig. 10).

Shiwalik foot hills which cover sizeable area of Ropar block have moderate to dense forest vegetation under Protected and Un-classed forest. Hence considerable area (88.85 km^2) of the block is under forest which is 22.76% of the total geographical area of the block. As compared to the percentage area under forests out of the total geographical area of Punjab State i.e. 6.12%, it is quite high and slightly less than the district's area

Figure - 9

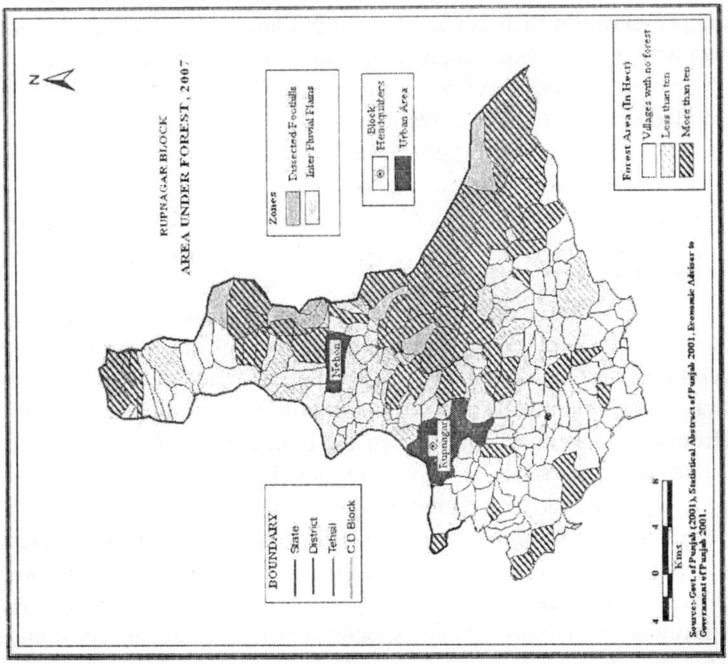

Figure-10

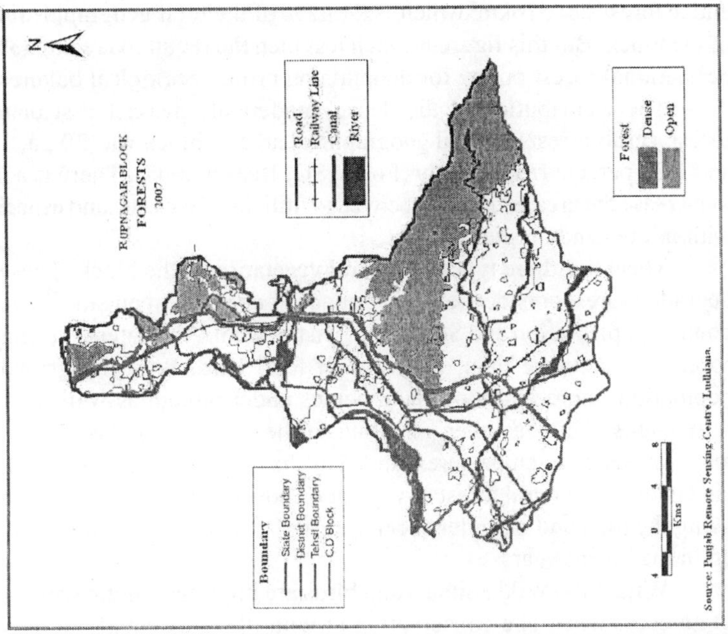

under forest i.e. 516km^2 which is 24.42% of the total geographical area of the block. But this figure is much less then the required area of 33% as per national forest policy for maintaining proper ecological balance.

The contribution of the dense, moderately dense forest or open forest, scrub forests to total geographical area of block was 20.25, 1.31, and 1.20 percent respectively (Forest SOI Report 2007). There is a need to increase the area under forest cover by utilizing the waste land especially with in the kandy region.

There are three types of natural vegetation in the block. These are degraded forest outside the forest boundary, open forest outside the forest boundary, plantation and Scrubland. On the whole, the natural vegetation spreads over more than one third of total land. With the growth of population; more and more land comes under plough as well as under settlements. This has been resulting of the clearing of shrubs, bushes, mangroves and even pastures land. This has been resulting in decline of land under the natural forest or vegetation cover. However tree population along the road and or under social forestry scheme has also been picking up in the recent years.

Wild Life-Wild animals and birds are important biotic component of an ecosystem. The conservation of wild life is essential not only for the preservation of Flora and Fauna but also for sustained agricultural productivity. Wild life preservation is therefore a dire necessity for stability of ecosystem and survival of mankind. The Rupnagar Block can be divided into three main zones:

1. The Dissected foothill Region.
2. Wet Land Region
3. Plains Region

1. The Dissected foothill Region

This is one of the major forested areas of the state, which otherwise has 84% land under agriculture. Since it harbors several species of economic and medicinal value, special efforts need to be initiated by the state government to conserve this important natural resource of state. However it has been noticed that the biodiversity of this area is under threat due to various developmental activities. The region is inhabited by a large number of endangered species like *pangolin, python, barking deer, hog deer, sambar*. These species require more effective protection. The population of some other species of wild animals and birds like wild Boar, Nilgai, Jackal, Hares, Mongoose, Rhesus Monkey, is also rich in this region. Population of these species is increasing in this region steadily. This area has many check dams which provide fresh water storage for

Fauna also increases variety of flora. Aquatic vegetation like water Hyacinths, Nelumbo, Hydreilla, Trapa etc. exist is abundance in this water bodies.

2. Wet land region

The major wet-land here is Ropar Wetland, which constitutes three villages namely Katli, Mianpur Bandalia, Mianpur Raina which are very rich in aquatic Fauna and Flora. It is an artificial wet lands but gradually acquiring natural specialty and natural view. The Ropar wet-land is situated near Rupnagar town. It is created out of compromising of water of Sutlej River over which as a result of construction of barrage and head works. The water accumulated in the shape of a lake. The depth of the lake varies from half meter to six meter. A few shallow water bodies feature along both sided of the river are located within the wet land area. The Sadavarat forest (28.33 ha.) is also included in the wetland. This wetland area is a house for many special wild life creatures and is a paradise for nature lovers.

The influx of migratory birds in this lake takes place annually during months of October, November and December. The vegetation of this area consist of *Shisam, Kikar, Banyan, Pipal, Neem, Amaltas, Mulberry, Siris, Jamun, Mango, Willow, Semul, Khajoor* and others

3. Plains region

The plain region comprises cultivated plain area, road sides, canal sides, river side's etc,. Some wild animals and birds like wild Boar, Hares, Porcupine, and Quails etc. are found in this region which requires protection to maintain a balanced ecosystem. However, certain areas of Ropar shiwaliks and Ropar city show poor diversity due to erosion and land degradation.

Conclusions

In this paper an attempt has been made to prepare an inventory of physical resources and map these upon RS and GIS for the effectiveness and usefulness of RS and GIS technology as a tool for micro level planning. Though some part of the study area was a part of the foothills of the lower Himalayas and other area was very near to this zone still have very less forest area in this zone, which can be picked as serious research agenda. Most of the area under foothill zone has coarse eroded sandy soils mixed with pebbles. Against this, soils in inter-fluvial plains vary from coarse loamy to fine loamy as we move nearer to the natural

With the water getting polluted this view of Sutlej might be a thing of the past.

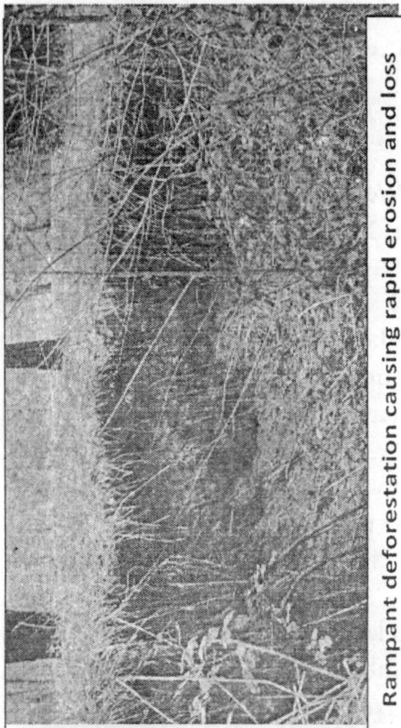

Rampant deforestation causing rapid erosion and loss of fertile soil.

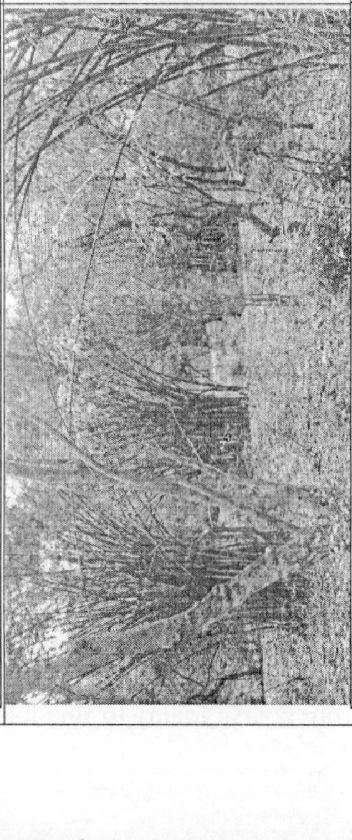

Forest area being reduced to dry lands.

Rapid deforestation might soon lead to the extinction of these rare bird species

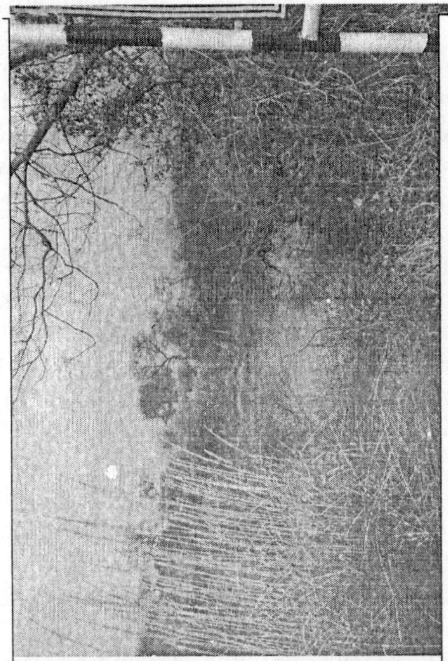

Vegetation surrounding wetlands which acts as a barrier for soil erosion

One of the wetlands of national importance. When its gone so would go most of the eco-system in the area.

drainage. The river Satluj forms the main drainage system in the area and roughly flows in northeast-west direction. There are numerous choes (seasonal rivulets) which join the river Satluj or die out in the piedmont areas, indicative of high permeability in piedmont zone. The central part of the block experiences low ground water level of 5-10 meters below ground level because of excessive extraction of the underground water for the cultivation of high water demanding crops in the pre monsoon season but it gets recharged again with the advent of monsoon rains. Post monsoon period experiences 2-5 meter below ground level in the entire Block. In the study area, considerable area (88.85 km^2) of the block is under forest which is 22.76% of the total geographical area of the block. As compared to the percentage share under forests out of the total geographical area of Punjab State i.e. 6.12%, it is quite high and slightly less than the district's area under forest i.e. 516km^2 which is 24.42% of the total geographical area of the block. But this figure is much less than the required area of 33% as per national forest policy for maintaining proper ecological balance. Most of the villages under and near foothill zone have a sizable forest area of 10 hectares and above with few exceptions, for example Dhaloh, Majri, parts of Bara Pind have no forest area because of sandy soils and non availability of seasonal choes and river. There are very few villages with very little area under forest i.e., less than ten hectares, Allowal, Tarf, *Cherian*, Allowal, Alampur, Bholukalewal to name a few. 137 villages of the total have no forest area, and 130 out of these fall in Inter-fluvial plains. Because of the presence of Ropar wetland and being a part of the shiwalik foothill zone, rupnagar block has rich diversity of flora and fauna. The results may indicate that many endangered and vulnerable species are found in this area. Many of the faunal species are listed under different categories of conservation status of IUCN, CITES, CAMP and IW (P) Act. are found in this region which requires protection to maintain a balance ecosystem.

References

- Anderson, J.R., et al., 1971, "Land use and Land cover changes. A Framework for Monitoring", *Journal of Research by the Geological Survey*, 5: 143-153. U.S.G.C, Reston, Virginia.
- Bhat, L.S. 1996, "Decentralized Planning: Lessons from Kerela" in Kant et al. (eds) *Reinventing Regional Development*, Rawat Publications, New Delhi.
- Burrough, P.A. 1986, Principles of GIS for Land Resource Assessment, Clarendon Press, Oxford.

- Gautam, N.C. 1983, "Satellite Remote Sensing Techniques for Natural Resources Survey", in Singh et. al. (eds) Environmental Management, Allahabad Geographical Society, Allahabad.
- Ginsburg, N. 1957, "Natural Resources and Economic Development", *Annals of Association of American Geographèrs*, 47(3):197-212.
- Government of India, 2007, *Groundwater Information Booklet: Ropar District*, Punjab. New Delhi.
- Government of India, 1980, *Handbook of Agriculture*, Allied Publishers, New Delhi.
- Jerath, N. 2006, *Biodiversity in the Shiwalik Ecosystems of Punjab.* Punjab State Council for Science and Technology, Chandigarh.
- Kumar, P. 2001, *Working Plan for Strip and Block Forests Ropar Forest Division,* Govt. of Punjab, Rupnagar.
- Lillesand. M.T and W.R. Kiefer, 2007, *Remote Sensing and Image Interpretation*, Wiley India Publications, New Delhi.
- Lo. C.P.and A.K. Yeung. 2008, *Concepts and Techniques of Geographic Information System,* PHI Learning Private Limited, New Delhi.
- Rawal, J. K. 2004, *An Understanding of Land-use/ Land-cover with particular reference to forests in India: Abstract.* Forest Survey of India, Dehradun.
- Zimmermann, E.W. 1951, *World Resources and Industries*, Harper and Brothers, New York.

36. Management of Natural Resources for Sustainability

Sudhakar Dwivedi
M.C. Dwivedi

The management of natural resources has become more integrated through its hundred year's history. In this chapter we attempt to integrate ecological, economic and policy factors into a functional as well as applied framework. It is true that the natural resource management requires an understanding of the interaction between natural and social processes. The natural resources are the raw materials used to satisfy human wants. These natural resources may be classified into two major basis i.e. biotic natural resources and abiotic natural resources. The judicious use of these resources to fulfill the human needs is essential. Many natural resources are renewable and they can use again and again by the human being. Planning, policy and administration are the main elements to identify and solve the problems of natural resources. The natural resource management organizations have to identify the problems and means for solving the existing problems of the resources.

Key words: Natural resources, Renewable resource, Ecosystem, Sustainability.

Introduction

This is the era of increasing social and ecological awareness. The natural resources are the raw materials used to satisfy human needs. These resources are divided into basic categories like, soil, water, air, minerals,

The basic components of natural resource management system

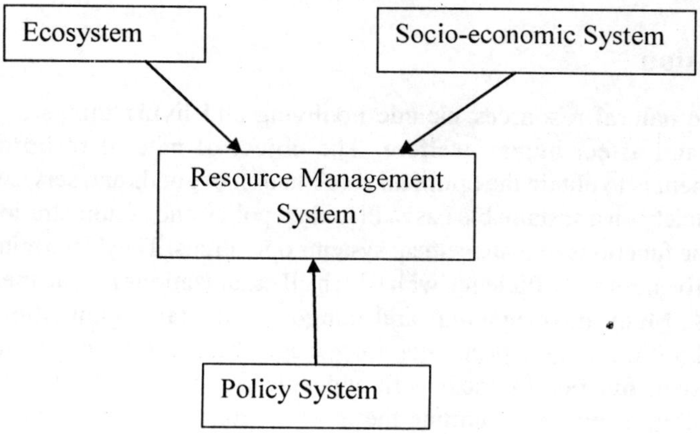

geographically based reference system. Management unit often are split up on a regional basis and management plans often define the limits / boundaries of each separate area to be affected by management and to what degree.

Public or private managers operate with the intent of serving the interests of people who use the resources. The private sector manages natural resources to sell to customers or to use themselves in manufacture of refined products and then sold. The governmental agencies are mandated through their authorizing low to serve resource users, called customers. However it is always difficult to determine exactly who is the customer? For example when the forest service sells timber to the company then they said company is the customer and company is the tax payers who are also affected by the income generated from the sale. Similarly many large businesses, agencies often produce many different products for different groups of resource consumers. However, because of their typical broad mandates, governmental agencies have no choice but to serve all customers uniformly well while private businesses can specialize or diversity as they choose within broad regularity limits.

The effective management system monitors opportunity provision and use. The system also focuses on public and customer satisfaction to evaluate management success. Such type information is always used by management to recognize goals and strategies for accomplishing organizational purpose. This information also may be reported to the concerned public groups. The private firms can measure success by the income and profit they sustain while monitoring goods and services production for improved efficiency

Conclusion

The natural resources include nonliving and living things that interact and affect human welfare. The object of natural resource management is to obtain the optimum combination of goods and services for the society on a sustainable basis. Planning, policy and administration control the functions of management systems operations. They determine the effectiveness and efficiency with which all organizations pursue their purposes. Many governmental and nongovernmental organizations involved with common principles to manage these resources. Many organizations function for the benefit of people in some way. They use a variety of strategies to determine the most appropriate distribution of benefits among society.

References

- Freeman A.M. (1993). *The measurement of environmental and resource values*. Resources for the future. Baltimore.
- Hackett, S.C. (2001). *Environmental and natural resource economics*. Theory, policy and the sustainable society. M.E. Sharpe, Armonk.
- Sankar, U. (2001). *Environmental Economics*, Oxford University Press.
- Sengupta, R. (2000 an). *Ecology d Economy: an Indian Perspective*. Oxford University Press.
- Schiller, B.R. (1994). *The economy today*, 8th edition, McGraw – Hills
- Tietenberg T. (2003). *Environmental and Natural Resource Economics*, 8th Ed.

37. Consequences of Environmental Protection through Resource Conservation

Neha Sharma and Manoj Sharma

Environment-man interface defines the root of anything that has operational comprehension. The operational whole resembles the outcome of a series of interrelated forces of interaction that originate from nature, culture and individual human action. Paper shows inter relation between Human and Environment. The environmental presence and the human resources are highly interdependent. Environment gives meaning and value to a world of development, No living species no society or no development can afford to survive without drawing sustenance from its environment. Environmental protection needs the agenda of resource conservation the most. Resources represent the soul of environment. The agenda of environment protection is appeared on agenda in 1960s. When the ideals and realities of environmental crusade were elaborated in the perspective of impending environmental crisis.

Key words: Environmental protection, Resource conservation, Geography, Environment

Introduction

Environmental protection and resource conservation makes a comprehensive and inclusive theme. The theme draws its meaning and contents from the philosophy and practices contained in human induced environment based development. The contemporary face of development has intended to inscribe varied cultural accomplishments, which are the outcome of environmental presence and human skills of resource manipulation/ways of exploitation and management. The theme of resource managements views, in its comprehensive, rationale of practical use cf environment as absolute physical space resource and relative economic resources as part of ensuring sustainable human and eco-worlds (Sharma, 2008). The theme may be explained by a concept diagram (Fig. 1)

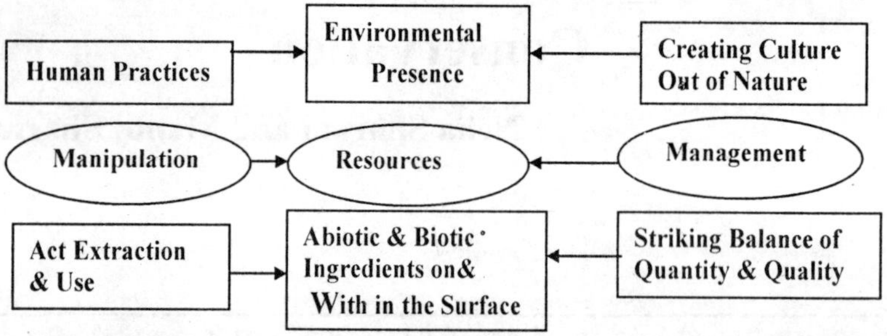

Fig. 1

Environmental Protection Through Resource Conservation

Let's ask ourselves-Is there anything other than environment / resource which concern us, one and all, more commonly and in greater extent and intensity? The answer speaks certainly not. The environment matters and it matters with in and out. The human identity and also the total biotic identity rest with the biosphere part of Earth's environment. It matters much for us for the simple reason that it provides us with a habitat to live within and to act upon to create in good many senses. The earth's environment matters also because it scripts and prescripts changes as every living thing produces changes in its environment (Taylor & Humpstone, 1975). This generates concerns in view of the relation that non-human changes are naturally negotiable through critical links of the environmental system, but human induced changes, significantly, differ in impact by disturbing the natural state of dynamic equilibrium.

The theme of EP and RC has generated common concerns for the sheer reasons that both environment and resources hold meanings to dynamic and complex relationships between people / human worlds and habitats / eco-worlds. These relationships have produced, what I would prefer to call, the FEEL and FEAR CONCERN, explained under the concept diagram (Fig. 2)

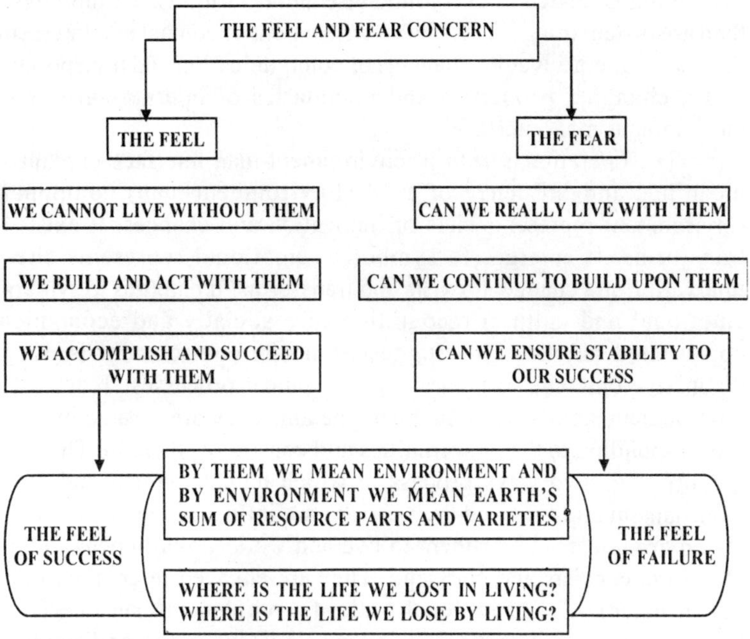

Fig. 2

An appreciation of the diversity and variety of environment / resources forms a theme that runs through the entire span of the spatial organization / human world and its relationships with the world of environment. There are changes to establish "an intellectual foundation" enabling the feel of a sustainable "geographical imagination" (Knox and Monston, 2001). The environmental presence and the human resources are highly interdependent. Each fills specified roles in complex and ever-changing networks of interactions and change in the face ordained and desired environment man interface.

Environment– Man Interface: Operational Truth

Environment-man interface defines the root of anything that has operational comprehension. The operational whole resembles the outcome of a series of interrelated forces of interaction that originate from nature, culture and individual human action. The comprehension of environment-man interface is substantiated by the first law of geography that states-everything is related to everything else and near things are more related than are distant things. This law views element of contact and interaction, dependent themselves on channels of communication and transportation, to fix changing properties and boundaries of operational space of environment-man interface.

The operational truth of environment-man interface explains the meanings and influence of E-M (Environment-man) components. Environment represents sites of innovation and changes. It exists and transforms. It emerges to symbolize emotional represents sites of innovation and change. It exists and transforms. It emerges to symbolize emotional and cultural recognition of a socially and economically constructed habitat. This suggests that the environment is not just distinctive outcome of natural / geographical processes. Tenets of the environment signify dynamic phenomenon. They are created by people too, responding to the opportunities and constraints there in. There exist a continuous two ways process in which humans create and modify environment and they and their actions are influenced by the conditions the environment sets for them to live and work. Environment and man, therefore, coexists and coextend. They are engaged in creating culture out of nature. The operational truth of environment-man interface is, however, always incomplete and on constant move and occurring simultaneously at different scales. The theme of environment-man interface establishes the basis of EP and RC. This may be conceptualized as follows (Fig. 3).

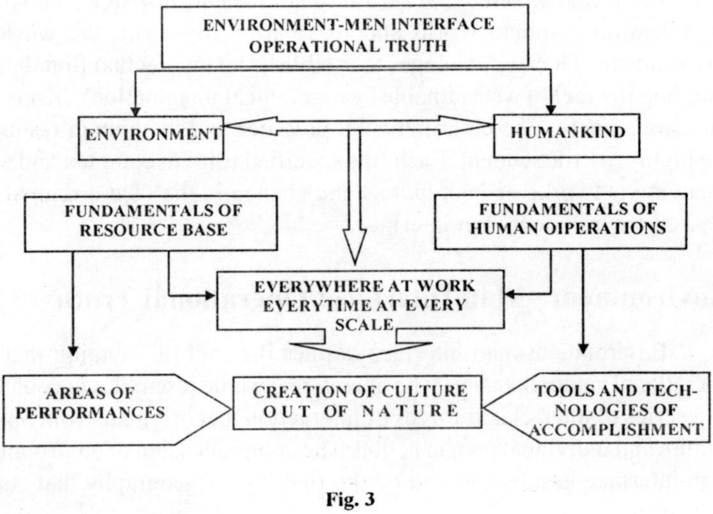

Fig. 3

The operational meaning of environment-man interface seeks substance in the comprehension that geographical elements of the environment are static in limited sense only, they become dynamic as human element as soon as man interacts with them. The interface does, however, not remain corresponsive. Human excesses have been found to excel the acceptable pressure on the environment. This has strained the relationship.

Environment : In A World of Development

It is now agreed upon that the world is fast moving the development way and the environment is being forced to meet thresholds and frontier. It is convincing too. No living species no society or no development can afford to survive without drawing sustenance from its environment (Hereding, 1997). Environment, in this sense, giving meaning and value to a world of development. Environment draws, thus the bottom line of the imperative of ecological concerns. The world of development has intended to witness the defining movement of the modern age based on anthropogenic performances and human accomplishment. The environment has, however, remained at the centre stage. The performers of development have seen accomplishments as partnership with the environment. The world of development has grown under compulsions and limits having their roots in environment.

The Homocentric Paradigm of development has defined the environments role keeping man at the centre of the world. The environment, in a world of development, has been subjected to a new kind of its treatment by men in which man has explored New Frontiers, reached out to New Environments and learnt to understand and master nature in new ways. Humans have impacted the world of environment in a bigger way and they have impacted it to the level of putting it into stress. The homocentric paradigm perspective perceives the idea of development out of inevitable injuries to environment (Fig. 4)

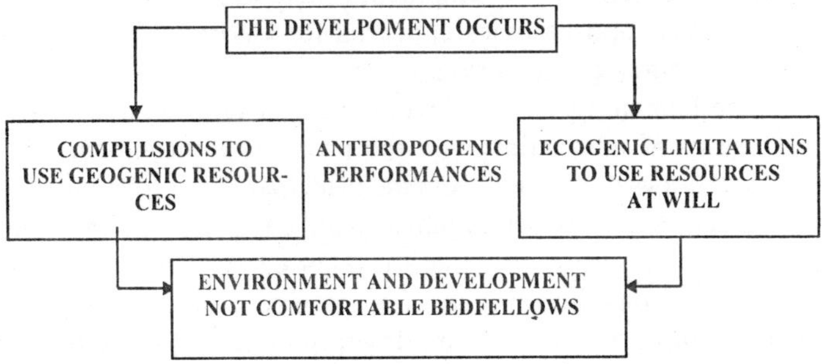

Fig. 4

Development has been considered to signify the outcome of the highest form of interaction between life in general and the environment (Fedorov, 1980). In this comprehension, any event of human action tends to correspond to event of development and all events of development draw flesh and blood from environment. All these happened / have been happening under forced and desired human agenda. The agenda of development has formed the essential part of human inheritance within the environment necessarily a protected environment.

Environment and Human Agenda: That Have Existed

Human's natural agenda has been to pursue with development. Development, too, is naturally resource-based. Environmental resources, in developmental context, open windows of opportunities to pursue the agenda of development. The most common experience says that development means no accident and implies its beginning and continuance in resource use. Development also implies that it depends on use of apt technology. The agenda of development, that has continued to pursue, brings out the comprehension that:

(i) Development represents need-based human agenda.

(ii) The agenda of development continues to prompt mankind to learn lessons of performances and cultural accomplishments with and within the environment.

(iii) Passion for achievements drives development and this passion leads to intensification of environment/resource harness.

(iv) Passion for achievements drives development and this passion leads to intensification of environment/resource harness.

(v) The agenda of development interferes with the critical balance of the nature and extent between **Abiotic Control** and **Biotic Response.**

(vi) The agenda of development generates necessary environmental issues and assumes concerns for environment protection and resource conservation.

The human agenda of development is said to have established a long history of interaction with the physical environment (Bradshaw, 1979) and draws conviction from the Principal of interdependent or inter-functional effects between organisms and environments (Kumar, 1977).

The relationship of humans to their earthy environment has been considered to constitute the core concept of development (Sharma, 1979). The contemporary agenda of development and its environmental implications are subjected to following mechanism (Fig. 5). Everything has a cost and so has the development. Likewise, something is there to pay the price and the environment pays the price of development. The environment is subjected to pay the price on account of homocentric agenda of development that has put development at par with and likely to overtake the environment. The human kind has had commitments to environment and agenda of development. Human agenda of development has remained influenced by two relevant paradigms:

(i) **Neo-Franciscan Paradigm:** Cemented the man's belief that nature and biotic things are God's creation for serving men. The agenda of development committed itself to use and enjoyed the environment at men's whimsical will and act.

(ii) **Neo-Luddite Paradigm:** Cautions men against use of development technologies as a Frankenstein monster, a runway more powerful than its creator. This paradigm suggests reorientation of human agenda of development based on quantity power with virtue and wealth with well-being.

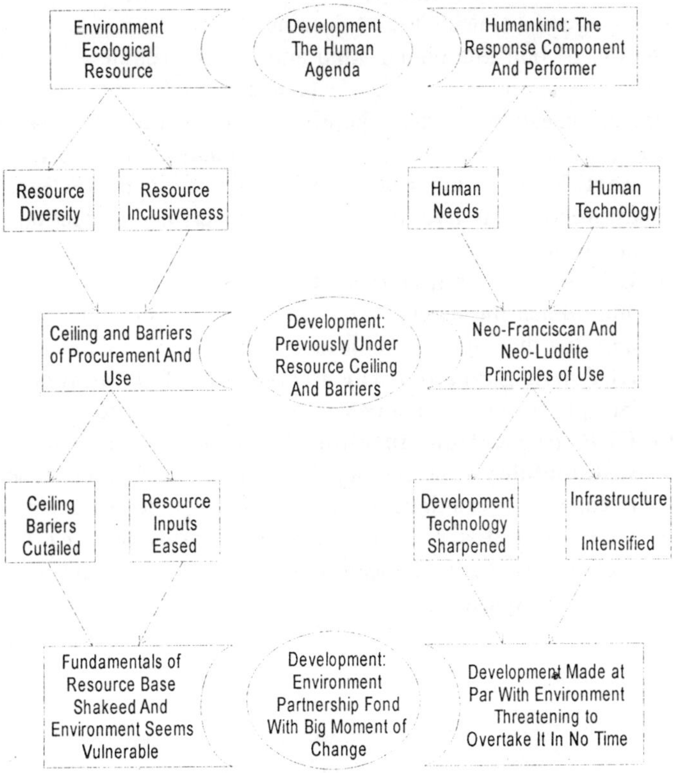

Fig 5

Earth's environment marks its geographic identity of an integrated material system conditioned in a state of dynamic equilibrium (Ryabchikov, 1975). Man's march to progress embraces, however, disturbing consequences on the environment. The agenda of development that we have been pursuing has the implications where the technological power at the command of each is greater than ever before and is growing rapidly. Besides, the agenda has generated a host of environmental issues every one of which has been found to have characteristics that makes simples solutions unrealistic (Luoma, 1984).

EP Through RC : The Agenda that we Need

The earth lives – It breathes: Ours are among the cells that give it life. Like its other cells, our take form, grow, die and are replaced as par of the earth's life. If it sickens, we will be afflicted, if it dies so too will we (Sarinen, 1974). The expression compliments humans to represent building ingredients of Earth's environment and warns against our continuing agenda of using the environment as resources. It is apparent that identity of both the human kind and the environment is mirrored through each one's and their coexistent and coextensive relationship. The fundamental question emerges, therefore, Why we need agenda for EP. There are, however, reasons/logics behind pursuance of agenda of EP.

(i) **Earth's Environment Protection means our protection.** If we aspire to stay to live we must ensure safety and protection of our environment.

(ii) **EEP ensures maintenance of regional ecology.** Environmental and human ingredients are found in actual situations or places defining the regional and local ecologies.

(iii) **EEP will guarantee the protection of Biosphere's life:** Biospheric environment ensures life support system.

(iv) **EEP protection intends to aim at environmental sustainability:** The essence of environment-man existence lies in capacity to continue to stay against threat to sustainability.

(v) **EEP provides credibility to the relevance of human technology:** Technological supremacy proves relevant only when it intends to usher environmental solution.

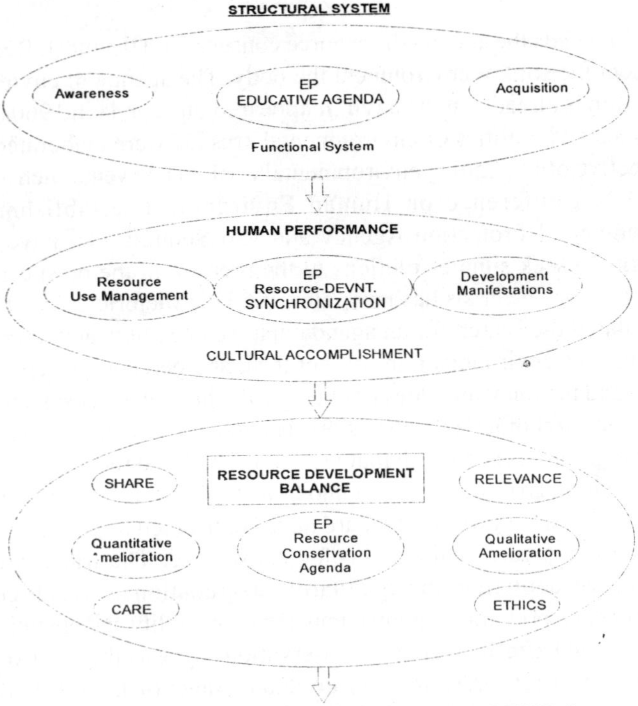

Fig. 6(i) SYSTEM BEADS OF EP AGENDA

System Beads of Ep agenda

- EP is necessarily an act of doing.
- The practical part of EP implies that what one does is goods but what is even more important is how one does it and how good he does it.
- Though development is credited to human actions. It is resources that act surrogate for it.
- No environment is more wide spread and ecologically demanding than resources.
- Concerns for resources and interest are using them and associated problems mark the fundamentals of EP.
- A good as resource management are they could be made still better.
- RC explains the control on direction of human development as co-existent and co-extensive with the environment.
- RC proves a perfect solution for all those fundamentals that ensure practically effective and technologically secure EP.

Fig. 6(i)

EP needs the agenda of resource conservation the most. Resources represent the soul of environment the body. The agenda of environment protection is already on the anvil. It appeared on agenda in 1960s. When the ideals and realities of environmental crusade were elaborated in the perspective of impending environmental crisis. The events such as Earth Day, UN Conference on Human Environment, establishment of Environmental Protection Agency and Rio Summit, etc. have shown concerns to seek simple solutions to the rhetoric in the perspectives of single issues. The crisis has now intensified the rhetoric crisis. This has necessitated the search for an agenda that we need now and ahead.

Earth's environment, in present form and perception, represents a system and the humanity does not lie outside the global ecosystem, but is an intrinsic part of it (Haggett, 1979). The issue of resource conservation comes high on agenda of EP that we need. The three recent instances of (a) Population Explosion, (b) Techno Industrial evolution and (c) Intense material philosophy of life standard have contributed to putting pressure on finite resource container / ecological reserve. The trio has caused quantitative depletion and qualitative degradation of resources and environmental pollution making it imperative to modify and operationalize the EP agenda afresh. Resource conservation suggests the mechanism of desired use of resources to meet optimum wants of largest number of men for longest possible period. Resource conservation means, in this comprehension, protection of resources against human acts of misuse, wastage and unwanted over extraction.

The fundamentals of EP pin in to the comprehension that the world of humanity coincides with the world of development (Sharma, 2006-07). Such coincidence comes out of human agenda which inevitably articulates performances and accomplishments or the system of doing with the environment. EP has, thus, always been on human agenda. EP agenda has taken different forms in time and area perspectives. Realizing the nature and extent of environment as the largest habitat system, it is pudent to think of an agenda of EP as a system model (Fig. 6).

Environment Protection defines a system of agenda basics of which implies relevance and ethics. Resources relevance and human ethics relate not so much to the extent of human accomplishment as to the sense of commitment to society and environment as well EP. Through RC provides a system agenda base on philosophy and practice of relevance and ethics.

References

- Bradshaw, M.J. (1979): Earth the Living Planet, ELBS Low Priced Edition, p. 159.
- Feadrav, E. (1980): Man and Nature, Progressive Publishers, Moscow, p. 6.
- Haggett, p. (1979): Geography: A Modern Synthesis, Herper and Row, Publishers, N.Y. p. 47.
- Ali, S.M. (1999): The Geography of the Paranas, Peoples Publishing House, New Delhi, 134.
- Mathur, R.N. (1961): Some Characteristic Features of Water Table in Meerut District, U.P. N.G. I.I. VII, 269.
- Rashmi Shikha, Nandeshwar Sharma, Regional Science Association India, p. 66.

market demand, and promoting broad based sustainable development that reduces poverty" (Scherr, White and Kaimowitz, 2003).

Adivasis or Tribals

Adivasis or tribals as they are called in India are defined by the elected house of the Indian Parliament (Lok Sabha) to be the ethnic groups that are, "the aborigines of India..." (Lok Sabha, 1995). They have been living in and around the forests from time immemorial. The tribals constitute 8.2% of the nation's population amounting to 84 million people as per the 2001 census. Out of the 5653 distinct communities, 635 are considered to be tribals and out of these 577 are the Scheduled Tribes as they are provided constitutional safeguards by the schedules of the constitution (Bijoy, 2003, 2008). They have been further provided protection by the enactment of Panchayats (Extension to Scheduled Areas) Act (PESA) 1996, which empowers the Gram Sabha (decentralized units of governance in Scheduled areas) to control the resources and the right to, "customary law, social and religious practices and traditional management practices of community resources" (PESA, 1996). Unfortunately this Act remains poorly enforced at many places and evictions continue.

Of the numerous problems that inflict the tribal people, the biggest problem facing them is the land alienation whose origins can be traced to the British colonialism whose forest policy objective was inclined towards commercial considerations rather than human. Ironically, it was exacerbated in independent India with the government forcibly evicting poor tribal communities in the name of conservation. It was because the government saw them as a major threat to the already debilitating forest cover. This pushed them into abject poverty and indebtedness, further deteriorating their health and nutritional status, as it was the only means of their subsistence. Their resistance to such measures was met with brutal repression at the hands of the state.

One of the major causes of the dependence of tribals on forest produce is the lack of alternatives for fuel needs, money and food. It is amply clear that in the absence of any government support for these necessities, they have nothing else to fall back upon but the forests, or else they won't survive. So, rather than blaming the tribals, we should try to look for the alternatives to cater for their needs for survival, which will automatically take the stress off the forests.

Forest Policy and Historical Perspective

In order to understand the nature, enormity and the implications of this problem and in order to find a suitable way forward, we shall have to go back in time to analyze the origins and genesis of this disputed legacy. Prior to the advent of the British to India, the peasants and the rural communities, "enjoyed untrammeled use of forests and wastes in their vicinity" (Guha, 1983). Most of the forests were managed by communal property use and this right of was recognised by custom rather than any law (Sarin, 2005). So, it was right from the days of pre-Raj oriental rulers that the forests were managed on communal basis by separate villages as a joint property of its members. This 'Community Ownership' meant that because the forests were vital to the village economy, villages evolved boundaries between them to regulate their use. Within the boundary of a village, forests were the common property of inhabitants who had exclusive rights to their use and, "these boundaries were premier institutions in traditional forest management, a necessary condition for enabling the inhabitants of a village to evolve a mutually acceptable and sustainable pattern of forest use" (Somanathan, 1991).

The British brought about a change not only in the political setup but also in the socio-economic fabric of the closely knitted Indian society. In early years of the Raj, the newly arrived British rulers were quite indifferent to the forest property rights and the use of their products. Most probably this attitude was due to the, "seemingly immeasurable extent of its natural resources" (Whitcombe, E. 1971 cited in Guha, 1983, p.1883). India was the most prized possession of the British Empire as it was the biggest reservoir of natural resources and the greatest potential supplier of raw materials for the booming British economy. And due to this there was, "a fierce onslaught on Indian forests" in the early and middle nineteenth century. To the extent that, the Indian timber saved the British against Napoleon, when the British had too many irons in the fire.

The period after the revolt of 1857 was marked by unabated deforestation. It was driven by a need for better means of communication required for better troop deployment, draining of raw materials and import of finished goods etc. This need gave birth to Railways in India and it consumed enormous amount sleepers both for railway track construction and fuel. As a result, there was a drastic decline in the overall forest cover, both accessible and remote. As the potential of the coal mines was not fully exploited, forests provided for the domestic fuel, locomotive needs in addition to the above stated services. (Guha, 1983)

But soon the British administration realized the limits of their policy and that the Indian resources were in fact exhaustible. So, to secure their interests, effective control and supervision became imperative. It was under these conditions and with the help of German foresters, the Forest Department (FD) of India was set up in 1864 (Guha, 1983). Forest Department (FD) was used as a tool to regularize state control on the forests hitherto managed by the tribals and peasants on community ownership basis for centuries. The establishment of the FD was followed by the Indian Forest Act 1878, which "empowered the government to reserve for the state's use all forests not on privately owned land. In fact, the British government simply refused to recognise traditional notions of community ownership and declared all non-private land to be state property" (Somanathan, 1991).

Ironically, there were major historical continuities between the pre and post-independence forest policies. Though, the driving interests behind forest policy formulation changed from the imperial needs for railways and wars in colonial times, to industrial and urban interests in post-independence era. Yet their successful execution has been achieved only at the expense of tribals and their habitats in both the periods (Guha, 1983). In fact, the post-independence policy of the government in tackling the unrest and discontent among the tribals and peasants has been far more brutal, involving the use of force, even killing of the protesters.

As mentioned before, in the pre-British days forests were managed by villagers on the principle of community ownership and it is this pattern of decentralized village ownership that Mahatama Gandhi championed for. Jawahar Lal Nehru, the first Prime Minister of India and the deputy of Gandhi in pre-independence days, followed him publically but Gandhian ideas didn't reflect in his post-independent policies (Swain, 1997). He put India on the path of development by investing in heavy industry, and national planning making India the first non-communist country to have a comprehensive central planning, thus neglecting the Gandhian philosophy of decentralized village community and his concern for social environment and the protection of the community (Swain, 1997; Guha, 1994). In the biomass and fossil fuel dependent Indian economy, this post-independence development strategy has clearly taken its toll on the forests. This made the FD to take the exclusive control of the forests in order to check its further degradation.

In the process of exercising exclusive control over forests, the FD has denied the historical right of the tribals of managing the forests on communal basis and deriving a sustainable livelihood out of it. They have been alienated, marginalized and forcibly evicted from their forests that they have been managing for centuries. They have been branded as 'encroachers' upon the land they used to own by historical custom, though not by any formal law of the land. There is little doubt that the major cause of the decimation of the forests and the biodiversity which they hold has been brought about by the acts of commercial forestry to meet the industrial demands of fuel and timber. The industrialists have up to now heavily influenced the policy and decision making of the FD in their favour and derived huge profits out of it. Forests were viewed mainly as a material source of providing raw material for their industries which would enable them to make huge profits. But they don't miss a chance to criticize the FD for phasing out the subsidies on the raw materials in wake of the fast depleting resources that it is meant to protect. Seeing this change in the attitude of the FD, they start lobbying for the release of degraded forests for their use as 'captive plantations' (Guha, 1994). So, the State becomes captive to the opportunistic interests, opening the door to rent seeking and corruption.

These tribals have been until now oppressed and condemned unheard, because they don't have a voice that can echo in the corridors of the big offices and legislative houses where policies are framed. The wild life conservationists and FD on the one hand are pressing hard for the exclusive state control of the forests. The social activists on the other hand are well averse to this practice and advocate a 'hands off' policy (Guha, 1994) and a decentralized village control setup. In between these extreme positions, the aspirations of the tribals or Adivasis, as they are locally called, get churned and remain unheard.

Current Government Policies

India takes pride in the rate of its economic growth and development over the past two decades. In the past few years its G.D.P has been growing at a staggering rate of nearly 9%. But this development has come at a price and has taken its toll on the resources. Earlier, the forest land was cleared to make way for agriculture but now even agricultural land is being diverted for industries and the newly created Special Economic Zones (SEZs). The FD has left no stone unturned in evicting the tribals. Ironically, on the other hand the diversion of forest land for non-forestry purposes, which is strictly prohibited by the Forest Conservation Act 1980, except with the concurrence of the Central Government, has gone up (Bijoy, 2008).

It is beyond all doubts that no society can become a post-food society, though societies can become post-fossil fuel or bio-mass societies and India can surely show the way because it has got the resources. But, unfortunately it lacks the political will. Where will we find the food to feed the booming billion? By creating the ludicrously incongruous jungles of concrete on prime agricultural land, which at many places have been created by clearing pristine forests? If the unabated exploitation of resources especially forests goes on, it can be visualized that burgeoning industries which are the hallmark of economic development of any country, will themselves act as a limiting factor on Indian economy and eventually cripple it. It is true that at some places damages have been caused to the forests by the tribals who derive their subsistence from it. But the symptoms should not be mistaken with the malaise. (Sekhsaria, 2007)

Imperial and industrial interests have dictated the policies of the FD in the pre and post-independence periods respectively, showing utter disregard for sustainability and social justice (Guha, 1994). In fact, the Indian Forest Act of 1927 which enabled the government to proclaim any piece of land as a forest and claim it as a government land, was meant to safeguard the interest of commercial forestry and ensure the regular supply of timber. As a result of blanket notifications, the land inhabited by Adivasis (tribals) for generations was converted into reserved forests. And, as the record of rights did not exist in the tribal areas, the settlement was done perfunctorily by officials, thus converting them into encroachers upon their own lands that they have historically looked after (Bijoy, 2008).

As the pressure from wildlife conservationists and green groups piled up on the Ministry of Environment and Forests (MoEF), in 1972 the Wildlife Protection Act was enacted as a first solid step towards wildlife and forest conservation. It was further strengthened by passing of the Forest conservation Act 1980. This act was meant to check the diversion of pristine forest land for non-forestry purposes except for the concurrence of the central government. It was envisioned to provide legal remedy for the settlement of the rights of the tribals. But instead, it aggravated the plight of tribal communities by causing further evictions in the name of conservation and denying their legal rights (Bijoy, 2008). Because of the fact that their rights were not recorded by the officials, tribals fell prey to forced evictions. Instead of taking action against the officials for dereliction of their duty, the government victimized the victims themselves (Bijoy, 2008). All this was done to achieve the arbitrary target of 33% forest cover set by the Forest Policy of 1952, despite the fact that there were great discrepancies between the forests on paper and the forests on ground, as large tracts of non-forest lands were converted into forest land by 'blanket legislations' (Swain, 1997).

But this is just a minor example of the injustices meted out to the tribals in the name of conservation. The atrocities that are to follow are far more horrendous. The MoEF issued, what turned out to be a circular of carnage in May 2002, outlining a 'time bound action plan' to evict the encroachers by 30 Sep. 2002 (Bijoy, 2008). About 3,00,000 forest dwellers were evicted from their habitats in the two years that followed the circular. These evictions were accompanied by serious human rights abuses. Their houses were gutted, their women were molested and raped and youth were killed. Scores of villages were set on fire, leading to further clashes and even more deaths (Bijoy, 2008).

The MoEF, in 2004, acknowledged that tribals were subjected to historical injustices, the violation of the laws by the government and also the fact that they have been denied their traditional rights to the forest lands (Bijoy, 2008). This did not, however, result in a cessation of hostilities. In 2005, 180 houses were burnt in M.P state. The enforcers of the law were so keen to abide by it that they burnt a hut after pulling a pregnant woman who was in labour, out of her house and she had to deliver her child in the open (Sarin, 2005). This inhuman repression, other atrocities and forced mass evictions led to nationwide protests, boycotts spearheaded by a group called Campaign for Survival and Dignity (CSD). Thousands across India were injured and gave court arrests (Bijoy, 2008).

The government succumbed under pressure and asked the Ministry of Tribal Affairs (MoTA) (established with an aim to safeguard the interests of the Scheduled Tribes) to formulate the 'Scheduled Tribes and Forest Dwellers (Recognition of Forest Rights) Act. This Act also popularly known as Forest Rights Act or Tribal Act came into force in Jan. 2007 despite staunch criticism from the MoEF.

For the tribals it has come as a sigh of relief and MoTA has seen it as a victory for the cause of tribals. They see it as a victory because it is for the very first time that the tribal rights to forests have been given formal recognition. The biggest positive that has come out of it is that it has protected the tribals from being labelled as 'encroachers' on their own lands by giving them the security of land tenure thus safeguarding them any further forced evictions. This act further paves the way for the sustainable use of forests so that the tribals can sustain their livelihood by giving them the right to access to Minor Forest Produce.

But, this act has been criticized by the MoEF, wildlife conservationists and green groups who feel that this act would be the last nail in the coffin for the already debilitating forests in India and will undo all that has been achieved over the past. A realistic concern they have, is that land and timber mafia in collaboration with political forces might misuse the provisions of the act (Bhargav, 2007; Mazoomdar, 2006). They also might, ruse the illiterate tribals vested with the land rights, to part ways with their prime forests and exploit them later.

Another imminent threat is the commercialization of forests that might result from the sale of forest based produce meant for the household needs of the tribals. These are very genuine concerns that need to be addressed rationally. But overall, the Forest Rights Act, 2007 is a great step forward towards the emancipation of the suffering tribal communities. Unfortunately, this issue has turned out to be a matter of prestige for MoEF and MoTA, with both taking extreme stands.

The problem can be viewed from a different perspective. It is not just the MoEF and MoTA that are confronting each other. MoEF represents the cause of Physical Geography i.e. forest conservation where as MoTA represents the cause of Human Geography i.e. tribal populations inhabiting the forests. So, in some way, it is Physical Geography itself that has come into confrontation with its sibling, Human Geography because their interests have been brought into conflict.

National Forest Policy for the Preservation of Forest Rights of Tribals

Tribals age-old symbiotic relationship with forests is well known. Recognising this fact, even the National Forest Policy committed itself to the close association of tribals with the protection, preservation and development of forests and envisaged their customary rights in forests. It is, however, a matter of serious concern that about 5000 forest villages do not have minimum basic living conditions and face a constant threat of eviction.

The National Policy suggests that any forceful displacement should be avoided. Human beings move on their own to places with better opportunities. The forest villages may be converted into revenue villages or forest villages may be developed on par with revenue villages to enable the forest villagers enjoy at least the minimum amenities and services that are available in revenue villages. The main highlights of National Forest Policy are—

1. The National Policy, therefore, mandates that -Educational and medical facilities, electricity and communication, approach roads and such other basic amenities be provided to forest villages.
2. Public Distribution System (PDS) and Grain Banks be established to prevent food problems.

3. Advanced agriculture and animal husbandry technologies be introduced so that forest villagers raise their production, incomes and economic standards.
4. Bank and other institutional loans be made available for entrepreneurs with viable projects of income generation
5. Tribals be given opportunities to partake in community forest management and encouraged to form cooperatives and corporations for major forest related operations
6. Integrated area development programmes be taken up in and around forest areas
7. Tribals rights in protection, regeneration and collection of minor forest produce (MFP) be recognised and institutional arrangements made for marketing such produce
8. Efforts be made to eliminate exploitation by middlemen in cooperatives like Tribal Development Cooperative Corporations (TDCCs), Large Sized Multi Purpose Societies (LAMPS) and Forest Development Cooperatives by introducing minimum support prices for non agricultural produce on the lines of minimum support prices for agricultural produce.
9. The Scheduled Tribes and Forest Dwellers (Recognition of Forest Rights) Act 2005 when it comes into force will be strictly implemented.

Conclusion

India's tribal communities have remained marginalized both under colonial and post colonial administrations due to discounting of their time honoured customary rights over the forests. This denial of their historical rights has led to conflicts and further alienation. This conflict can't be reconciled by taking extreme stands but by having a holistic approach that accommodates both the perspectives. Effective conservation efforts will only bear fruit if both the government and the tribals come together, co-operate and act as partners. If they share not only the onus of responsibility, but also the power to make decisions. Governments reluctant to devolve power to the tribals because of the power implications must learn from their increasing environmental infirmities and ineffective resource management. What tribals really need is not upliftment, but empowerment. And empowerment is only possible if they are actively involved in decision-making process and given their due in resource allocation. The global models of Protected Areas and National Parks must be adapted and acclimatized to the needs of the local inhabitants, so as to take their concerns and interests into account. This joint effort to conserve forests will make the hitherto incompatible approaches of state and community controls seem compatible. Thus, simultaneously paving the way for effective long term forest conservation and sustainable livelihood for tribal communities.

References

- Bhargav, P. (2007). Four Hectares of Forests: Correcting history or destroying collective future. SANDEE News. 14:19-10.
- Bijoy, C. R. (2008) Forests Rights Struggle: The Adivasis Now Await a Settlement. American Behavioural Scientist 2008; 51; 1775
- Bijoy, C. R, (2003) The Adivasis of India- A History of Discrimination, Conflicts and Resistance. PUCL Bulletin, (http:// www.pucl.org/Topics/Dalit-tribal/2003/adivasi.htm) Accessed on 09-01-2009.
- Census, (2001). State wise Tribal Population Percentage and Distribution in India, (http://tribal.nic.in/indiamap.html)
- Forest Rights Act (2007), Gazette of India, 02-01-2007, Ministry of Law and Justice, New Delhi (http://www.scribd.com/doc/933438/Forest-Rights-Act) Accessed on 27/12/08.
- Guha, R. (1983) Forestry in British and Post-British India: A Historical analysis. Economic and Political Weekly, Vol. 18, No. 44 (Oct. 29, 1983), pp. 1882-1896.
- Guha, R. (1994) Forestry Debate and Draft Forest Act: Who Wins, Who Loses? Economic and Political Weekly,, Vol.29, No. 34 (Aug. 20, 1994), pp. 2192-2196.
- Lok Sabha Debates ser. 10 Jun 41-42 1995 v.42 no. 41-42.
- Mazoomdar, J. (2006) It is a jungle out there. Indian Express, 19 June 2006. Panchayats (Extension to Scheduled Areas) Act (PESA) Dec. 24 1996, No. 40 of 1996. (http://hppanchayat.nic.in/pdf%20files/Pesa.pdf) Accessed on 16-01-09.
- Sarin, M. (2005) Laws, Lore and logjams: Critical Issues in Indian Forest Conservation. GATEKEEPER SERIES 116, pp. 3-19.
- Sarin, M. (2005) Scheduled Tribes Bill 2005: A Comment. Economic and Political Weekly, May 21, 2005, pp. 2131-2134.
- Scherr, S. J., White, A and Kaimowitz, D. (2003) A New Agenda for Forests: Conservation and Poverty Reduction. Washington, D.C, ISBN 0-9713606-6-9. (http://www.forest-trends.org)
- Sekhsaria, P. (2007) Conservation in India and Need to Think Beyond 'Tiger vs. Tribal'. BIOTROPICA 39(5): 575-577
- Somanathan, E. (1991) Deforestation, property rights and Incentives in Central Himalaya. Economic and Political Weekly, Vol. 26, No. 4 (Jan. 26, 1991), pp. PE37-PE39+PE41-PE46.
- Swain, A (1997) Democratic Consolidation? Environmental Movements in India. Asian Survey, Vol. 37, No. 9, pp. 818-832, University of California Press.

Index